AF327231

Baltic Amber
– a Palaeobiological Study

Sven Gisle Larsson

1978

SCANDINAVIAN SCIENCE PRESS LTD.

Klampenborg - Denmark

Edited by
Leif Lyneborg

Text composed by
City Composing & Typesetting Services
Kuala Lumpur, Malaysia

Text and plates printed by
Vinderup Bogtrykkeri A/S
DK-7830 Vinderup, Denmark

Publication date
1 February, 1978

ISBN 87-87491-16-8

Contents

Introduction

Amber has always been a familiar object to the peoples who have lived around the Baltic Sea. Since men first wandered this way, it has been picked up around these coasts, and during the ensuing centuries, so much amber was collected from the early Oligocene deposits on the Samland peninsula near Kaliningrad, earlier Königsberg, that it became of great importance to the chemical industry, as well as acquiring fame all over the world as the "Baltic amber" of jewellery. Finally, as a result of its rich content of very well-preserved fossils, in particular small terrestrial arthropods and plants, it has provided the biological sciences with an excellent opportunity to learn something of the earlier history of plant and animal life. The problems presented by the latter aspect of the study of amber are the theme of the present work.

Amber is plant resin of very varying degrees of fossilization, secreted by plants which occupy scattered positions within the botanical system. This phenomenon is not exclusively associated with the Baltic, since amber is found almost all over the world. Nor is it especially associated with the early Tertiary, but is known at any rate from Mesozoic times right up to the present, when it occurs in the form of its raw material, the fresh-flowing resin of plants. Amber-like minerals exist which are of greater age than the Mesozoic, but their nature is uncertain and they may be the result of accumulations of plant pollen, compressed under the action of great heat.

Amber is thus a heterogeneous material, and its full significance for a more general understanding of the past cannot be appreciated until the results of the research on the many local deposits can be coordinated. Each separate region can provide significant information, leading to an overall understanding. The present study is meant as a contribution to this orientation, especially as concerns the Western Baltic. Even though we assume that the amber deposits of the Western and the Eastern Baltic originate from one and the same continuous wooded territory, the observations presented in what follows would nevertheless suggest the likelihood that such extensive territorial regions have been involved, that from the point of view of fauna, at any rate, there have been differences between East and West. Unfortunately, no comparative botanical material is available. Not the least value of the Copenhagen collection is that it supports the likelihood of this difference.

The study of the literature, which constitutes the basis for the systematic sections in the following pages, has raised a number of fundamental problems in coordinating the results obtained in the various contributions to amber research. No matter how attractive and well-defined these fossils may appear when seen embedded in the amber, they can never fully satisfy the taxonomist; there will always be more or less important characteristics which remain concealed. The taxonomist must therefore often make guesses on the basis of his knowledge of living forms. The amber fossils, which in general differ only slightly from present-day fauna, at any rate if the oldest deposits are excluded, may in the case of Baltic amber show a maximum of systematic relationships with present living forms from almost anywhere over the globe. A few of the species can be identified among recent fauna, and many genera are found both in the past and the present, but numerous genera and also higher groups have become extinct. The working scientist must therefore be familiar with representatives of his group from all over the world. This, however, has not always been the case. Very often, and especially formerly, entire fossil faunas have been dealt with by one and the same person. The result has been incomplete descriptions, of poor benefit to later research. Later generations of students have found it necessary to describe the species anew, but all too often the types and other irreplaceable classic material have been lost in troubled times, not least in the Baltic regions, making final identification impossible. Similar difficulties have resulted from innumerable *nomina nuda* in older literature, empty names without descriptions.

The significance attached by the individual scientist to the concept *genus* has varied considerably. There has, however, been an increasing tendency to sharpen the morphological requirements, and fewer and fewer species have been placed in living genera. This lack of uniformity has made it difficult to draw general conclusions, not least in a statistical analysis of the material. The author would like to appeal to those taxonomic specialists who have already become familiar with the problems of one amber fauna, to attempt to get into touch with other amber collections, so that their experience might contribute to a uniform analysis of the same systematic group from various periods of the earth's history.

The collection of amber fossils in Copenhagen is of younger date. The actual reason for the origin of the collection was a visit by the late Professor Alexander Petrunkevitch in the summer of 1949. He was on a trip through Europe to collect material for his 1958-book "Amber spiders in European collections." The results of his collections in Denmark appeared rather meagre to us: 42 pieces of Danish amber and 62 pieces of Prussian amber, from 3 different collections, one of them private. Recognizing the quite special demands made by fossil terrestrial arthropods on investigators' taxonomic familiarity with living species, the entire public material of amber fossils was therefore collected together in the Zoological Museum, Copenhagen, and today this collection comprises more than 8000 specimens, including about 500 spiders. This collection of amber originates almost exclusively from Danish coasts. About 700 pieces are Prussian amber, a number of them very old and more or less vitrified after having been lying for generations in different museums.

The Zoological Museum wishes to express its thank to the then leadership of the Geological Museum for the permission to have that museum's collection of about 800 pieces as a loan collection in the Zoological Museum collection. We also wish to thank Augustinus Fondet and Copenhagen University for important economic support in making purchases, and not least Carlsberg Fondet and Statens almindelige Videnskabsfond, which made it possible for us to give a considerable part of the material the provisional technical treatment (grinding) necessary for the scientific investigation. We would also like to thank those amber dealers whose interest and pecuniary restraint made it possible for us to increase the collection so considerably. Messrs. A. Henningsen, A.K. Andersen and the late J. Flauensgaard are hereby extended our particular thanks, the last-named in particular having carried out the technical pre-treatment of the material. The author is an entomologist, but the biological data published nevertheless require a geological framework within which they can be understood. For guidance in this field, I am grateful to a number of Danish geologists, and in particular to Dr. Niels Bonde, Institute for Historic Geology and Palaeontology, Copenhagen. Also the special chemistry of the resin makes a scientific control desirably, and in this respect I am greatly indebted to Dr. Martin Etlinger, H.C. Ørsted Institute, Copenhagen. I would also like to thank Dr. Børge Petersen, curator of the Zoological Museum, Copenhagen, for invaluable economic aid with the translation (H. Cowan, B.Sc.). Finally, the author would like to express his great debt of gratitude for the understanding and support he has always met in the Zoological Museum, quite especially through the instrumentality of Dr. S.L. Tuxen, leader of the Entomological Department.

SECTION 1

Amber and Resin

Amber is fossilized resin, and is thus of vegetable origin. It has a hardness varying from 1—3. The specific gravity is from 1 to 1.3, most often around 1.06. The melting point .varies from about 120ºC to 400ºC, being between 350º and 400ºC for Baltic amber. Amber is inflammable; on heating, some amber volatilizes without melting. These great variations in characteristics appear to depend in part on the systematic position of the source plant but also on the conditions of fossilization. The solubility varies; some amber is almost insoluble, some is almost completely soluble in chloroform, benzene, ether and acetone, and also in many cases in various alcohols and alkalis. However, the degree of solubility in these substances is often very slight, exposure only resulting in swelling and softening of the material, so that it can be cut with a knife. If amber is re-exposed to drying after this softening, it does not recover its earlier shape, but cracks and crumbles very rapidly and is completely destroyed. Baltic amber is distinguished from most other petrified resins by its high content of succinic acid (3—8%). Amber which has a high content of succinic acid is called succinite, while amber without succinic acid is called retinite. Sub-fossil amber and amber of angiospermic origin are often described as copals.

Baltic amber varies very considerably in appearance, and is often very beautiful. It may be quite clear, varying from light golden to more or less reddish-brown, at times with a greenish or bluish hue. Other pieces may be opaque as a result of containing a milky-white or greyish-golden mass, and pieces are often encountered in which opaque and transparent amber meet along a relatively sharp boundary. Many of these variants have their own designation, but are generally without scientific interest. It may often be difficult to decide the cause of the ' milk amber," but it seems to have arisen in more than one way. It is not unusual to find insects in amber which are milky on one side of a boundary layer in the amber, but not on the other side. In such cases, the side of the insect which has first come into contact with the fluid resin is the one to become milky. It must be assumed that at any rate the prerequisites for the formation of this milkiness in the amber existed while the insect was alive, and that it is due to reactions in the insect; the subsequent resin flow, which has maintained its clearness, presumably did not take place until the death of the insect. Milky amber is also often found at the anus of the insect and especially at the mouth, while only occasionally and in connection with more widespread opacity it is found to involve the wings; beetles are sometimes seen to have completely clear pronotum and elytra, while the remainder of the insect is enclosed in the whitish mass. In cases where the phenomenon has developed without the presence of insects, it is conceivably a consequence of raindrops beating on trunks or branches. Something similar may be found in the case of the present-day modest flow of resin on spruce and pine in our woods. Moisture in the form of animal excreta, and also in the form of the tree's own sap (Conwentz, 1890, p.99), probably plays an important part in the development of the opacity, since resin and water are on the whole incompatible, but as long as the surface of the resin has not hardened, emulsification of the two components is possible. On the other hand, it is characteristic that the "resin" of the cherry tree, which is a gum, a carbohydrate, does not become "milky" in the rain, but slowly goes into solution. Opacity in connection with enclosed insects forms in most cases a massive and quite sharply defined layer, which it has been impossible to remove so far, and those fossils which are surrounded by this layer cannot be made use of with our present technical aids.

Milky amber does not always appear to have developed in connection with an insect's secretions or other forms of fluid. It is conceivable that under the influence of a high temperature, for example strong solar irradiation, the fresh resin gave up some of its less permanent constituents in the form of vapour over a short period of time, and as a result of the high frictional resistance in the resin this vapour has persisted as a cloud of microscopic bubbles, which have "frozen fast" during the hardening, so that the amber became opaque. It is stated that with this type of milky amber, which is independent of the presence of insects, it should be possible to produce translucency by boiling in certain oils (Durham, 1957, p.3). Hommerberg (1947, p.11) recommends careful warming in amber oil. As a result, the air in the small bubbles is pressed out, and the oil penetrates during the subsequent cooling. Opacities of this kind are often more or less greyish in shade and without sharp boundaries. In other amber pieces, clouds of inorganic or organic detritus are the cause of the opacity (Schubert, 1939, p.38).

A great number of pieces of amber show a laminated structure, in which it is possible to see how the individual parts have arisen as a result of a number of resin flows at brief intervals (in German "Schlauben"). One can imagine the resin being warmed up by the sun, and in a relatively fluid form flowing down the tree trunk, followed by an inhibition of the flow by cooling during the night, so that the surface partly solidified. Then a fresh flow occurred over that of the previous day, without being able to dissolve completely the hardened skin already formed. As evidence of the low viscosity possessed by many of these flows, parts of spiders' webs can be seen which have been engulfed by the resin, without the threads having broken.

These fine spaces between two flows have of course nothing directly to do with the insects, but many of the fossils are found in fact in these cracks; they have been captured by the first flow and have struggled to become free, as may be seen from the whirls raised in the amber by legs and wings; only then has the subsequent flow closed over their transparent tomb. Whether the insects have remained hanging to the surface of the resin, or have sunk into the mass, seems to vary with the time of their activity and their relative surface area. Night-flying and dusk-flying forms usually remain stuck to the resin, and the same is the case with delicately winged insects such as midges. One often finds that crickets and other powerful insects which have struck the surface skin with relatively great force have given it a dent; it was only broken where bristles and spines penetrated, but enough to hold the victim firmly until the next resin flow formed. Small wingless animals such as mites, worms and many spring-tails, probably driven out of the moss by the freshly flowing resin itself, are most often enclosed in a single resin flow, and the same is the case with larger animals having a relatively small surface.

In many of the larger insects which have been caught in the resin, parts of the body have been eaten, probably by small birds, during the brief period in which the insect was in its open prison. The attachment between those layers which have developed as a result of the presumed diurnal rhythm is often faulty, and exceedingly fine clefts are formed which are often almost invisible. However, it is important to be aware of their existance, as the amber is very brittle at these sites and splits easily, and this will usually cleave any fossil present. There is hardly any doubt that the existence of these clefts is one main reason for the smallness of most of the pieces of amber collected around our coasts today. Those pieces of amber which do not contain diurnal clefts have probably very often been produced from a single flow only, and are fragments of larger lumps which have been broken against stones during their transport in streams and along shores.

However, Conwentz states (1890, p.96) that "Im Uebrigen blieb noch eine beträchtliche Menge Succinit im Innern des todten Holzes zurück - - -. In den meisten Fällen wurden aber die Harzstücke später frei, nachdem dass die umgebende Holz, vielleicht erst im Verlauf von Jahrhunderten, zu Mulm zerfallen war". Of course, such internal amber formations do not show any laminated structure, nor insect fossils.

The resin production, however, was hardly attached solely to the tree trunks. Stalactite-like formations, likewise with diurnal clefts, and amber drops (Text-fig. 1) suggest that the resin has also originated from knots and large branches. Some pieces of amber are witness to falls to the forest floor from great heights. There are more or less flattened pieces containing earth and other materials from the forest floor in the more flattened side. Other pieces, again, bear impressions of the surface structure of palm leaves, impressions they may have received during a fall to the floor of the forest.

In addition, much amber contains innumerable fine, air filled clefts which do not appear to have been formed in the resin, but later on during the fossilization process (splitting as a result of inner tension on evaporation of volatile constituents). They result in reflections and other optical phenomena of surprising effect, and they often increase the attractiveness of the specimen. Numerous drop- or pearl-shaped bubbles of very varying size and of unknown origin usually contain air, but in some cases are partly filled with solid bodies or fluid particles, presumably remains of organic origin. Many of these clefts and bladders contain coloured deposits (often black or smoky, at times with a play of rainbow colours), possibly traces of dried plant juices. These coloured clefts often give the amber the most remarkable appearance. The impression may be of fluid drops which have penetrated through the clefts while forming rhizome-like patterns. At times they have misled amber investigators, who have seen in them the impressions of corals or other marine animals, which in reality have never been found in amber (Kirchner, 1944 and 1950). These patterns may be due to the amber starting to break down.

Stability

Amber is in fact nothing like so stable a substance as one might be led to believe in view of its great age, and in particular it is in varying degree sensitive of poly-

Text-fig.1. A drop of amber collected on the shore of Kattegat west of Elsinore, Zealand, Denmark. Length: 23 mm. Coll. Holger Andreassen. (G. Brovad phot.).

merization, oxidation and the drying-out influence of the air. In those cases where the amber is lying in the soil, immersed in the sea or in the water-table, nothing particular happens to it; a vitrified surface layer is formed, which throughout the intervals of geological time protects its internal structure against further vitrification. Where the coastal surf releases the buried amber, this layer is rapidly rubbed off, and it then undergoes a coarse grinding by sand and gravel, so that one can observe something of what is hidden in the interior. If the amber is not collected, it will itself become gravel in due course, and innumerable tons of amber must have disappeared in this way in the course of time, e.g. the Åhus-sediments containing much microscopic amber sand (Cleve-Euler & Hessland, 1948).

Jewellery of amber was used even at the time of the first inhabitants of Denmark, and pieces were often formed to represent symbolic figures of varying significance. Many of these articles are among the treasures in our museums. Two categories of finds in the archeological collections are of particular interest. First are the bog sacrifices, bodies which have been under water from earliest times, and in these finds the pearls are only slightly vitrified, the surface being still relatively clear even today, about the same as that of amber on the beach. Then there are the grave finds, which have been lying in ancient graves, where they have been exposed to the action of the atmosphere. These pearls are strongly vitrified when found, and the surface is encrusted and quite opaque.

This relatively rapid breakdown of the amber in those cases where nature itself has not preserved it, must necessarily play a role in the local placing of recent natural deposits. Nowhere in the Baltic region is amber found in its primary site, the forest floor where it lay originally; at all times it has been transported by rivers or by ice; some is found today in fresh-water deposits (for example in the Alnarp Valley (Text-fig.7) in Skåne and in the lignite deposits in Jutland), most

occurs in marine layers near the coast (as in the "blue earth" of the Samland Peninsula), and much is scattered throughout the depths of our moraines. Only where the water-table or the sea water has protected it throughout geological periods, has it "survived." Similar conditions appear to have held for all fossil plant resins of great age, for example the Canadian amber from the Upper Cretaceous period (McAlpine & Martin, 1969b), whereas it is still possible to collect sub-fossil amber (copals) of much younger age in the forest floor, as in the Philippines and northern New Zealand.

As we are thus only familiar with amber from trees that have grown in the vicinity of fresh water, which has been able to transport the amber to suitable preservation sites, but not from trees in other biotopes, the nature of the amber fossils must give the impression that the resin-exuding trees can only have grown in especially damp climates. This may indeed have been the case, but there is actually no evidence available of this.

The worst fate which can befall a piece of amber is to lie in an open box in a show case which is not air-tight, and be exposed to the dry air of the museum. Our museums have many specimens of amber which have been lying in this way for far more than a century, for example specimens from King Frederik III's "Royal collection" from the 17th century. A considerable part of this amber was excavated from the foundations of Copenhagen's defences when they were being rebuilt during the reigns of Frederik III and Christian V. Some of these pieces were of considerable size, up to about 1½ kg in weight (Werlauff, 1836, p.215). Pieces which were formerly golden in colour have now a dull, reddish-brown tint, and their ground surfaces have in the course of time become crackled to a dense, fine network, which admittedly does not penetrate very deeply into the material. However, a fresh grinding of these pieces requires great care, if the piece is not to crumble away between one's fingers; they feel dry and are exceedingly fragile. Jewellery of amber, which is used frequently, appears to be protected against this drying by its repeated contact with the human skin.

This "death" and destruction of amber, which is taking place at a relatively rapid rate, has naturally caused museum staffs great concern, and has given rise to quite an extensive literature, going as far back as the 18th century, when serious attempts were first made to understand the scientific significance of the well preserved amber enclosures, particularly in the form of terrestrial arthropods and plant remains. It soon became clear that one of the factors leading to the destruction of the material was the drying-out as a result of the evaporation of volatile substances, resulting in strains between the exposed surface layer and the temporarily protected interior. It was these strains which among other things resulted in the crackling of the surface. This process appeared to be encouraged when the amber was exposed to direct sunlight.

It was in particular in the museums of the coastal towns of Prussia that interest was aroused in the preservation of amber (Tornquist, 1910b). This was quite natural, since these were museums with the possibility of extensive collections, especially from the excavations in Samland.

One of the simplest and most obvious attempts was to store amber in water, possibly slightly saline. This was relatively easy to carry out, also on a large scale. It was a copy of the natural process, but the small containers lacked the self-cleaning power of the natural state, and antiseptics had to be added on account of bacterial growth and similar contamination. These agents, however, had a harmful influence on the amber. Storage in alcohol was also tried, and valuable material was completely destroyed in this way, since the alcohol softens and partly dissolves amber, thus adding to the consequences of the drying by removing important constituents. Storage in mineral oil has been tried (Petrunkevitch, 1958), and this has the advantage that by penetrating cracks and fissures (possibly with the aid of a vacuum), disturbing refractive effects are reduced to quite a considerable degree, thus facilitating the scientific study of the fossils; the mineral oil has in fact approximately the same refractive index as amber and resin. Also mineral oil, however, is found to have an effect on the resin, so that after the lapse of a few years there is a dark staining which spreads from the cracks into which the oil has penetrated. There is also the further disadvantage that where the oil has penetrated into the spaces within the fossils, the absence of refraction which ensues causes these to become more or less invisible. A satisfactory solution to this problem has so far not been found. Nowadays, about the best way to preserve amber is by storing it as airtight as possible, and in particular by protecting it against the risk of air movements. Particularly valuable pieces should probably be embedded like microscopic preparations. It is also possible that some of the many synthetic substances developed during recent years could be used as a protective skin or for embedding (Schlee, 1970, p.5).

The nature of resin production

Resin is produced in spaces in the parenchymatous tissue (vesicles or channels). These spaces are formed in various ways, each resin-producing plant species having its own pattern. In some species the active cells separate to form an intercellular space, and the cells represent an active epithelium around a cavity which collects the secretion (schizogenic development). In other species, the collecting cavity is formed by the destruction of the producing cells (lysigenic development). In many cases the process is a combination of these two types. Conifers show a well-developed system of schizogenous resin channels, lying longitudinally in the xylem and transversely in the medullary rays. The flow of the resin often appears to take place from cracks which develop as a result of tensions in the bark at times when the growth is particularly pronounced. For example, in spruce this may take place in the spring, when the cambium is particularly thick and swollen. The resin then exudes on trunks and branches in amounts depending on the productivity of the particular species. In some species of plants the amounts may be very considerable indeed. However, large quantities of resin are stored, more or less permanently, in major lysigenous fissures in wood and bark (Conwentz, 1890, p.95).

There are only a few existing groups of plants which produce resin in such amounts that they can be regarded as possible sources of the copal and amber formations of the past, but they are found among both conifers and deciduous trees. Among conifers might be mentioned in particular species of Araucariaceae, but also among Pinaceae and Taxodiaceae a considerable excretion of resin may be found, but not in general in such large and coherent masses. Among the deciduous trees, the resin-secreting forms may be found in particular among Leguminales; but also among Burseraceae, Guttiferae, Anacardiaceae, Dipterocarpaceae and several others, species are found which are rich producers of resin. A feature common to these particularly high-yield plants is that they are found in the tropics and sub-tropics, and are only weakly represented in the warmest temperate regions.

In modern times, really heavy secretion of resin is found in tropical South America *(Hymenaea, Bursera* and *Protium)*, in the central and southern parts of Africa *(Daniellia, Copaifera, Hymenaea (Trachylobium)* and *Canarium)*, in tropical and sub-tropical Asia especially in Indonesia and the Philippines (the dipterocarp *Shorea*, from which dammar resin is obtained, as well as the araucarian *Agathis alba* (= *dammara)*, which gives manila copal). In the Auckland district of northern New Zealand, kauri copal is obtained from *Agathis australis (Dammara australis)*. Most of these resins flow naturally in plentiful amounts, but the flow is further encouraged for industrial purposes by tapping (artificial wound formation). The secretion often occurs in large clumps (up to about 15 kg), and plant parts and small animals are often found engulfed in the mass. The reason for this natural production of resin is unknown, but according to Langenheim (1967, 1969), the production, at any rate in *Hymenaea* and some species of pine, appears to be greatest in trees of pronounced vitality, in other words trees which have a particularly high sap tension in their tissues.

In many of the localities mentioned, and not least in the *Agathis* forests in the East, similar masses of resin are found in the soil, often in such amounts that they are collected for industrial purposes. Here it is a case of both recent resin and more or less fossil copal. Fossil copal can also be found where the forests have long since died out. Accumulations of copal can furthermore be found at the mouth of rivers where deltas are formed or in estuaries, in other words material which these rivers have washed out of recent or earlier forest floors, and which has then collected at sites where the current was weak enough to allow the light copal to sink to the bottom.

There is no essential difference between copal and amber, and scientists use a variety of definitions for the two concepts. For some, amber and succinite are synonymous concepts, but inconsistency is demonstrated by not including for example fossil *Agathis* resins under amber, although in their chemical structure they are very closely related. On the other hand, the chemically different retinites are nevertheless most often designated amber. The purely physical differences, for example in melting point, hardness and solubility, are probably due to variations in the degree of fossilization of the original material and to differences in botanical origin. Copals are most often more recent and more easily affected in various ways than other forms of amber. In the case of *Hymenaea*, both resin and copal and amber are known, considered from the physical aspects.

As already mentioned, members of the pine family secrete considerable amounts of resin, in fact so much that various species of pine are cultivated particularly for tapping the resin for industrial purposes.

Under normal conditions in the temperate zone, as for example in present-day Denmark, the resin production of the coniferous forest is quite insignificant compared with that of the various tropical copal plants. However, Bachofen-Echt (1930, p.39; 1949, p.26) mentions an example from the island of Arbe in the Adriatic. This was a growth of old pine trees (species not mentioned) which had only reached a height of 3—5 metres. The harsh climatic conditions on the island, with severe seasonal fluctuations, had turned the growth into a crippled "goblin wood." There were large resin deposits on the trunks, and on the ground around the trees the resin had collected in lumps which could attain the size of a man's head. This shows that under certain natural conditions, pine trees can exude such large amounts of resin that it might be considered possible to demonstrate its presence in the soil over a rather long period of time. However, this period does in fact appear to be strongly limited, as resin of pine may be sensitive to attacks by microorganisms and oxidation by air (Langenheim, 1969). It is possible that the resin exudate on Arbe is a reaction to unfavourable conditions of existence, but it may also be under the control of unknown hereditary factors, and finally, the chemistry of this resin is unknown.

Where pines are cultivated with the practical purpose of collecting resin for industry, a considerable effort to increase the yield has been made in recent years. North American studies on the species *Pinus palustris, P. caribaea* and *P. elliottii* have shown that under uniform conditions, some individuals yield more than twice as much as others, and that to a pronounced degree this characteristic is inherited. It has thus been possible to establish a plantation with twice as great a production as an average plantation of the same species under corresponding conditions. The yield appears to be clearly dependent on the rate of growth and is therefore likewise proportional to the size of the tree crown. As a result, the practical purpose resulted in the further gain that trees with a large production of resin also produced a wood mass which was about 12% greater than average.

Factors influencing the secretion of resin are the number of resin channels and their calibre, the fluid pressure and the viscosity. The pressure and viscosity are determined to a high degree by the hereditary characteristics of the tree, while the number and calibre of the channels are also dependent on the age of the tree. It is above all the hereditary characteristics which determine the yield. However, the sap pressure also varies with the conditions, and depends on the supply of water.

These investigations (Snow, 1949, 1954; Boudreau & Schopmeyer, 1958; Squillace & Dorman, 1961; Dyer, 1963; Squillace, 1966), quoted here from Langenheim (1967, p.222) show that healthy and vigorous trees have a considerably greater production of resin than sickly trees, in which all the physiological processes are weakened.

Originally, it was considered that very special conditions must have been present for such a heavy production of resin as is suggested by the large amounts of amber that could be found over such a small area as the Samland Peninsula, often in large pieces. No species of pines known today produces resin in such large amounts. It was Conwentz (1890, p.81) who suggested that the phenomenon was due to pathological circumstances, namely that a disease, *succinosis*, must have attacked all the pine trees in the forest, so that literally not one healthy tree remained. This idea has fascinated most amber investigators for many years. The possible reason for the disease has been discussed, for example attacks by wood-destroying insects such as bark beetles, but this explanation must be considered quite inadequate. Schubert (1961) suggests that an abnormally increased resin production might be due to a strong disturbance of the physiological equilibrium, as trees are especially sensitive to harmful effects wherever they grow in regions which bound their natural distribution. It must be considered, too, whether any bacterial (or fungal) disease might have had a chance of injuring the trees during such a period of physiological disturbances. Has there possibly been a parallel to the recent plant disease, the gumflow of cherries and prunes, during which a surplus of gum strongly and often suddenly reduces the water transport in the tissues of the plant?

Even though attacks by Polyporaceae and insects can hardly be denied some degree of significance, the above-mentioned experiments in breeding of races of American *Pinus* with a high resin yield suggest that the plentiful resin production of the amber tree was a normal phenomenon. It was not a sign of disease, but if anything, evidence of favourable conditions. For this reason alone, it cannot be expected that the amber tree can be identified with any now living *Pinus* species.

The chemical and physical characteristics

In the course of time, mineralogists have provided designations for many kinds of amber on the basis of general physical and chemical properties. Primarily it was classified into succinite with up to 8% succinic acid, and retinite without succinic acid. Chemical analyses were based in the first instance on Helm's classic studies. However, the method employed by Helm involves several sources of error and requires quite a considerable amount of material for the analyses, which is a special inconvenience when it involves smaller pieces of amber, whose scientific or artistic value demands that they should be spared as much as possible. This inconvenient state of affairs has been overcome, however, as a result of the extensive studies carried out in recent decades on the chemistry of amber.

The most recent fundamental chemical investigation of Baltic amber, with consideration of its botanical origin, is that briefly reported by Gough & Mills (1972). The substance is agreed to have originated from conifers and has long been known to contain a few tenths of a percent of the monoterpene alcohol borneol, free or esterified, with about a 10% excess of the dextrorotatory over the levorotatory isomer. Gough & Mills identified small amounts of nine diterpene resins acids with skeletons (Text-fig.2) of the usual labdane (I), pimarane (II) and abietane (III) types in the ether-soluble fraction. The bulk of succinite is presumably derived from polymerization of diterpenes, which are the major constituents of conifer resins. Since the source is more narrowly thought to have been extinct *Pinus*, and since the resins of modern European pines consist mainly of abietanes, exemplified by abietic acid, a scheme has been put forward (see Rottlander, 1974, p.78) for conversion of abietic acid by ultraviolet light to dimeric, partially decarboxylated molecules that might be assembled into a polyester perhaps resembling amber.

Evidence that this sequence of processes actually takes place to a significant degree, however, is nonexistent, and the abietane residues in dimerized resin of the kind proposed as the amber precursor are clearly revealed on dehydrogenation (Gough & Mills, 1972; see also Beck, 1972; Beck et al., 1975). Abietic acid in buried wood is well known to decarboxy-

Plate 1.
Ichneumonid wasp. The great number of stellate hairs from oak shows that the amber resin flow in question took place in the spring or early summer when the oaks were still flowering. Below the wasp a male of a ceratopogonid biting midge is visible.

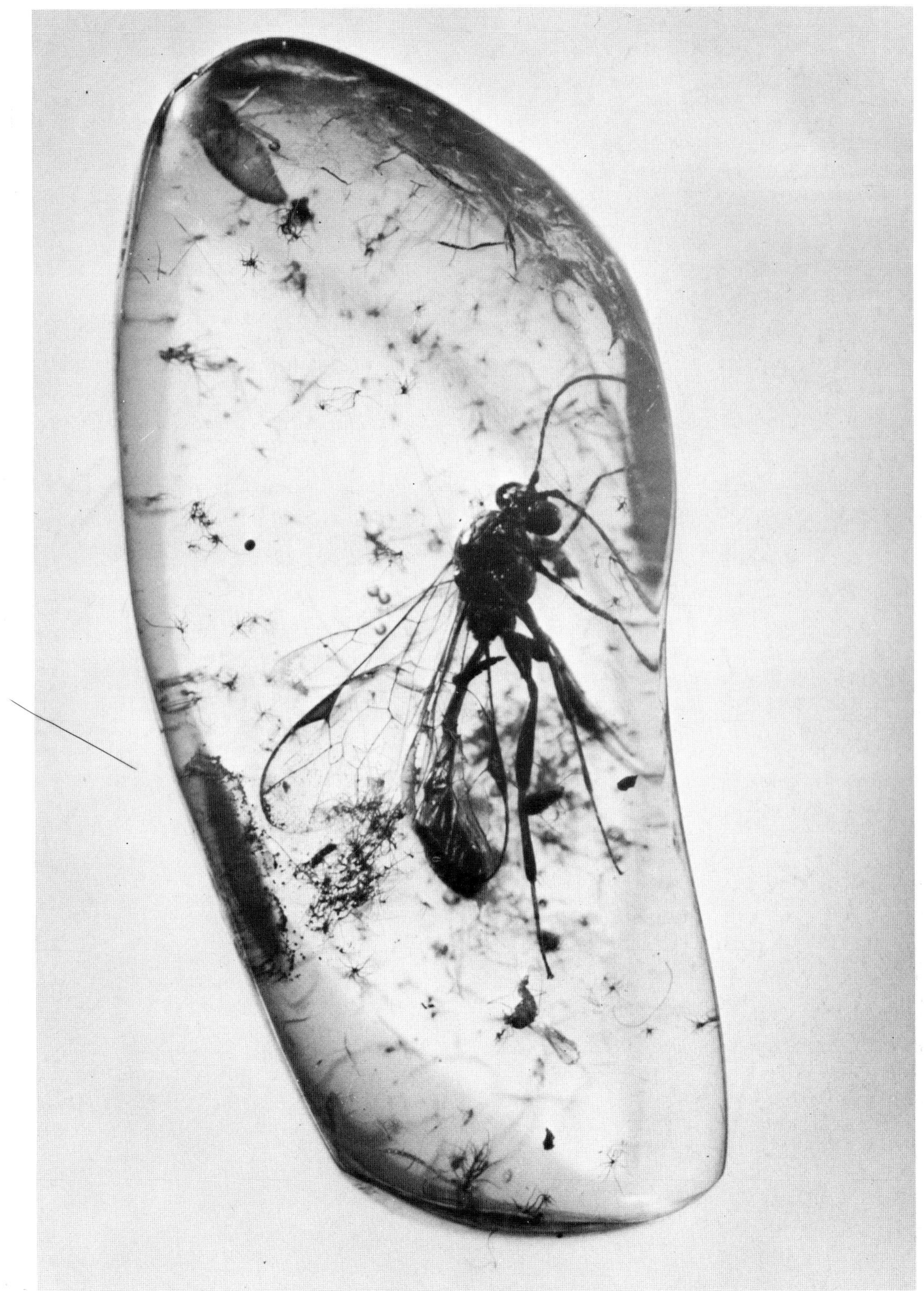

late and disproportionate as the mono-
mer, giving hydrocarbons like fichtelite
and retene (7-isopropyl-1-methylphenan-
threne, VII), which form no appreciable
part of amber. Fossil resins composed
principally of abietane polymers are
accordingly believed to be unknown
(Langenheim, 1969; Thomas, 1970, p.59;
Gough & Mills, 1972). The skeletons
of diterpene monomers can be distinguished by means of selenium dehydrogenation, which converts labdanes (Carman & Craig, 1971a, b) to products including agathalene (1,2,5-trimethylnaph-
thalene, IV) and in some cases 1,1,4,7-
tetramethylphenalan (V) or pimanthrene
(1,7-dimethylphenanthrene, VI), pima-
ranes to pimanthrene and abietanes to
retene. The pioneer work of Schmid and
collaborators (see Beck, 1972; Rottländer, 1974) established that the ethanol-
insoluble portion of succinite was dehydrogenated to agathalene and pimanthrene and that ethanol-soluble fractions
yielded the same products plus in one
instance retene. By quantitative evaluation of similar experiments Gough &
Mills estimated that 75-80% of the
diterpene residues in Baltic amber belonged to the labdane type, 15-20% to
the pimarane type and 5% or less to the
abietane type.

The ability to harden by polymerization
appears characteristic of resins containing a large proportion of labdanes with a
conjugated, terminal diene unit in the
side chain. A prime example among
conifer resins is the recent and fossil
copals from the genus *Agathis* (Thomas,
1969, p.599) of Araucariaceae. The
Agathis monomers, represented by the
cis and *trans* communols (elliotinols) and
communic (elliotinoic) acids, are labda-

trienes which besides the diene grouping
contain an isolated, 1, 1-disubstituted
double bond attached to the ring system
(exocyclic). Polymerization of communic acid, methyl communate and
communyl acetate (Carman et al., 1970,
p.1655) proceeds through the terminal
double bond of the side chain and leaves
intact the exocyclic double bond, which
gives rise to infrared absorption also
characteristic of Baltic amber. The infrared spectra of Baltic amber and
Agathis australis (kauri) copal are in fact
notably similar (Langenheim, 1969;
Thomas, 1970), the differences being
mainly ascribable to the succinic acid
residues of the former. Thus removal of
succinic acid from the insoluble portion
of succinite by alkaline hydrolysis leaves
a remainder whose infrared spectrum is
practically identical with that of poly-
meric kauri resin, and conversely succiny-
lation of kauri polymer gives a product
with infrared spectrum close to that of
Baltic amber (Gough & Mills, 1972).
The monomers of Baltic amber hence
appear to have been chiefly labdatrienes
of the same general sort as the communols and communic acids. These
compounds, though not quantitatively
conspicuous in recent pines, occur in
both principal divisions of *Pinus. P.
elliottii* oleoresin contains about 5% each
of communol and communic acid, and
the latter constitutes up to 10% of the
resin acids of *P. strobus* (Joye & Lawrence, 1963, p.3274; Joye et al., 1965,
p.429; Santamour, 1967, p.82). The
source of Baltic amber can thus scarcely
be excluded from an ancient range of
chemical variation of *Pinus.*

Succinic acid is a universal cellular
metabolite, but its presence in succinite

Plate 2.
A. Bud-scale of oak, densely covered with stellate hairs, some of
which are about to fall off (G. Brovad phot.).
B. Amber containing many fine fossils. A silverfish is seen at
lower left, but other views of the same piece of amber would also
have shown excellent specimens, among others an earwig, a centipede, a caddis fly and some gnats. (Privately owned and phot.).
C. Fossilized wood of an araucarian tree found in the same deep
layers of the Limfjord Mo-clay as the few known Mo-clay amber
pieces. The annual rings of the wood are very densely arranged,
and the piece is astonishingly heavy. (G. Brovad phot.).

at the beginning is uncertain. Oleoresins of modern Pinaceae are reported to furnish succinic acid on boiling with water or slow pyrolysis without exclusion of oxygen. The recorded yields are less than 0.2%, the higher figures quoted (Thomas, 1970) comprising total water-solubles or loss. Even crude abietic acid on fusion with alkali or pyrolysis has been said to give detectable succinic acid. Unsaturated labdane esters of acetic acid occur in *Agathis* (Thomas, 1969) and in *Larix* among Pinaceae (Mills, 1973, p.2407). Similar natural diesters of succinic acid have not been observed, but possibly an extinct pine produced such compounds, suitable for entering into an exceptionally resistant cross-linked polymer.

During the last few decades, various techniques have been developed for the study of the botanical relationships of ambers and resins. Not least, work with *infrared spectroscopy*, which requires less than 3 mg of substance, has been a major advance in amber research. These methods are due above all to Hummel (1958), Moenke (1962), Beck et al. (1964, 1966) and Langenheim & Beck (1965). What follows is mainly cited from Beck (in press). See also, however, the quite recent publication by Broughton (1974, p.583), whose description of the infrared spectrum of Baltic amber coincides in the main with that of Beck.

The spectrum of the commonly occurring Baltic amber (succinite) may be used (Text-fig. 3) to explain the infrared spectrum. Especially characteristic of Baltic amber is an almost horizontal section between 8.0 and 8.5 μ (1250-1180 cm^{-1}), followed by an especially strong absorption at around 8.7 μ (1150 cm^{-1}), after which the absorption decreases rapidly (f in Text-fig. 3, the "Baltic shoulder"). This section is due to carbon-oxygen single bonds. Similar carbon-oxygen bands are also found in some North American forms of amber (e.g. amber from Maryland, ambrosine from South Carolina and a small proportion of the Cedar Lake amber). The explanation for this is the content of succinic acid in these types of amber.

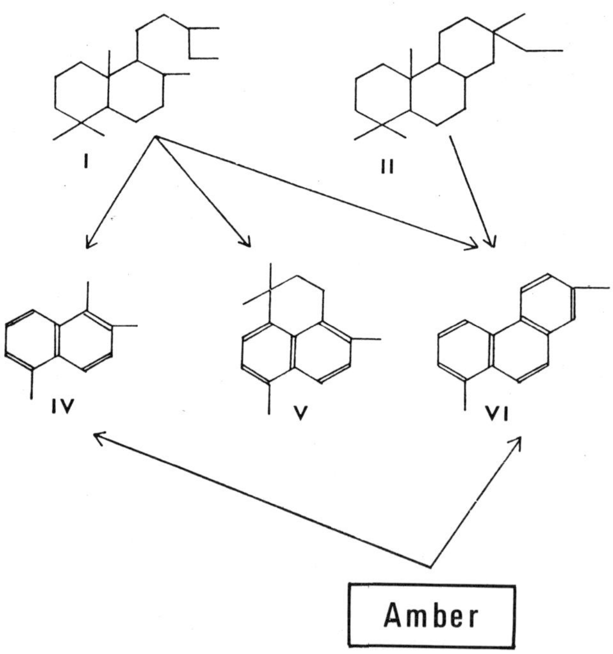

Text-fig.2. Dehydrogenation pathways of resin diterpenes. See text for explanation. (Redrawn from Beck, 1972).

An absorption band at about 2.9 μ (3450 cm^{-1}) (a in Text-fig. 3) marks hydroxyl groups, including alcohols, carboxylic acids, hydroperoxides and likewise free water. The strength of the absorption varies with the water content of the amber. Simultaneously and proportional to this there is a rise or fall in the intensity of an absorption band at about 6.1 μ (1640 cm^{-1}) (b in Text-fig. 3), and this is therefore also considered to show the uptake of water. This short-wavelength section of the spectrum is without special significance for comparisons with other spectra of fossil or recent resins, but it shows, which is valuable in problems of preservation, that amber is not completely unaffected by water, and it explains why water-filled bubbles in fresh amber gradually dry out during museum storage.

The absorption bands at about 6.9 μ (1450 cm^{-1}) and 7.25 μ (1375 cm^{-1}) (c_1 and c_2 in Text-fig. 3) show that methyl and methylene groups are present in amber.

The absorption at 11.3 μ (885 cm^{-1}) (d in Text-fig. 3) is among the most valuable of the carbon-hydrogen bands. It is undoubtedly due to out-of-plane wagging vibration of 2 H-atoms on a 1,1-disubstituted double bond ($RR'C=CH_2$). The intensity of the absorption varies considerably, possible owing to such double bonds having a tendency to oxidation, but it can be recognized even in strongly vitrified amber samples. This absorption band makes it possible to distinguish Baltic amber from many other forms of amber, including glessite, the "sogenannte Pseudo-Succinit Frankreichs" (Beck, in press), retinite from North America and true burmite from Burma. Among those forms of amber showing the 11.3 μ absorption band might be mentioned gedanite, South American copalite, ambrite from New Zealand and ambrosine from North America. It is also present in a part of the Sicilian simetite, but may be completely lacking.

All fossil resins show strong variations in absorption by carboxyl and ester groups between 5.65 and 5.90 μ (1770 and 1695 cm^{-1}) (e in Text-fig. 3). The absorption band is always very wide and often partially disappears.

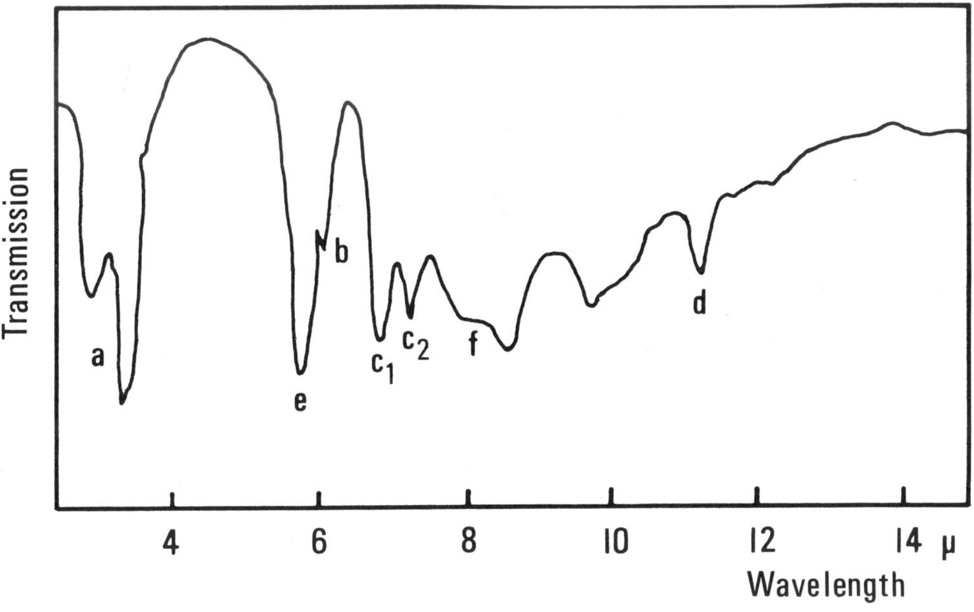

Text-fig.3. Infra-red absorption spectrum of Baltic amber. For explanation of the lettering, see the text. (Redrawn from Beck, 1972).

Two identical infra-red spectra of pure substances indicate that the substances are identical, but for resins, which consist of a mixture of substances with in general a high molecular weight, it is only the dominant constituents which can be expected to give strong absorption bands. Even though absolute identity of the spectra can never be expected, similarity makes it likely that a corresponding similarity exists for the most important constituents. When the resins polymerize, the spectra of recent and fossil resins from the same plant species maintain such a degree of similarity that corresponding peaks can be seen, even though they are most often weaker in the fossil than in the recent material. At greater wavelengths, some bands may be missing in the fossil resin on account of polymerization or oxidation (Langenheim, 1969).

In interpreting the infra-red spectra, therefore, some idea is obtained as to which species of plants may have given rise to the various deposits of amber and copal. However, in the case of Baltic amber these results differ so much from the well-established view of orthodox palaeobotany, that as already mentioned, there are, for the time being at any rate, grounds for interpreting the results with some reserve. In the case of Baltic amber, it has been considered from early times that the mother plant was without doubt that conifer which Goeppert (1836) called *Pinites*. Goeppert found, however, that the wood of these *Pinites* differed somewhat from that of recent *Pinus* species, in that he had not succeeded "eine Conifere im fossilen Zustande zu entdecken, welche hinsichtlich der grossen Poren der Markstrahlen mit *Pinus sylvestris*, oder mit der Gattung *Pinus* Lk. übereinkommt." (Goeppert & Berendt, 1845, p.88). Since then, Conwentz (1890) and other authors have referred it to the genus *Pinus*, even though Conwentz (1890, p.63) writes; "dass der Bau der Bernsteinhölzer nicht allein den Typus der Kiefern, sondern auch den der Fichten zeigt, wenngleich er vornehmlich zu ersteren hinneigt,"

and on p.145: "Zu diesen Bernsteinbäumen gehören vornehmlich vier Kiefern-Arten, von welchen aber keine einzige unserem nordischen Charakterbaum der Gegenwart, *Pinus silvestris* L., nahe steht." These dendrological studies were followed up most recently by Schubert (1961), and also he classifies the Baltic amber tree as a pine, *Pinus succinifera* Conwentz, but he too was unable to find any recent *Pinus* which could reasonably be considered a descendant of the amber pine. Baltic amber might possibly have come from the subgenus *Haploxylon*, the group to which among others *Pinus strobus* belongs, if, indeed, it derives from a *Pinus* species at all, which is now being questioned by an increasing number of workers (e.g. Gough & Mills, 1972, p.527).

Judged from the infra-red spectra, no impression is gained of a close relationship, but rather the contrary. The cleaned pine tree resin (colophonium) contains as mentioned abietic acid (Type III, Text-fig. 2), as is usual among Pinaceae. This has the effect of splitting up the absorption band at 7.25 μ (c_2 in Text-fig. 3) in the ratio 1:3, but it is just this which is not the case with Baltic amber. In addition, resin acids of the abietic type, in contrast to succinite, have no "Baltic shoulder" at 8—9 μ and no 1, 1-disubstituted double bonds (absorption band at 11.3 μ) (Beck, in press). There is thus a quite considerable difference between the two spectra. In order to be certain that this difference could not have arisen during the process of petrification, an attempt was made to hasten artificially the polymerization associated with petrification, and it proved possible to do this in the course of about 3 months, by means of ultraviolet rays from a powerful mercury vapour lamp using a special technique (Beck, in press). Following this "petrification process", neither pure abietic acid, colophonium nor fresh resin of *Pinus resinosa* gave spectra which were related to the spectrum of succinite, but on the other hand two samples of manila copal *(Agathis alba)* did, together with a Congo copal.

This might encourage one to assume that the Baltic amber tree was an araucariacean, but even though we know that the Araucariaceae were of wide distribution in the northern hemisphere in the late Cretaceous and the early Tertiary (found for example in the early part of the Jutland Mo-clay horizon; M. Breiner, personal communication), no trace of them is known from Baltic amber, either as wood, needles or pollen. On the other hand, remains of Goeppert's *Pinites* are common, at times in the form of wood fragments enclosing amber. A few present day Pinaceae are known, however, in which the main resin acids are not of the abietic acid group, but belong to the labdane type and have the "Baltic shoulder" and the double bonds characteristic of succinite (absorption band at 11.3 μ), namely, the North American *Pinus lambertiana*.

Pinites succinifer thus shows apparently a greater relationship with the Araucariaceae than with the great majority of Pinaceae. On the basis of the overall chemical and morphological conditions, therefore, the tendency is to consider it as a rather primitive type, representing an early stage of the developmental history of the Pinaceae, which in their chemistry (just like the *Pinus lambertiana)* still retained archaic characteristics in common with the Araucariaceae. One might also speculate whether at that time the genus *Pinus* as a whole was of such a young degree of development that, like its forefathers among the Araucariaceae-like plants, it still produced resin of the labdane type. This might explain why amber containing abietic acid is unknown.

Using infra-red spectra analysis, an attempt has been made to establish more or less sharply demarcated types comprising both recent and fossil resins, which individually are characteristic of a definite plant group (Langenheim & Beck, 1965; Langenheim, 1969; Beck, in press). There is thus the type containing abietic acid, characterizing Pinaceae except for the species already mentioned, which together with *Agathis* species,

some North American amber forms and a few dicotyledonous types have the characteristic absorption bands at 8—9 μ (carbon-oxygen single bonds) and 11.3 μ (1, 1-disubstituted double bonds). Among the Araucariaceae, *Araucaria* differs somewhat from *Agathis*, and there also appears to be some variation within the latter. Within the Taxodiaceae there appear to be two patterns at least, one for *Sequoia* and *Taxodium* and another common to *Metasequoia* and *Sequoiadendron*. Among angiosperms the spectra are easily recognizable for Burseraceae *(Bursera, Protium* and *Canarium)*. Among Leguminales, the African genera can be divided into two groups, one including *Daniellia* and *Copaifera* while the other includes only *Hymenaea*, common to the American tropics. Styracaceae, Rutaceae, Anacardiaceae, Hamàmelidaceae, Zygophyllaceae and other families also give recognizable spectra (Langenheim, 1969, p.1161).

Mischer et al. (1970, p.111), likewise using physical methods of analysis, have attempted to separate the various types of amber so as to follow the distribution of the individual types, i.e. of the presumed source plants. Among these methods, *mass spectrometry* in particular would appear to offer possibilities as a useful tool in the systematization of amber. The published study includes analyses of only 81 amber samples, however, together with a few details on other amber sources. The results obtained from these samples appear to supplement and confirm the results from infra-red spectroscopy.

The starting point for the investigations was amber from Samland (40 samples), which was found to give a very uniform picture. Clearly deviating from this was amber from Lebanon (12 samples), Romania (6 samples), Canada (15 samples) and Sicily (6 samples). The Sicilian material could be divided into two groups, however, one of which had points of similarity with typical Baltic amber, and one of the pieces from the Romanian collection could not be distinguished from this with certainty.

21

Samples were also studied from the Caucasus, the Kiev region and Switzerland. "Die Bernsteine aus den beiden ersten Gebieten konnten nicht vom "baltischen" Bernstein unterschieden werden und werden deshalb diesem Gebiet zugeordnet" (p.119). Infra-red spectroscopy showed the presence in these samples of the "Baltic shoulder." Also in these analyses, the consumption of material was very low, ≤ 1 mg.

In some cases, it is considered that X-rays help to confirm agreement between crystalline constituents in amber and resin. Further, it has been possible in some of them to recognise specific constituents, as for example triterpenoid alcohols, characteristic of definite angiosperms, in both recent and fossil resin (Frondel, 1967). Using X-rays it has been possible for example to establish Burseraceae as sources of both the Baltic glessite and the Highgate resin from the London basin.

That there is still some lack of complete certainty in the evaluation of the biological character of the individual amber types is seen among other things from the following quotation from Beck (1972, p.297), referring to Rottländer & Mischer (1970, p.668): "Bisher erschienene Ergebnisse stellen die Verwandtschaft des Bernsteins aus der Unterkreide des Libanon mit dem gleichartigen der Rossfeldschichten Österreichs fest. Beide werden botanisch den Araucariaceen (Diterpen-Typ I) zugeordnet, aber die Massenzahl 604 im Rossfeld-Bernstein wird trotzdem der Diabietinsäure (Diterpen-Typ III) zugesprochen."

It is obvious that from the botanical side great weight has been laid on these analytical investigations, but they have been just as important for archaeologists, as they have contributed important indications with regard to early trade routes. For example, one might mention the Mycenaean culture, which was destroyed around 1100 B.C. Beck writes on this (1972, p.296—297):

"Wir haben feststellen können, dass die meisten Bernstein-Artefakte des mykenischen Kulturkreises aus Succinit gefertigt wurden; lokale Fossilharze wurden gleichzeitig und gelegentlich mitverwandt. - - - - Funde aus dem Hagenauer Forst im Elsass - - - erwiesen sich als baltisch. Neue Funde von baltischen Bernstein-Schiebern aus Südfrankreich zeigen den Weg, den die mykenischen Bernsteinschätze genommen haben können. Ein besonders wichtiger Einzelfund baltischer Bernstein bezeugte erstmalig menschliche Ansiedlung in pleistozänen Schichten einer Höhle bei Bristol in Südwestengland."

The oldest trade routes thus appear to have passed through France and adjacent European regions, and it may well be permissible to assume with Werlauff (1836) that the earliest suppliers of amber were Nordic people from the coasts of the Bay of Helgoland, as at that early stage the Eastern Baltic was hardly included in the culture region of the peoples of the Mediterranean. It must however be remembered that succinite is also (although uncommonly) found in other sites in Europe than the North Sea coasts and the Baltic coasts, mixed with amber of other botanical origin.

References

Andrée, K., 1927: Vom "Ostpreussischen Gold." Dem Bernstein, im allgemeinen und von der Klebsschen Bernsteinsammlung und ihrer Bedeutung für Königsberg und die Bernsteinforschung in besonderen. — Jahresber. Königsb. Univ. 1926-27, p.18-36.
— 1929: Bernsteinforschung einst und jetzt. — Bernstein-Forsch. 1, p.I-XXXII.
— 1937a: The scientific importance of amber and recent research in this field. — Research and Progress 3, p.76-83.
— 1937b: Der Bernstein und seine Bedeutung in Natur und Gesteinwissenschaften, Kunst und Kunstgewerbe, Technik, Industrie und Handel. — Königsberg. 219 pp.
— 1939: Über Inklusen im allgemeinen und über Bernsteininklusen und Bernsteininklusenfälschungen im besonderen. — Bernstein-Forsch. 4, p.52-77.
Bachofen-Echt, A.v., 1930: Der Bernstein und seine Einschlüsse. — Verh. zool. bot. Ges. Wien 80, p.(35)-(44).

—   1949: Der Bernstein und seine Einschlüsse. — Wien. 204 pp.

Barry, T.H., 1932: Natural varnish resins. — London. 294 pp.

Beck, C.W., 1972: Aus der Bernstein-Forschung. — Naturwiss. 59, p.294-298.

—   in press: Die Infrarot-Spektren des Bernsteins. — (Archaeo-Physika?), 29 manus.-pp.

Beck, C.W., C.A. Fellows & E. MacKennan, 1975: Nuclear magnetic resonance spectrometry in archaeology. — Adv. Chem. Ser. 138, Archaeol. Chem. 12, p.226-235.

Beck, C.W., E. Wilbur & S. Meret, 1964: Infrared spectra and origin of amber. — Nature 201, p.256-257.

Beck, C.W., E. Wilbur, S. Merit, D. Kossove & K. Kermani, 1966: The infrared spectra of amber and the identification of the Baltic amber. — Archaeometry 9, p.96-109.

Berry, E.W., 1927: The Baltic amber deposits. — Sci. Monthly 24, p.268-278.

Berthelot & Buignet, 1861: Recherches sur le camphre de succin. — Ann. Chim. Phys. (ser.3) 61, p.471-480.

Black, G, 1919: Amber and its origin. — Amer. Mineral. 4, p.83-85, 97-99, 118-120, 130-131.

Boudreau, P.F. & C.S. Schopmeyer, 1958: The inheritance of exudation pressure and viscosity of resin in slash pine. — Pl. Physiol. 33 (suppl.).

Broughton, P.L., 1974: Conceptual frameworks for geographic-botanical affinities of fossil resins. — Canad. J. Earth Sci. 11, p.583-594.

Brüning, E., 1900: Über die Harzbalsame von *Abies canadensis* (L.) Miller, *Picea vulgaris* Link und *Pinus pinaster* Solander. — Inaug.-Dissert., Bern (Publ. with A.Tschirch, Arch. Pharm. 238, p.487-504, 616-630, 630-648).

Carman, R.M., D.E. Cowley & R.A. Marty, 1970: Diterpenoids XXV. Dundathic acid and polycommunic acid. — Austral. J. Chem. 23, p.1655-1665.

Carman, R.M. & W. Craig, 1971a: Diterpenoids XXVII. The selenium dehydrogenation of manool. — Austral. J. Chem. 24, p.361-370.

—   1971b: Diterpenoids XXX. The selenium dehydrogenation of agathic acid. — Austral. J.Chem. 24, p.2379-2388.

C.H.: see Hommerberg.

Cleve-Euler, A. & I. Hessland, 1948: Vorläufige Mitteilung über eine neuentdeckte Tertiärablagerung in Süd-Schweden. — Bull. Geol. Inst. Uppsala 32, p.155-182.

Conwentz, H., 1890: Monographie der baltischen Bernsteinbäume. — Danzig. 151 pp.

—   1896: On English amber and amber generally. — Nat. Sci. 9, p.99-106, 161-167.

Dahms, P., 1914-1921: Mineralogische Untersuchungen über Bernstein. — Schr. Nat. Ges. Danzig (N.F.) 13, p.175-243; 15, p.1-42, 57-68, 78-88.

—   1920: Über rumänischen Bernstein. — Zentralbl. Mineral. 1920, p.102-118.

—   1922: Hohlräume und Wassereinschlüsse im Bernstein. — Zentralbl. Mineral. 1922, p.327-335, 353-363.

—   1925: Über die künstliche Klärung des Bernsteins. — Schr. phys. ökon. Ges. Königsb. 64, p.17-29.

—   1926: Ambra und Bernstein. — Schr. Nat. Ges. Danzig (N.F.) 17, p.15-29.

Durham, J.W., 1956: Insect bearing amber in Indonesia and the Philippine Islands. — Pan. Pac. Entom. 32, p.51-53.

—   1957: Amber through the ages. — Pacif. Discov. 10 (2), p.3-5.

Dyer, C.D., 1963: Naval stores production. — Agr. Ext. Serv. Bull. Univ. Georgia Coll. Agr. 593, Athens. 28 pp.

Eichhoff, H.J. & G. Mischer, 1972: Massenspectrometrische und emissionsspektroskopische Untersuchungen an Bernstein zur Herkunftsbestimmung. — Z. Naturforsch. 27b, p.380-386.

Farrington, O.C., 1923: Amber, its physical properties and geological occurrence. — Field Mus. Nat. Hist. Geol. Chicago 1923, p.25-31.

Frondel, J.W., 1967: X-ray diffraction study of some fossil and modern resins. — Science 155, p.1411-1413.

Goeppert, H.R., 1836: Fossile Pflanzenreste des Eisensandes von Aachen. — N. Acta Acad. C.L.C. Nat. Cur. 19 (2), p.150.

Goeppert, H.R. & G.C. Berendt, 1845: Der Bernstein und die in ihm befindlichen Pflanzenreste der Vorwelt. — In Berendt: Die im Bernstein befindtlichen organischen Reste der Vorwelt I, p.61-126.

Gough, L.J., 1964: Conifer resin constituents. — Chemistry and Industry 1964, p.2059-2060.

Gough, L.J. & J.S. Mills, 1972: The composition of succinite (Baltic amber). — Nature 239, p.527-528.

Haller, A., 1892: Contribution a l'étude des camphols et des camphres. — Ann. Chim. Phys. (ser. 6) 27, p.392-432.

Hampton, B.L., 1956: A new primary resin from *Pinus caribaea*. — J. Org. Chem. 21, p.918.

Harrington, B.J., 1891: On the so-called amber of Cedar Lake, N. Saskatchevan, Canada. — Amer. J. Sci. 42 (3), p.332-338.

Helm, O., 1877: Notizen über die chemische und physikalische Beshaffenheit des Bernsteins. — Reichardt's Archiv der Pharm. 8 (3), p.229.

—   1878a: Über die mikroskopische Beschaffenheit und den Schwefelgehalt des Bernsteins. — Schr. Nat. Ges. Danzig (N.F.) 4 (3), p.209.

—   1878b: Gedanit, ein neues fossiles Harz. — Schr. Nat. Ges. Danzig (N.F.) 4 (3), p.214.

—   1881: Mitteilungen über Bernstein. II. Glessit; III. Über sicilianischen und rumänischen Bernstein. — Schr. Nat. Ges. Danzig (N.F.) 5, p.291-296.

— 1882: Mitteilungen über Bernstein. VI. Über die elementare Zusammensetzung des Ostsee-Bernsteins. — Schr. Nat. Ges. Danzig (N.F.) 5 (3), p.9.

— 1884c: Mitteilungen über Bernstein X. Über blaugefärbten und fluorescirenden Bernstein. — Schr. Nat. Ges. Danzig (N.F.) 6, p.133-134.

— 1884d: Mitteilungen über Bernstein XI. Über knochenfarbigen und bunten Bernstein. — Schr. Nat. Ges. Danzig (N.F.) 6, p.134-138.

— 1891: Mitteilungen über Bernstein XIV. Über Rumanit; XV. Über den Succinit und die ihm verwandten fossilen Harze. — Schr. Nat. Ges. Danzig (N.F.) 7, p.189-208.

— 1896a: Mitteilungen über Bernstein XVII. Über den Gedanit, Succinit und eine Abart des letzteren, den sogenannten Mürber Bernstein. — Schr. Nat. Ges. Danzig (N.F.) 9, p.52-57.

Hommerberg, C., 1947: Bärnstenar. — Malmö. 14 pp.

Howes, F.N. 1949: Vegetable gums and resins. — Chronica Botanica, Waltham, Mass. 188 pp.

Hummel, D., 1958: Kunststoff-, Lack-, und Gummi–Analyse. Chemische und infrarot-spektroskopische Methoden, II. — München.

Jensen, J., 1965: Bernsteinfunde und Bernsteinhandel der jüngeren Bronzezeit Dänemarks. — Acta Archaeol. 36, p.43-86.

Joye, N.M. & R.V. Lawrence, 1963: The isolation of a new diterpene acid. — J. org. Chem. 28, p.3274.

— 1967: Resin acid composition of pine oleoresins. — J. Chem. Engin. Data 12, p.279-282.

Joye, N.M., E.M. Roberts, R.V. Lawrence, L.J. Gough, M.D. Soffer & O. Korman, 1965: The structure of the dicyclic diterpenoids of slash pine. The identity of elliotinoic acid and communic acid. — J. org. Chem. 30, p.429.

Kaunhowen, F., 1913: Der Bernstein in Ostpreussen. — Jahrb. Kgl. Preuss. Geol. Landesanst. 34 (27), p.1-80.

Kirchner, G., 1944: Korallen im Bernstein. — Umschau Wiss. Techn. 48, p.113-115.

— 1950: Amber inclusions. — Endeavour 9 (34), p.70-75.

Klebs, R., 1880: Der Bernstein. Seine Gewinnung, Geschichte und geologische Bedeutung. — Königsberg.

— 1882: Der Bernsteinschmuck der Steinzeit von der Baggerei bei Schwartzort und anderen Lokalitäten Preussens. — Beitr. Naturkd. Preuss. 5, Königsb. 75 pp.

— 1888: Über die Farbe und Imitation des Bernsteins. — Schr. phys. ökon. Ges. Königsb. 28, p.20-25.

— 1890a: Aufstellung und Katalog des Bern-

stein-Museums von Santien und Becker, Königsberg Pr., nebst einer kurzen Geschichte des Bernsteins. — Königsberg.

Koritschoner, F., 1902: Untersuchungen über das Russische Pech und das Harz von *Pinus palustris*. — Inaug.-Dissert., Wien (publ. with A. Tschirch, Arch. Pharm. 240, p.568-584).

Langenheim, J.H., 1967: Preliminary investigations of *Hymenaea curbaril* as a resin producer. — J. Arnold Arbor. 48(3), p.203-227.

— 1969: Amber: A botanical inquiry. Amber provides an evolutionary framework for interdisciplinary studies of resin-secreting plants. — Science 163, p.1157-1169.

Langenheim, J.H. & C.W. Beck, 1965: Infrared spectra as a means of determining botanical source of amber. — Science 149, p.52-55.

— 1968: Catalog of infrared spectra of fossil resins (ambers) in North and South America. — Bot. Mus. Leaflets, Harv. Univ. 22 (3), p.65-120.

Langenheim, R.L., J.D. Buddhue & G. Jelinek, 1965: Age and occurrence of the fossil resins Bacalite, Kansasite, and Jelinite. — J. Paleont. 39 (2), p.283-287.

Langenheim, R.L., C.J. Smiley & J. Gray, 1960: Cretaceous amber from the arctic coastal plain of Alaska. — Bull. Geol. Soc. Amer. 71, p.1345-1356.

Lee, Y.-T. & J.H. Langenheim, 1975: Systematics of the genus *Hymenaea* L. (Leguminosae, Caesalpinioideae, Detarieae). — Univ. Calif. Publ. Bot. 69, 110 pp.

McAlpine, J.F. & J.E.H. Martin, 1969a: Canadian amber. — The Beaver, 1969, p.28-37.

— 1969b: Canadian amber — a paleontological treasure-chest. — Canad. Entom. 101, p.819-838.

Mills, J.S., 1973: Diterpenes of *Larix* oleoresins. — Phytochemistry 12, p.2407-2412.

Mischer, G., H.-J. Eichhoff & T.E. Haevernick, 1970: Herkunftsuntersuchungen an Bernstein mit physikalischen Analysenmethoden. — Jahrb. Röm.-Germ. Zentralmus. Mainz 17, p.111-122.

Mischer, G. & R.C.A. Rottländer, 1971: Diabietinsäure durch UV-Bestrahlung von Kolophonium. — Tetrahedron Letters, no. 17, p.1295-1298.

Moenke, H., 1962: Spectralanalyse von Mineralien und Gesteinen. — Leipzig.

Niederstadt, B., 1901: Über den neuseeländischen Kauri-Busch-Copal von *Dammara australis* und über das Harz von *Pinus silvestris*. — Inaug.-Dissert., Berlin (publ. with A. Tschirch, Arch. Pharm., 239, p.145-181.

Petrunkevitch, A., 1958: Amber spiders in European collections. — Trans. Conn. Acad. Arts Sci. 41, p.41-400.

Pieszczek, E., 1880: Über einige neue harzähnliche Fossilien des ostpreussischen Samlandes. — Reichardt's Archiv der Pharmacie, 14 (6), p.433.

Rottländer, R.C.A., 1969a: Dehydroabieten durch UV-Bestrahlung von Harzsäuren. — Tetrahedron Letters, no.47, p.4127-4128.
— 1969b: Bernstein durch Dimerisierung von Abietinsäure. — Tetrahedron Letters, no.47, p.4129-4130.
— 1970: Identifizierung von Ir-Banden des Succinits und einiger Derivate. — Tetrahedron Letters, no.24, p.2127-2128.
— 1974: Die Chemie des Bernsteins. — Chemie unserer Zeit 8, p.78-83.
Rottländer, R.C.A. & G. Mischer, 1970: Chemische Untersuchungen an libanesischem Unterkreidebernstein. — N. Jb. Geol. Paläont. Monatsh. 1970 (11), p.668.
Santamour, F.S., 1967: Notes on pine resins and pest resistance. — Morris Arboretum Bull. 18, p.82-86.
Schlee, D., 1970: Insektenfossilien aus der unteren Kreide. — 1. Verwandtschaftsforschung an fossilen und rezenten Aleyrodina (Insecta, Hemiptera). — Stuttg. Beitr. Naturkd. 213, 72 pp.
Schmid, L., 1939: Im Bernstein vorkommende Säuren. — Monatsh. Chem. 72, p.311-321.
Schmid, L. & A. Erdös, 1933: Chemische Untersuchung des Bernsteins. — Liebigs Ann. Chem. 503, p.269-276.
Schmid, L. & W. Hosse, 1939: Über Bernstein. — Monatsh. Chem. 72, p.290-302.
Schmid, L. & H. Körperth, 1935: Über Bernstein. — Monatsh. Chem. 65, p.348-350.
Schmid, L. & F. Tadros, 1933: Chemische Untersuchung des Bernsteins. — Monatsh. Chem. 63, p.210-212.
Schmidt, G., 1903: Ueber den Harzbalsam von *Pinus laricio* (oesterreichischer Terpentin). — Inaug.-Dissert., Bern (publ. with A. Tschirch, Arch. Pharm. 241, p.570-588).
Schubert, K., 1939: Mikroskopische Untersuchungen pflanzlicher Einschlüsse des Bernsteins. I.Holz. — Bernstein-Forsch. 4, p.23-44.
— 1961: Neue untersuchungen über Bau und leben der Bernsteinkiefern (*Pinus succinifera* (Conw.) emend.). — Beih. Geol. Jahrb. 45, Hannover, 143 pp.

Snow, A.G., 1949: Research on the improvement of turpentine practices. — Econ. Bot.3, p.375-394.
— 1954: Progress in development of efficient turpentining methods. — U.S. Dep. Agr. Forest Serv. Southeast. Forest Exp. Sta. Pap. 32.
Squillace, A.E., 1966: Planning tree improvement research of Olustee. — Amer. Turpent. Farm. Ass. Coop. J.28 (5), p.11-13.
Squillace, A.E. & K.W. Dorman, 1961: Selective breeding of slash pine for high oleoresin yield and other characters. — Rec. Adv. Bot. Univ. Toronto Press 2, p.1616-1621.
Thomas, B.R., 1966: The chemistry of the order Araucariales. Part 4. The bled resins of *Agathis australis*. — Acta Chem. Scand. 20, p.1074-1081.
— 1969: Kauri resins — modern and fossil. — Organic Geochemistry (ed. G.Eglinton and M.T.J. Murphy, Springer, Berlin), ch.25, p.599-618.
— 1970: Modern and fossil plant resins. — Phytochem. Phylogeny (ed. J.B. Harborne, Academic Press, London), ch.4, p.59-79.
Tornquist, A., 1910a: Geologie von Ostpreussen. — Berlin.
— 1910b: Die in der königlichen Universitäts-Bernsteinsammlung eingeführte Konservierungsmethode für Bernsteineinschlüsse. — Schr. phys. ökon. Ges. Königsb. 51, p.243-247.

Unverdorben, O., 1827: Ueber die Harze. — Ann. Physik. Chem. 11, p.27-59.
Weigel, G., 1900: Über die Harzbalsame von *Larix decidua* und *Abies pectinata*. — Inaug.-Dissert., Leipzig, (publ. with A. Tschirch, Arch. Pharm. 238, p.387-410, 411-427).
Werlauff, E.C., Bidrag til den nordiske Ravhandels Historie. — Vid. Selsk. hist. phil. Skr. 5, p.150-316. København.
Zinkel, D.F. & B.B. Evans, 1972: Terpenoids of *Pinus strobus* cortex tissue. — Phytochemistry 11, p.3387-3389.

The Origin of Baltic Amber

During the greater part of the Eocene, the land masses of Northern and Central Europe were relatively quiescent, while Southern Europe, in fact the Mediterranian regions as a whole, were in continual movement, a "wagging" up and down of changing land masses. Just as it does today, the North Sea (Text-fig. 4) lay between England and the European continent, but to the south, the Netherlands and at least the western part of North Germany were submersed, and the same was the case for almost the whole of Denmark. The Scandinavian continent stretched to the most northerly tip of Jutland, and the Kattegat did not have its present appearance. How far the "Baltic Bay" of the North Sea extended eastward is uncertain; this has probably varied. The eastern boundary of the European continent was the Ural Mountains, and immediately east of these the Sea of Ob extended from the Arctic Ocean to Tethys. Extensive regions of Southern Russia comprising both the Black Sea and the Caspian were under water. The ancient Central European mountains, the Hercynian mountains, constituted a permanent part of the central regions of the continent.

There is no doubt that the European continent has been covered with forest wherever this could grow, and under the conditions which then existed this will have been wherever the climate was not too dry. Nor can it be doubted that the resin-producing plants have spread just as extensively. Theoretically, therefore, one might expect to be able to find amber everywhere within this area. There were several factors, however, which hindered this from being the case. First, there is no type of amber which is particularly resistant to vitrification of various kinds. As a result, amber from the Mesozoic and the older Tertiary is never found nor will in the future be found in its primary deposits in the original forest floor, but always where it has been deposited after transport by streams, in sites where the water-table or the sea has provided a quite special protection against oxidation and drying. As a result, only those resins with a relatively great power of resistance, and which are secreted in particularly great amounts, have had a sufficiently high capacity for survival until now. Adequately large amounts of resin are secreted at the present day only by plants growing in the tropics or subtropics, this being the case for both deciduous and non-deciduous trees, and it is most likely that this has always been the case.

As the preservation of amber has been dependent on water transport, nothing precise can be said about the position of the site of production, based on the site at which it was found. Some amber has had a prolonged transport, other amber has had a brief transport, and in many cases the transport has been repeated again and again, so that finally, the amber has been deposited in sites which are far younger than itself. Typically the amber is deposited in bays with weak streams, in deltas or at the mouth of rivers, in other words in a continental coast zone; here, the light material is given the chance of sinking to the bottom. At this site there may be a century-long collection of material within a narrowly delimited region, so that the impression is one of a disproportionately rich production. Part of the amber, however, probably most of it, is carried out to sea where, as in the case of the tertiary clay at Åhus, Skåne (Cleve-Euler & Hessland, 1948, p.157), for example, it may be common as microscopic particles, but where in the form of larger pieces (as for example in the North Jutland Mo-clay) it is widely scattered and apparently a rarity. Where amber is produced in regions near a continental water-shed, it may be due to chance whether it is recovered on the south or the north coast of the old territory. The production from quite a small region may be spread over very great distances.

Pretertiary ambers

The Baltic amber is no isolated phenomenon, but part of a world-wide picture (Text-fig. 6), but undoubtedly that part we know most about. As early as the coal deposits of the Carboniferous, an amber-like mineral occurs, e.g. in England, but this is considered to be a formation arising from pollen which has been exposed to a high temperature and pressure, and which is therefore not a resin derivative. Particularly in the younger Palaeozoic, the extinct plant group Cordaitales was widely distri-

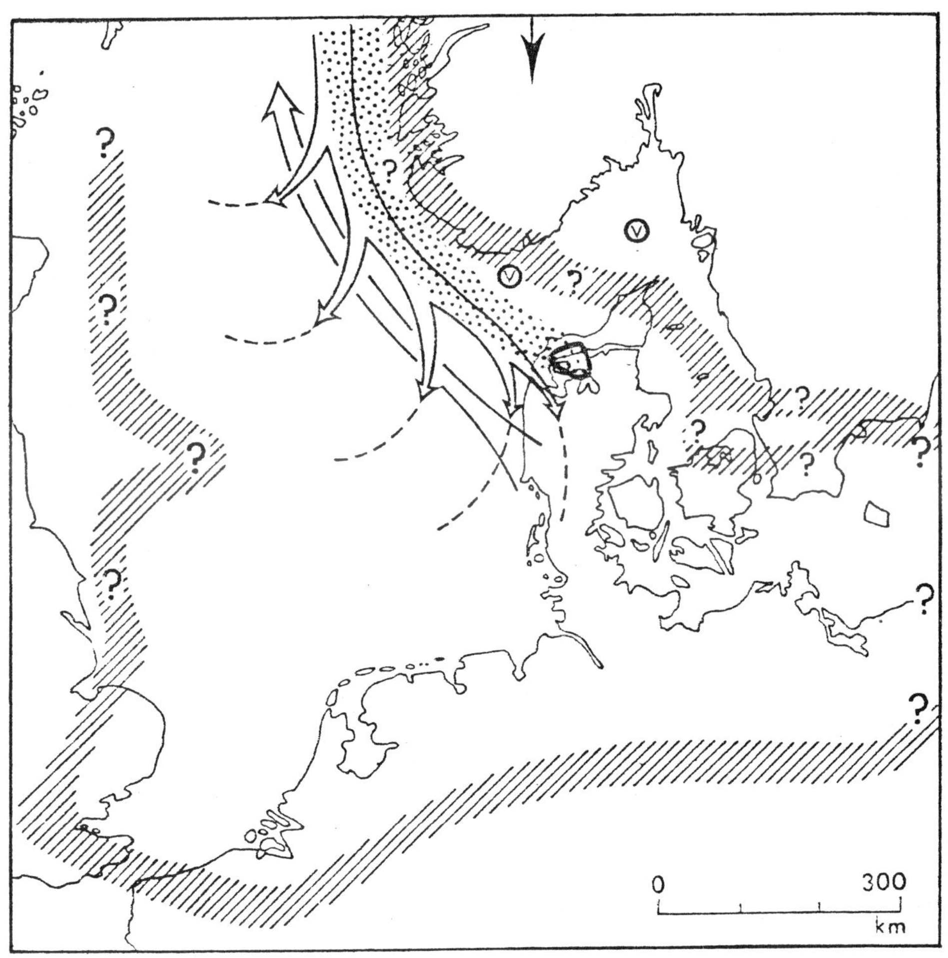

Text-fig.4. Model of the palaeogeography of the North Sea and western Baltic in Early Eocene when the Mo-clay was deposited. The approximate position of the coast line is shaded. The Mo-clay area in Jutland is encircled by a solid line. Encircled letters "V" indicate possible position of volcanic centers. Arrows and stipling lines mark supposed dominating winds and currents. (Bonde, 1974).

buted, and anatomical features in the fossil wood suggest that at any rate some of them might have been resin-producers.

In the Mesozoic and Palaeogenic, the Protopinaceae together with *Protodammara* and other very primitive Araucariaceae were widely distributed over the entire Holarctic region. The determination of the wood of these plants has often caused palaeobotanists great difficulties, sometimes involving considerable uncertainty; not least, it has in many cases been difficult to distinguish them from Cordaitales. The first more or less definite evidence of the development of specialized resin-producing plant-cells is known from the Triassic, from the petrified bark of *Araucarioxylon arizonicum* Knowlton (Langenheim, 1969, p.1161), referred by Kräusel (1919, p.189) to the lower Cretaceous or upper Jurassic. This plant was placed by Kräusel in the Protopinaceae. After that period, resin-secreting plants were general, at first exclusively conifers, later also angiosperms in steadily increasing numbers.

In a newspaper note (Harald Steinert, St. Galler Tagblatt, 7.November 1976) is mentioned a century-old find of three diminutive pieces of amber from the Keuper formation (Triassic) in Switzerland. The dating of this find can probably be correct, as resin-producing plants have existed in these surroundings at that time.

The oldest with certainty known amber is from the Jurassic (about 160 million years old), and comes from Bornholm, Denmark (Text-fig. 6) (Langenheim, 1969, p.1160, personal communication from B. Eske Koch, Denmark). It is found in insignificant amounts, and no fossils are known from it. Nor has infrared spectroscopy been performed, but it probably originates from a conifer. The primitive araucariacean *Pagiophyllum* was found on Bornholm at the same time (Seward & Ford, 1906, p.377), and it is possible — admittedly only theoretically so — that this might be the mother plant.

The transition between the Jurassic and the Cretaceous (about 130 million years ago) is represented by "Libanesischer Bernstein" (Schlee & Dietrich, 1970). Also apart from its age, this is a very remarkable find, because it contains Mesozoic insect fauna from the coast regions of the Tethys Sea, i.e., either from the south coast of Laurasia or from the tropical north coast of Gondwanaland, Schlee & Dietrich have most confidence in the latter view. The analysis of this material is only starting, but the descriptions which are available so far of two insect species belonging to the Aleuroidea (Schlee, 1970) promise much for the future. It appears as if phytophagous insects, like the majority of Hemiptera, were still at the commencement of that violent development which was encouraged by the flourishing of the angiosperms, and which practically speaking terminated in the Baltic amber 80—90 million years later. Botanically, the Lebanesian amber (diterpene-type I) is referred to the Araucariaceae (Rottländer & Mischer, 1971, according to Beck, 1972, p.297).

Many amber deposits from the Cretaceous are known in North America, Siberia and Europe (Langenheim, 1969,

Period			Million years passed
Cenozoic	Quaternary	Present	2-3
		Glacial	
	Tertiary (Neogene)	Pliocene	12
		Miocene	23
	Tertiary (Palaeogene)	Oligocene	35
		Eocene	55
		Palaeocene	70
Mesozoic	Cretaceous		136
	Jurassic		200
	Triassic		225
Palaeozoic	Permian		280
	Carbonic		345

Text-fig.5. Division of the earth's younger history.

p.1160; McAlpine & Martin, 1969b; Jerichin & Sukatscheva, 1971; Schlüter, 1975). Only a few of them belong with certainty to the lower Cretaceous, and a number of them are not more closely dated within the period. Among the oldest ones are some European fossiliferous localities from north-western France, Older Cenoman, about 100 million years old (Schlüter, 1975, p.151). Jerichin & Sukatscheva (1971, p.3) published amber from the Coniac period from northern Middle Siberia (about 85 million years old). In America, the most significant localities lie in the Alaskan Arctic Coastal Plain, Ceder Lake in Manitoba and various sites in the region of the Atlantic Coastal Plain (from Massachusetts to Maryland). Many fossils are known, most of them from the Cedar Lake region, but the study of the material is far from completed (McAlpine & Martin, 1969b). Most of the European findings are associated with the Alpine regions (Switzerland, Austria and Hungary). They probably originate from forests which have grown in the Hercynian region, and are deposited in the southern coastal region, which has since been elevated and become part of the Alps. There are no fossils and only few definite analyses (Mischer et al., 1970) from this region, but for example the protopinacean *Araucarioxylon* sp. Jacobsohn has grown in the region (Kräusel, 1919, p.190). There are

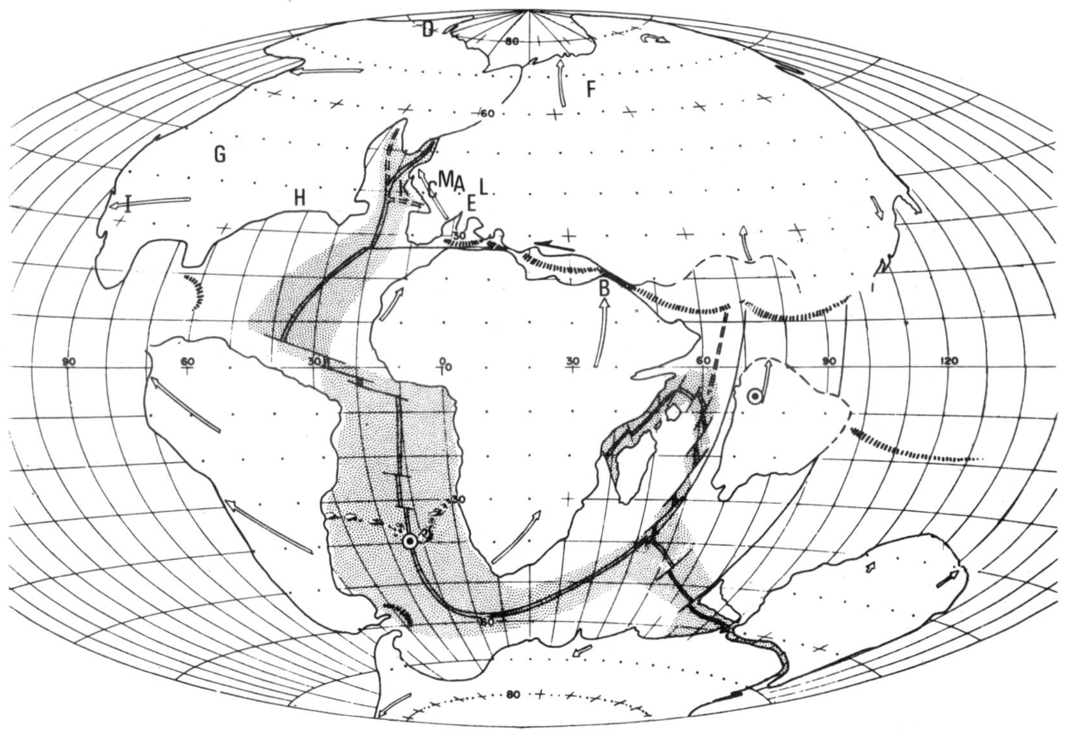

Text-fig.6. Position of the continents at the end of the Cretaceous, with locations of ambers, as old as, or older than, the Baltic amber. — A: Bornholm (Jurassic); B: Lebanon (Earliest Cretaceous); C: Northwestern France (Early Cretaceous); D: Alaskan Arctic Coastal Plain (Early Cretaceous); E: Central Europe (Cretaceous of different ages); F: Central North Siberia (Coniac, Cretaceous); G: Cedar Lake, Manitoba, Canada (Late Cretaceous); H: Atlantic Coastal Plain (Late Cretaceous); I: Simi Valley, California (Eocene); K: Southeast Coast of England (Eocene); L: Baltic Coast (Eo-Oligocene); M: Rhine Valley (Eo-Miocene). The English amber seems, however, to have a double origin, partly being Baltic. (Based on map in Dietz & Holden, 1970b).

spectroscopic studies from most North American localities (Langenheim & Beck, 1968), and there are grounds for believing that the Taxodiaceae have been dominating resin-producing plants. Amber containing succinic acid is known, however, mostly from the Atlantic Coastal Plain, but also to some extent from Cedar Lake (presumably from Protopinaceae and primitive Araucariaceae). Finds of amber are also reported, however, originating from *Liquidambar* (Hamamelidaceae) (Langenheim, 1969, p.1160).

In the **Tertiary,** the number of amber deposits is found to increase quite considerably everywhere, and it is now that the angiosperms acquire their great significance alongside the primitive Pinaceae (of diterpene-type I) and the Araucariaceae. Also amber from Taxodiaceae plays a considerable role. On the other hand, amber of typical pinacean character is completely lacking, both here as well as in all older deposits. The Baltic amber originates from the beginning of this period. Fossils are also found in Burma amber and Sicilian amber. Great interest is attached to the rich deposits of "Chiapas amber" (Mexico) from the Oligo-Miocene, as it belongs to a region where the climate has more or less remained unchanged since, and where the source plant *Hymenaea* is still active as a resin producer.

Amber from northern Jutland

The oldest Tertiary amber known within the Danish region originates from the earliest Eocene. Only quite a few pieces are concerned, found deposited in cementstone in the North Jutland Mo-clay (Text-fig. 7), in the one case found placed naturally in a piece of fossil wood, in another case together with typical insect fossils of the Mo-clay, in both these cases within the older layers of the formation. A third specimen is mentioned by Palmgren (1941, p.1).

Mo-clay is a marine deposit, consisting almost exclusively of diatomaceous shells. Geologically it has been described by Bøggild (1918), Gry (1940 and 1965) and Bonde (1973 and 1974). It is very rich in fossils, in particular large amounts of fish (Bonde, 1966), and in quite definite layers there are strikingly many land plants and fresh-water plants and insects (Andersen, 1947; Henriksen 1922, 1929; Koch, 1960; Heie, 1967, 1970a, 1970b, 1976; Larsson, 1975). The plant fossils are evidence of a double origin; some belong to a river-bank flora from quiet streams and estuaries or from tidal estuaries, apparently lowlands near coasts, while other plant material, a forest flora, carries traces of long and hard transport. Mo-clay amber has apparently the same origin as this wood flora, which must have constituted the most western part of the Baltic amber forests.

At the time (Text-fig. 4) the Mo-clay was deposited, most of Denmark was a part of the North Sea, perhaps with the exception of the most northerly and most easterly regions, which together with the drained areas of the Kattegat have constituted part of the South Scandinavian territory, and the sea floor has probably sloped rather steadily WSW from this coast. It is exceedingly likely that both the fossil land insects and plants, together with the contemporary amber, originate from near Scandinavia, transported respectively through the air and by the river system.

It might be considered that these amber finds were of no significance on account of their scarcity, but this would be incorrect. Where amber has collected in an amber-twig-layer, this is the result of many years of concentration of material within strongly limited regions. Where it, forced by the rivers, pass these collecting sites out into the open sea, there will be a further spread of the already scanty material, and the finds will to an even lesser degree correspond to the actual amounts. It is possible that the Mo-clay amber should even be considered as evidence that in the region along the then coastline of North Jutland there is or has been one or several amber-twig-layers with a richer content.

Other factors support such an argument. Along the west coast of the Kattegat, quite considerable amounts of amber are collected on the stretch from Sæby to Randers Fjord. Much is found at the estuary of the Limfjord, especially when the entrance at Hals Barre has just been dredged, and much is also collected at Læsø. In addition, numerous pieces of amber are ploughed up or dug up from the ground near the coast (personal communication from the goldsmith A. Henningsen, Thisted). The conditions at Hals Barre in particular, but also the whole local profusion of the amber, appear in my opinion to suggest the possibility that there might be hidden amber-twig-layers in these regions, and that the incidence of deposits in the superficial layers on land is due to subsequent secondary transport. It is exceedingly difficult to conceive this amber as having been transported from East Prussia, but against the background of the Mo-clay amber, the idea seems obvious to me that it may have been brought by the same routes as the Mo-clay amber and collected in an amber-twig-layer. Should these suppositions be correct, the underwater layers containing amber must be from Eocene, perhaps older. At any rate they cannot be younger than the content of the excavations in the Samland Peninsula.

Professor Curt W. Beck, Vassar College, U.S.A., has made infra-red spectra of the first-mentioned specimen of these Mo-clay ambers, and the present author is very grateful for the following information. Professor Beck concludes (letter of

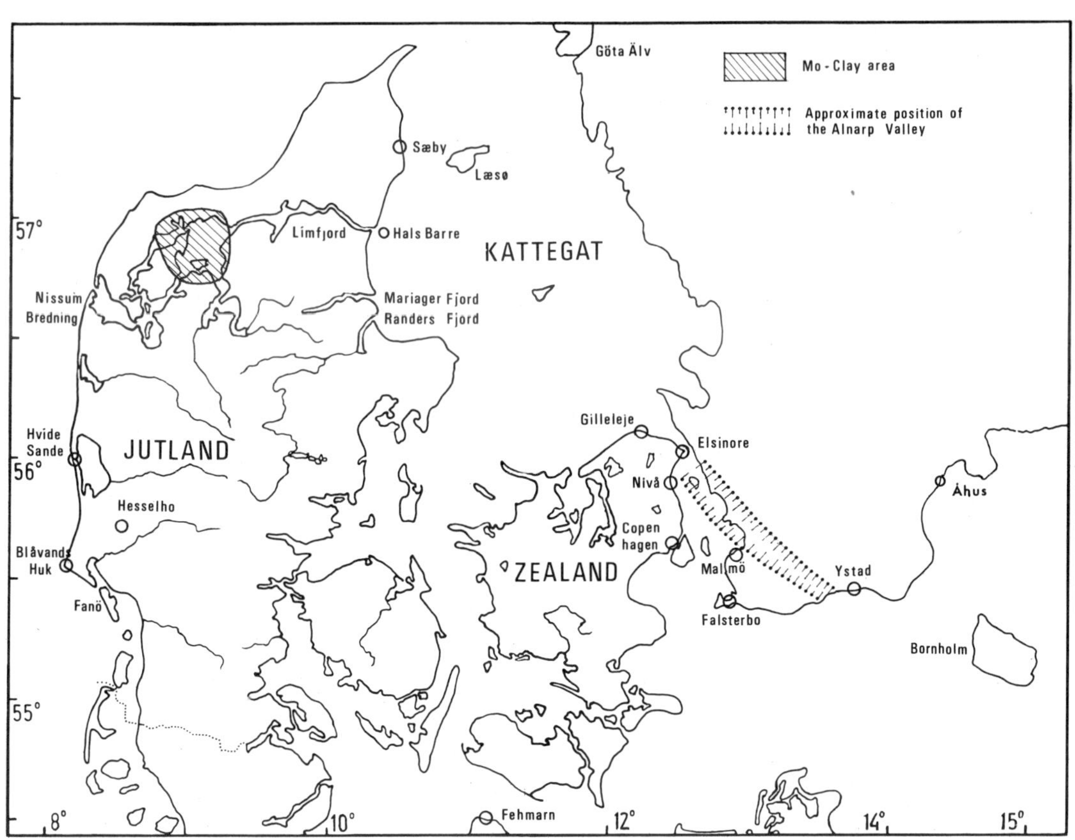

Text-fig.7. Recent coast-lines in Denmark and southwestern Sweden. Localities mentioned in the text are mapped.

May 18, 1975): "I have run two infra-red spectra on separate samples. The two spectra show excellent reproducibi-lity, i.e. they are identical except for small variations in intensities and resolu-tion.

From these spectra I can say with cer-tainty that the resin of the Mo-clay is not succinite or so-called Baltic amber. However, it is much more difficult to say what the resin is, than what it is not. The spectra agree quite well with those of the large and vague group of resins that have been called retinites, in Dana's broad sense of that term. Free carboxy-lic acid groups are indicated by a weak absorption near 7.9 microns; this is characteristically absent in Baltic amber, which is an ester resin rather than an acid resin.

It is impossible to make any assignment of a botanical nature on this evidence, except to say that the Mo-clay resin was *not* formed by the same trees which pro-duced Baltic amber. Retinites have been assigned to a wide variety of botanical sources, often on insufficient evidence. Still, most of them seem to derive from conifers fairly closely related to modern European pines of the sub-genus *Diploxy-lon*, while Baltic amber might have come from the *Haploxylon* sub-genus, if in-deed, it derives from a *Pinus* species at all, which is now being doubted by an in-creasing number of workers (e.g. Gough & Mills, 1972; Thomas, 1970)."

This investigation indicates that the problems of the botanical taxonomy of the amber forest are far from finally solved, at least as far as the western Bal-tic areas are concerned.

In addition to the above, a considerable number of amber specimens have been found in Jutland from younger Tertiary deposits. A single specimen from Mariager Fjord (Nørregaard, 1903, p.67) can be mentioned as an example, though some uncertainty is attached to this. It is said to be succinite and is reported to have been found in plastic clay of Middle Oligocene origin. The specimen may be assumed to have reached Jutland along the same route as the Mo-clay material, that is, from Southern Scandinavia. However, as formation of amber resin cannot have taken place in Scandinavia since the earliest commencement of the Oligocene, on biological grounds, this specimen has presumably been redeposit-ed. The mother plant has not been de-termined, but as the specimen is de-scribed as succinite, it presumably ori-ginates from the traditional succinite-producing amber tree, *Pinites succinifer*.

Another amber specimen reported by Nørregaard in the same communication was found in Hesselho brick-field north of Varde in western Jutland (see Text-fig. 7). This is a translucent retinite with resinous and conchoidal fractures, and in general deviating from all other amber found at that time in Northern Europe. The specimen was found in upper Mio-cene micaceous clay. The mother plant is unknown, it has not been the usual amber tree of the Baltic region. The Hesselho amber specimen must have been redeposited more than once. It may have reached its final deposit from South Scandinavia in the same way as the Mariager amber specimen. However, the possibility cannot be excluded that it

Plate 3.
A. Cicada of the superfamily Fulgoroidea, seen from above. The Fulgoroidea are relatively common in amber, especially among the Mo-clay fossils. (G. Brovad phot.).
B. Cicadas of the family Issidae, seen from below. An adult and a larva, probably belonging to the same species. (G. Brovad phot.).
C. Stone fly (Plecoptera), probably belonging to the family Leuctridae.
D. Larva of a case-building caddis fly (Leptoceridae). The larva was obviously primarily caught by its shelter; when it left this the animal drowned in the flow of semifluid resin. The body's white appearance shows that the skin was humid when the larva was trapped. (G. Brovad phot.).

A
B
C
D

A
B
C

originates from a tree of Central European origin, and in that case it could have been produced somewhat later than the Eocene.

For Mid and West Jutland, the Miocene was a very disturbed period, where the coast line moved backward and forward; large areas were alternatingly shallow floodplains and coastal swamps, to some extent with forests. The rivers brought enormous amounts of clay, sand and gravel from Scandinavia, presumably via the same river systems which had brought both the Mo-clay amber and the Mariager amber here. The vegetation of the swamp and swamp forest became the basis for the brown coal later formed in Mid and West Jutland; this was a very variegated transitional flora with both boreal and mediterranean features. Quite considerable amounts of amber have been found in the lignite deposits, including quite large pieces (personal communication from amber grinder J. Flauensgaard, Copenhagen). This has not been examined spectroscopically, but at any rate if it contains any succinite this cannot have been formed at the same time and on the spot, but must have been redeposited. Also this amber may originate from the South Swedish forests of the Eocene, eroded time and again from earlier concentrations by the river systems during the changing coast profile, and once more deposited near new coastlines and with greater geographic spread.

If there are grounds for believing that the Miocene lignite amber originates from the South Swedish Eocene forests, there is in my opinion no less reason to believe that this is the case with the coastal amber of West Jutland. Amber is found here (see Text-fig. 7) in good amounts from Fanø to Nissum Bredning, but by far the greatest part is collected on the stretch Blåvandshuk-Hvidesande (personal communication from the goldsmith A. Henningsen, Thisted). Here, amber is washed out constantly from the tidal zone, and it must originate from hidden depots in the Miocene or later regions.

There is much to suggest that the early Tertiary forerunner of Götaelven and possible parallel streams have brought much amber from Sweden over the more or less drained Kattegat and down over the young and steadily enlarging Jutland. In the shifting coastal zones, concentrations must have been formed whose contents have been partly redeposited over and over again, and which can now be seen in the most distant amber deposits along the west coast of Jutland.

Perhaps not without reason, the Limfjord district was a region which became known very early on the international amber market, probably even before the East Prussian (Bachofen-Echt, 1930).

According to these ideas the amber in northern Jutland should have its origin in forests which have grown in South Sweden, but of necessity only from the western side of the South Scandinavian water-shed. However, amber found in quantity in the large Swedish lake Mälaren at Stockholm most probably has also originated from these Scandinavian amber forests, but secreted from trees on the eastern side of the same water-shed. Neither of these two amber materials appears to have come from Samland or any other eastern locality.

The Alnarp Amber

Much of the Danish amber, however, is found in far younger geological formations; it has been transported here in the

Plate 4.
A. Two caddis flies (Trichoptera), from ventral and right lateral views. (G. Brovad phot.).
B. Male of a Chironomid midge. Note the thoracic profile, which is characteristic of the family. (G. Brovad phot.).
C. Moth-fly (Psychodidae), a well known little midge from present-day's toilets. (G. Brovad phot.).

33

course of the Ice Age or later, but has originated from earlier concentrations. At any rate, by far the greater part of the amber which can be collected along the coast of the Øresund, the specimens which can be collected in the ground in Copenhagen and environs and certain parts of north-eastern Zealand, originate from the Alnarp Valley. This prehistoric valley was regarded by N.O. Holst (1911), who described it, as a preglacial river, the lower outlet of the Vistula (Weichsel). According to V. Milthers (1935) it is a graben, and the material deposited here is ascribed by this investigator, with some reservation, to the last but one interglacial period (ibid., p.50). Most recently, the age of these deposits has been determined by K. Nilsson (1973, p.6) as being about 32,000 — 37,000 years, using C^{14}-measurements, and they must be assumed to originate from interstadials within the Vistula ice age. In what follows, most of the details regarding this valley are quoted from Holst, but corrected according to Milthers and Nilsson. Knowledge of the valley has been obtained from a large number of borings, most of them carried out in connection with the water supplies for Malmö.

Today, the Alnarp Valley (Text-fig. 7) cannot be immediately recognized in the landscape of Skåne, for in addition to the deposits with which it is filled, it is hidden under an approximately 35 metre thick postglacial layer of earth, the upper surface of which is on a level with the surroundings. Under this layer lies the old valley, which has great dimensions. The floor together with the often steeply sloping sides are of chalk, and the depression in the chalk layer is on the average about 40 m deep. In this extensive ditch, material is deposited consisting mainly of quartz sand, generally very fine in the upper layers and coarser deeper down. In the floor of the valley there is a layer of cemented pebbles, mainly of limestone and flint from the existing underground; most of the other stones originate from the bed-rock. Some of these pebbles show

that the Alnarp Valley must have had an inflow from northern districts (Holst, 1911, p.11). Within the sand there are local deposits of clay, some of them of quite considerable thickness, others quite thin, corresponding to the changing rate of flow of a stream and to local bays and inlets.

One characteristic of the Alnarp Valley is its high content of plant remains, as well as a number of fossil molluscs and a few remains of beetles. What has in general attracted most attention, however, is its rich content of amber. Amber and fossils are found in all layers of the depression, but according to Holst are especially common at a depth of 45-65 m below the present surface.

It is mainly from the surface and from the west bank of this "amber river" that the last inland ice has obtained the material which it has carried further to Denmark. For a large part, this has consisted of enormous blocks, which in their solid frozen state have been carried in coherent blocks over to the east coast of Zealand, where they today form part of the foundations of Copenhagen. It is mainly the amber from this which is found in the Copenhagen harbour excavations and other large excavations, and which can be found washed up on the adjacent coasts of Køge Bugt.

The Alnarp Valley in Skåne has a width of about 5 km; it is narrowest, and its floor lies highest, furthest south near the small coastal town of Skifarp, a little west of Ystad. From here it runs to the north-west in between Malmö and Lund, takes in the west coast of Skåne over a long stretch to the north and extends to the island of Hven. At some underwater parts of this coast, the deposits are either lying partly free or very superficially, and it may be assumed that it is to some extent from here that the relatively large amounts of amber originate which are collected around Falsterbo in the south-west corner of Sweden, brought here by the southward, relatively salt undercurrent of the Sound. It is probable that also the Danish coast obtains some of its amber from here, especially around Copenhagen and in Nivå Bugt. The valley continues over Zealand from Humlebæk to Gilleleje, south-west of Elsinore. According to Milthers (1935, p.39), there seems to be some justification for believing that this long valley has been established at the transition from the Tertiary to the Quaternary, or during the early Quaternary.

The plant and animal remains found in the Alnarp Valley suggest a climate which can hardly have differed much from that of the present. Among the conifers mentioned in the older plant lists are spruce, pine and *Sequoia*. In addition, Sarauw (1897) mentions *Pinus succinifera*, but this is highly unlikely, as this tree must have long since ceased to exist, at any rate so far north. Among the deciduous trees mentioned are alder, birch, oak, hazel, willow and elder. Among the herbaceous plants, fresh water plants are very common, and many of them belong to submersed species. In addition there are a number of mosses and ordinary meadow plants. Molluscs are generally recognized as now-living fresh-water genera: *Limnaea*, *Planorbis*, *Succineum* and *Pisidium*. Among beetles, *Pterostichus* is specially common. It is remarkable that both the *Otiorrhynchus* species which are widespread today in the East Atlantic region at high latitudes, *O. arcticus* and *O. dubius (maurus)*, are the only weevils mentioned from these deposits. Most of this plant and animal life is obviously far younger than the amber; it is pronouncedly boreal and originates from a climatic region which in the "amber period" prevailed much further · north of the amber forest.

The question then is where did the Alnarp Valley obtain its water, and thereby its amber. Holst (1911, p.54) also believes that he has the solution to this, being convinced that all amber originates from the earlier East Prussia, especially from the very large amber deposits in the Samland Peninsula. He imagines that the Vistula, in the course of an upheaval, which he describes in more detail, has had a course generally

corresponding to the present, and that from its present outlet it has continued out over the drained Baltic northwards, where it has joined up with the main stream coming from the east. During this course, the Vistula is assumed to have passed close to the Samland Peninsula and received its amber there. It is also considered to have received water from the Oder and possibly other North German river systems. The existence of this enormous river system for the drainage of very large areas of the eastern parts of North and Central Europe has later been denied, and has been the subject of much debate.

Quite recently, Nilsson (1973, p.6) has commented on this: "The probable environment producing this picture is a large river with a shallow sloping profile and a low rate of flow. With this point of view, we can return to Holst's theory on the Alnarp river with the exception that the river is not preglacial and that it has not produced the Alnarp Valley by erosion.

As Holst pointed out, a river of this kind must have had a large drainage area. As far as we can see now, it is clear that the floor of the Alnarp Valley, at about 60 m below sea level, has presented the lowest point of passage to the Atlantic for the waters around the Southern Baltic during the interstadial period in question. It seems reasonable therefore, with Holst, to assume that the Alnarp river then drained at any rate the southern region of the Baltic, including such rivers as the Vistula."

However, some of the pebbles from the valley floor and the fossil *Otiorrhynchus* species mentioned seem to indicate that the gigantic river system has received part of its water also from the north, probably from the melting Scandinavian glaciers.

The South Swedish coast has probably received much of its amber by this Vistula route. However, considerable amounts of microscopic or even very small amber particles are found in the Tertiary clay at Åhus, Skåne (Cleve-Euler & Hessland, 1948). This is an

Eocene layer which these authors consider as sediments, deposited near the coast. The amber in this must therefore under these circumstances be regarded as originating from the same Swedish amber forest as probably the Jutlandic coast amber. However, the nature of the deposit is questioned from various sides (see report in Bonde, 1974).

The Samland amber

The most famous amber of all is that from the Samland Peninsula in the Eastern Baltic, a little north-west of Kaliningrad. This is a marine deposit, as can be seen from the admixture of fauna (shark's teeth, oysters, sea urchins, crabs and much else, including a crocodile). Very large amounts of amber are found here even today, although the deposit has been exploited by digging and mining for many years. The oldest amber-bearing layer in the Samland Peninsula, which however has only a relatively insignificant content of amber, lies about 13 m above the youngest chalk-bearing layer. This oldest amber-bearing layer is considered to be late Eocene, and according to Jentsch (1903, p.122) shows a considerable extension into Russia. It is not until a level of about 20 m above the chalk layers and about 13 m below the mean sea level that the blue earth begins, "Die blaue Erde", which is a typical "amber-twig-layer", and which has a very high content of amber. The blue earth lies below the northern and north-western part of the Samland Peninsula, approximately from Kurisches Haff and a little way out into the Baltic at Frisches Haff, where it is exposed in the sea floor. It is a greyish-green, micaceous sand layer of 1—10 m in thickness. Above this is a considerable series of layers from the middle and late Tertiary, in which the lignite layers which are terrestrial, alternate with deposits without coal.

It is almost certain that the production of amber resin ceased in the Baltic region at the commencement of the Oligocene, as the temperature at that time in these regions fell below the normal limit for the production of resin, as far as we know this from recent forests. There is no definite evidence to show how far the Baltic arm of the North Sea extended into what is now the Baltic region. It may be assumed that in the course of the Eocene its outline has varied somewhat, and that its coast zone has in fact passed through the Samland region during the two periods when the amber deposits mentioned have been produced. The possibility that amber was transported to the Baltic during these two late and relatively brief periods only is, on the other hand, unlikely. In my view, it seems most likely that amber resin (or amber) formed in the East Baltic-Russian drainage area throughout the greater part of the Eocene, has been carried with the main rivers of the time further west, out into that day's North Sea, there to constitute part of the amber found in Helgoland's Bay, on Fehmarn, on the south coast of Skåne and other parts of the south-westerly Baltic region. Only the very last production of the forests has had a chance of deposition in the Samland Peninsula. The fossils in the Samland amber most probably represent exclusively flora and fauna from the period when the amber production in the forests of the region was ceasing, in other words from the district near the northern boundary of the amber pine territory.

Tertiary amber has been found in amounts at Brest-Litowsk. Tertiary amber is likewise known from Zitomir, Kiev (upper Eocene), Jekaterinoslav, Charkow and Borislav as well as eastwards to the Urals (Pleistocene) (Bachofen-Echt, 1930, 1949) and Caucasus (Mischer et. al., 1970, p.119). These findings are in sites which periodically were coastal regions in the early and middle Tertiary, and some of the findings have been made in districts where the mangrove swamps of the *Nypa* flora (Takhtajan, 1969) have been particularly suitable biotopes to receive deposits from the river systems of the amber forest. These must have come in particular from the West and Central Russian subtropics, in other words from the same forests which may have produced the Samland amber, but from the south side of the watershed. It must be considered likely that at any rate part of the amber collected in these localities, whether in small or in large quantities, has been found under conditions similar to those in the Samland Peninsula. The material found in the late Tertiary layers, however, has undoubtedly been redeposited one or more times — corresponding to the amber deposits in the Miocene in Jutland.

Central European amber

Amber has also been found in the Tertiary layer at Schmiedeberg, just north of the northern boundary of Czechoslovakia, and at Geisselthal near Halle (middle Eocene). This moves the boundary of the amber forest westwards in Central Europe, and as the habitat of these forests here lies in the Oder-Neisse, Elbe, Vistula drainage areas, an important part of the West German Baltic amber and the Holstein Baltic amber probably originates from here, possibly also some of the South and West Danish (Hesselho amber?). Even further west, there is amber of early Tertiary origin at several sites in the Rhineland (Langenheim, 1969).

It seems as if the number of succinite-producing *Pinites* trees have been decreasing westwards, while the number of retinite-producing trees have been increasing. This is also the case for the Rhine valley, where amber has also been found which is ascribed to the Cupressaceae and *Liquidambar* (Hamamelidaceae) (Langenheim, 1964, p.277, and 1969). In the Rhineland, amber production appears to have continued right up to the Miocene, when the climate must long since have forced *Pinites* southwards.

Swjatoi Ness on the north-eastern coast of the Kola Peninsula is one of the most curious amber localities (among others mentioned by Bachofen-Echt, 1949, p.8,

fig. 7). It lies far away from any theoretically possible site of growth for an amber forest. The obvious idea that this amber should have been transported by waterway from Central Russian amber territories does not appear particularly likely. Amber which has drifted ashore on Iceland (already mentioned by Mohr, 1766, p.338) constitutes a similar puzzling question. These two problems, however, may have a common solution. I think it most probable that both these materials have been transported by the Gulf Stream from the amber territories of the Atlantic Coastal Plain of North America. Unfortunately, no analyses exist of these ambers.

We know from the Mo-clay that the Baltic amber forest has existed at any rate from earliest Eocene. The probability is that forest with resin-producing trees has been growing in these regions for many millions of years previously. We know that conifers with a strong production of resin, both forms with resin containing succinic acid and forms in which the resin has been without succinic acid, have been distributed over almost the entire Holarctis during the upper Cretaceous. The finds of similar plant types in the older Cretaceous, together with the mentioned Jurassic find on Bornholm, seem to show that they have also existed here far back in the Mesozoic. Even though genera and species have altered over the millions of years, and even though the forest boundaries have continually moved backwards and forwards depending on the circumstances of varying landscape, coastlines and climate, it is nevertheless, without any actual break in the continuity, the same forest which in the Jurassic left amber on Bornholm, and which at the same time left the remains of *Pagiophyllum* in many European countries and *Araucariopsis* in the vicinity of present-day Kaliningrad. It is the same forest which, in the great period of Baltic amber 55 to 35 million years ago, was populated by trees such as *Pinites succinifer* and various Taxodiaceae of recent type, direct descendents of known or unknown forms in the forests of the Cretaceous.

The forests have presumably spread extensively also further south. There are amber localities, generally younger, in Savoy, on Sicily and in Romania, but there are also sites of greater age (Cretaceous) in Austria, Hungary and Switzerland. Further North the amber production of the forest appears to have been limited. For biological reasons, it does not appear to have been able to exist very far up in Scandinavia. Throughout the subtropical parts of Russia, the amber production must have been of very great magnitude; the rich deposits in the Samland Peninsula are not the least evidence of this. Even from the Samland Peninsula, the amounts we know are probably only poor by comparison with what has been carried out to sea, where it has either been ground away in the gravel and sand of the coastal zones or buried in the mud of the sea floor, or by comparison with the amounts which have crumbled away in the forest floor, vitrified by the fluctuating climate through millions of years. It is impossible to know how many sites have provided opportunities for small and large collections of amber, under favourable conditions. Nor can anyone predict whether a new "Samland" will turn up in the floor of the Kattegat, along the North German coast at Fehmarn, in South Russia or quite other sites. Much amber has probably been preserved, scattered throughout the mud of the sea floor, and will not appear until new land upheavals take place, in the same manner as the scanty Mo-clay amber.

The European amber trees must presumably have grown over all this great region, as long as tropical or subtropical conditions held. The development of the amber resin has ceased with the commencing fall in temperature, first in the Northern regions. In the Baltic region, *Pinites succinifer* and many other amber plants appear to have become extinct early in the Oligocene, or else they have stopped their excessive secre-

tion of resin. *Sequoia* species seem to have retained this ability somewhat longer. In South and South-East Europe, the resin secretion appears to have persisted into the Miocene.

West Baltic amber (Jutlandish, perhaps also in part East Danish and South Swedish) must thus be assumed to originate from a vegetation which has developed within the western districts of the enormous European forest region, where Russian resinous plants among others have been the source of the East Baltic amber (primarily, that from the Samland Peninsula). Furthermore, the West Baltic amber appears to originate from the entire period of production, whereas the Samland Peninsula amber appears to originate from a brief period of time, geologically speaking, towards the end of the amber period. It is therefore not remarkable that investigators with great expertise are surprised at times at the difference in the biological contents of the two types of amber. The geographical region is very extensive, and a period of almost 20 million years is a great interval of time.

On the above question, Hennig (1969b, p.7) writes as follows: "Einen besonderen Hinweis verdient jedoch der dänische Bernstein, aus dem ich 1965 und 1967 einige Arten beschrieben habe. Bisher wurde das im Dänemark gefundene fossile Harz zusammen mit dem ostpreussischen einfach als "Baltischer Bernstein" bezeichnet. Es musste jedoch schon auffallen, dass im dänischen Bernstein Vertreter mehrerer Familien oder Gattungen (nicht nur der Acalyptratae, sondern z.B. auch der Bombyliiden) gefunden werden, die im ostpreussischen Bernstein niemals beobachtet worden waren, obwohl die aus Ostpreussen stammenden Bernsteineinschlüsse in den Sammlungen noch immer unvergleichlich viel zahlreicher sind als die dänische."

It is an established fact that both among the western amber and the eastern amber there are pieces which do not originate from any *Pinites*. The amber forest was a mixed forest, and also contained resin-secreting angiosperms, undoubtedly also species not yet identified. There is a single piece (containing the acrocerid fly *Prophilopota succinea* which is evidence of this, and on which Jean H. Langenheim writes in a letter to Hennig: "I have never seen a spectrum like this one. I certainly do not think that it is Baltic amber." (Hennig, 1969b, p.2). Remember also the Mo-clay amber studies by Professor Beck (see above).

References

Andersen, J.P., 1947: Den stribede cementsten i de danske eocæne molerlag. — Medd. Dansk Geol. Foren. 11 (2), p.189-196.

Andrée, K., 1942: Die Herkunft des Nordsee-Bernsteins. — Forsch. Fortschr. 18.

Bachofen-Echt, A.v., 1930: Der Bernstein und seine Einschlüsse. — Verh. zool. bot. Ges. Wien 80, p.(35)-(44).

— 1949: Der Bernstein und seine Einschlüsse. — Wien. 204 pp.

Beck, C.W., 1972: Aus der Bernstein-Forschung. — Naturwissenschaften 59, p.294-298.

Berry, E.W., 1927: The Baltic amber deposits. — Sci. Monthly 24, p.268-278.

Berthelsen, A., 1974: Weichselian ice advances and drift successions in Denmark. — Bull. Geol. Inst. Uppsala 5, p.21-29.

Bielefeld, Dr., 1920-21: Der Bernstein und seine Verbreitung im nordwestlichen Deutschland. — Niedersachsen 26, p.306-307.

Bonde, N., 1966: The fishes of the Mo-clay formation (Lower Eocene). A short review. — Medd. Dansk Geol. Foren. 16, p.198-202.

— 1973: Fiskefossiler, diatomiter og vulkanske askelag. — Dansk Geol. Foren. Årsskr. 1972, p.136-143.

— 1974: Palaeoenvironment as indicated by the "Mo-clay formation" (Lowermost Eocene of Denmark). — Tertiary Times 2 (1).

Breiner, M., 1972: Moleret i Limfjorden. — Fur Museum 1. 24 pp.

Bøggild, O.B., 1918: Den vulkanske Aske i Moleret. — Danm. Geol. Unders. (II) 33, 84 pp.

Cleve-Euler, A., 1941: Alttertiäre Diatomeen und Silicoflagellaten im Inneren Schwedens. — Palaeontographica 92 A.

Cleve-Euler, A. & I. Hessland, 1948: Vorläufige Mitteilung über eine neuentdeckte Tertiärablagerung in Süd-Schweden. — Bull. Geol. Inst. Uppsala 32, p.155-182.

Conwentz, H., 1896: On English amber and amber generally. — Nat. Sci. 9, p.99-106, 161-167.

Gough, L.J. & J.S. Mills, 1972: The composition of succinite (Baltic amber). — Nature 239, p.527-528.

Gripp, K. & W. Tufar, 1965: Pyrit-Fossilien aus dem Unter-Eozän von Johannistal bei Heiligenhafen (Ost-Holstein). — Meyniana, Kiel, 15, p.29-40.

Gry, H., 1940: De istektoniske Forhold i Moleromraadet. Med Bemærkninger om vore dislocerede Klinters Dannelse og om den negative Askeserie. — Medd. Dansk Geol. Foren. 9 (5), p.586-627.

— 1965: Furs geologi. — Danks Natur, Dansk Skole, 1964, p.45-55.

Heie, O.E., 1967: Studies on fossil aphids (Homoptera: Aphidoidea), especially in the Copenhagen collection of fossils in Baltic amber. — Spolia Zool. Mus. Haun. 26, 274 pp.

— 1970a: Lower Eocene aphids (Insecta) from Denmark. — Bull. Geol. Soc. Denmark 20, p.162-168.

— 1970b: Notes on six little known Tertiary aphids (Hem., Aphidoidea). — Ent. scand. 1, p.109-119.

— 1976: Taxonomy and phylogeny of the fossil family Elektraphididae Steffan, 1968 (Homoptera: Aphidoidea) with a key to the species and description of *Schizoneurites obliquus* n. sp. — Ent. scand. 7, p.53-58.

Hennig, W., 1969b: Neue Übersicht über die aus dem Baltischen Bernstein bekannten Acalyptratae (Diptera: Cyclorrhapha). — Stuttg. Beitr. Naturkd. 209, 42 pp.

Henriksen, K.L., 1922: Eocene insects from Denmark. — Danm. Geol. Unders. (2) 37, 36 pp.

— 1929: A new Eocene grasshopper, *Tettigonia (Locusta) amoena* n.sp., from Denmark. — Medd. Dansk Geol. Foren. 7, p.317-320.

Hiltermann, H., 1949: Funde von bernsteinartigen Harzen in der Unterkreide Nordwestdeutschlands. — Schr. Nat. Ver. Schleswig-Holstein 24, p.70-73.

Holst, N.C., 1911: Alnarpsfloden, en svensk "Cromer-flod." — Sveriges Geol. Unders. (Ser. C), Årsbok 4 (1910): N:o 9, p.3-64.

Holtedahl, O, 1968: Hvordan landet vårt ble til. — Oslo. 237 pp.

Jentzsch, A., 1903: Vertreitung der Baltischen blauen Erde. — Zschr. Dtsch. Geol. Ges. 55, p.122.

Jerichin, V.V. & J.D. Sukatscheva, 1971: O Melovych Nassekomonossych "Jantarjach" (Retinitach) Severa Sibiri. — Akad. Nauk SSSR, Leningrad, p.3-48 (in Russian).

Keilbach, R., 1959: Die Herkunft des Bernsteins und seine wissenschaftliche Bedeutung. — Urania, Berlin, 22 (9), p.334-338.

Koch, E., 1960: (Ref. from a lecture in Palaeontologisk Klub, 9 November 1959). — Medd. Dansk Geol. Foren. 14, p.283.

Kräusel, R., 1919: Die fossilen Koniferenhölzer (unter Ausschluss von *Araucarioxylon* Kraus). — Paläontographica 62, p.185-275.
— 1949: Die fossilen Koniferen-Hölzer (unter Ausschluss von *Araucarioxylon* Kraus). II. Teil. — Paläontographica 89 B, p.83-203.
Langenheim, J.H., 1964: Present status of botanical studies of ambers. — Bot. Mus. Leaflets, Harv. Univ. 20, p.225-287.
— 1969: Amber: A botanical inquiry. Etc. — Science 163, p.1157-1169.
Langenheim, J.H. & C.W. Beck, 1968: Catalog of infrared spectra of fossil resins (ambers) in North and South America. — Bot. Mus. Leaflets, Harv. Univ. 22 (3), p.65-120.
Larsson, S.G., 1975: Palaeobiology and mode of burial of the insects of the Lower Eocene Mo-clay of Denmark. — Bull. Geol. Soc. Denmark, 24, p.65-81.
Martell, P., 1920: Der Bernstein an der Küste Ostpreussens. — Schr. Zool. Stat. Büssum Meereskd., p.118-123.
McAlpine, J.F. & J.E.H. Martin, 1969b: Canadian amber — a paleontological treasure-chest. — Canad. Entom. 101, p.819-838.
Milthers, V., 1935: Nordøstsjællands Geologi. — Danm. Geol. Unders. (5) 3, 2.edit., 192 pp.
Mischer, G., H.-J. Eichhoff & T.E. Haevernick, 1970: Herkunftsuntersuchungen an Bernstein mit physikalischen Analysenmethoden. — Jahrb. Röm.-Germ. Zentralmus. Mainz 17, p.111-122.
Mohr, N., 1786: Forsøg til en islandsk Naturhistorie. — København.
Nilsson, K., 1966: Geological data from the Kristiansstad Plain, southern Sweden. — Sver. Geol. Unders. (C), 605, 27 pp.
— 1973: Glacialgeologiska problem I, Sydvestskåne. (Summary: Problems of glacial geology in southwestern Scania). — Univ. Lund Dept. quatern. Geol.
Nørregaard, M., 1903: Rav og Retinit fra danske Tertiæraflejringer. — Medd. Dansk Geol. Foren. 9, p.67-68.
Palmgren, J., 1941: Bernstein in der vulkanischen Asche der Moler-Formation Jütlands. — Bull. Geol. Uppsala 28, p.1-2.

Sahlin, C., 1938: Den skånska bärnstenen och dess tilgodogörande. — Med Hammare och Fackla, Årsbok Sancte Örjens Gille 1937, p.1-35.
Sarauw, G.F.L., 1897: Cromer-Skovlaget i Frihavnen og Trælevninger i de ravførende Sandlag ved København. — Medd. Dansk Geol. Foren. 4.
Schlee, D., 1970: Insektenfossilien aus der unteren Kreide — 1. Verwandtschaftsforschung an fossilen und rezenten Aleyrodina (Insecta, Hemiptera). — Stuttg. Beitr. Naturkd. 213, 72 pp.
Schlee, D. & H.G. Dietrich, 1970: Insektenführender Bernstein aus der Unterkreide des Libanon. — N. Jahrb. Geol. Paläont. Mh. 1970, p.40-50.
Schlüter, T., 1975: Nachweis verschiedener Insecta-Ordines in einem mittelkretazischen Harz Nordwestfrankreichs. — Entom. Germanica 1 (2), p.151-161.
Seger, H., 1931: Der Bernsteinfund von Hartlieb bei Bresslau, Altschles. — Mitt. Schles. Altertumsver. 3, p.171-184.
Seward, A.C. & S.O. Ford, 1906: The Araucariae, recent and extinct. — Phil. Trans. R. Soc. London (B) 198.
Sharma, P.V., 1970: Geophysical evidence for a buried volcanic mount in the Skagerrak. — Bull. Geol. Soc. Denmark 19, p.368-377.
Steinbrecher, H., 1935: Die fossilen Harze der Braunkohlen. — Anz. Chem. 48, p.608-610.
Takhtajan, A, 1969: Flowering plants. Origin and dispersal. — Edinburgh. 284 pp.
Thomas, B.R., 1970: Modern and fossil plant resins. — Phytochem. Phylogeny (ed. J.B. Harborne, Academic Press, London), ch. 4, p.59-79.
Tornquist, A., 1910a: Geologie von Ostpreussen. — Berlin.
Wetzel, W., 1939: Miozäner Bernstein im Westbaltikum. — Zschr. Dtsch. Geol. Ges. 91, p.815-822.
— 1953: Mikropaläontologische Untersuchungen des schleswigholsteinischen Bernsteins. — N. Jahrb. Geol. Paläont. Mitt. H.7, p.311-321.

The Flora of the Amber Territory

The early Tertiary forests cannot be new formations, but have undoubtedly originated far back in time, although also as far as their conifers are concerned they have altered much in the course of time, both with respect to genera and species. The amber which has been found in Switzerland, Austria and Hungary in the Cretaceous deposits was most probably produced in the Hercynian mountains of the region and deposited in that coastline which was later elevated into the young Alpine mountains. Provided the theories of continental drift remain valid (here quoted primarily from Dietz & Holden, 1970), according to which the separation between Europe, Greenland and Canada in the North Atlantic was not fully effectuated until the end of the Mesozoic (or even later?), the separation between Europe and Eastern North America has been relatively recent (see Text-fig. 6). As a consequence there must be a close relationship between the forests giving rise to the amber we know from the Atlantic Coastal Plain, and the forests which have produced the Central European amber. This is a relationship which not only appears to be a reality in the case of the Cretaceous, but received further confirmation from our far more modest material from the Jurassic.

The conifers originated in the Palaeozoic, but their time of greatest flourishing was in the Mesozoic, and one gets the impression "that the existing coniferous families have had their origins in the middle Mesozoic times" (Darrah, 1939, p.146). Among those plants which are likely to have belonged to the phylogenetic foundation is the undoubtedly very heterogeneous group which is assembled under the designation Protopinaceae, and a number of very primitive forms, generally regarded as Araucariaceae. Kräusel (1919, p.190) includes in the Protopinaceae "alle die Hölzer, die dem Bau nach einen Übergang von dem älteren Araucarioiden zum modernen Typus bilden," including *Araucarioxylon arizonicum* Knowlton (see above), as well as *Araucarioxylon* sp. Jacobsohn from the Cretaceous (locality Vienna).

Also Seward & Ford (1906) mention primitive Coniferae with more or less pronounced araucarioid affinity, p.377: "The genus *Pagiophyllum* illustrates the wide distribution of a Jurassic type, which though not proved to be Araucarian is, in all probability, correctly referred to that division of Gymnosperms; it is recorded from Britain, France, Germany, Portugal, Bornholm, North America, and elsewhere."- - - "A new genus *Araucariopsis* was instituted by Caspary for fossil wood found near Königsberg with Jurassic boulders, and characterized by the presence of resinparenchyma, in addition to the Araucarian type of tracheids." Doubtful agathoid material from the Cretaceous is also mentioned, from North America and Bohemia, among other places. In addition, *Protodammara* is mentioned from the Atlantic Coastal Plain, also known elsewhere, e.g. from Greenland.

These palaeobotanic notes are subject to much uncertainty, but they show that the Mesozoic material has contained numerous conifers with affinity with both Araucariaceae and Pinaceae, and in addition with very widespread distribution. It is most probable that the resinous acids from these ancient plants have been derived from diterpene Type I, and that amber from these is succinite. The abietic acid type among the recent pinacean resins has hardly been established at that time. The fact that Baltic amber most often nevertheless is succinite, although originating from one or more Pinaceae, is probably an expression of the above relationships, and this may well be one fundamental reason why amber from typical Pinaceae is apparently missing. Already at an early stage, the Taxodiaceae appear to have developed their special resin structures.

Theoretically, the amber can be contemporaneous with the plant material deposited together with it. This appears to be the most likely state of affairs in the older deposits, e.g. in the case of the scanty Mo-clay amber. It is probably also the case in the "blue earth" of the Samland Peninsula, where apparently all older material has been deposited further west. None of the organic constituents of these deposits appear to be essentially older than the layers in which they are found. In research on the composition of the flora of the time, therefore, there are good reasons for not concentrating exclusively on those plant remains which are imbedded in the amber, but for also (with the reservations which follow) giving some attention to the other vegetable constituents of the "amber-twig-layer." In Neogenic and Quarternary deposits, however, the Baltic amber is undoubtedly far older than the layers in which it is found, and undoubtedly the majority of the plant fragments with which it is mixed are of quite another origin. One need only recall the amber and plant fragments in the Alnarp Valley material, or the material in the lignite deposits in Jutland.

There is no guarantee, however, that those plant species found in the material of amber-twig-layers have grown within the region of an amber forest. The amber-twig-layer material has been assembled by the flow of water, often by river systems whose size has been on a par with that of the largest now known. Such rivers are widely branched, and

many of their sources are right up at the level of the water-sheds of the continent in question. In the case of the Samland Peninsula, it would be quite impossible to map the source regions of the river systems which have deposited the material. We only know that a number of the streams in question have traversed the amber forest along some of their length, but we know nothing as how many other landscape types and climates these rivers have carried material from. All the flows have become mixed in the Tertiary lagoons of the Samland Peninsula. Nor do we know with certainty whether the amber forest has existed exactly at the time the amber was washed out of the ground. These forests may have ceased to exist years before this. By means of the plant material, we obtain just a glimpse of the flora from approximately the time the deposits were made, but this must not be interpreted as a picture of an isolated plant society. Like the flora of the London Clay, it may contain everything from boreal to tropical flora, and originate from both mountain and lagoon, and a considerable factor of uncertainty is introduced when it is employed for evaluation of flora.

The world flora of the past

Prior to the period of the Baltic amber forest, the world flora consisted of two main elements, an old and a new. Already for millions of years, the old element had covered those areas of the continents to which it was suited. They were cryptogams, including all those groups we know from the present day. In addition there were a number of gymnosperms: Bennettiales, which were just about to become extinct, Cycadales and Ginkgoales, both rapidly decreasing in numbers, and finally Coniferales — merely to mention the most dominant types. Pteridospermae, the first of all known seed plants, and Cordaitales, existed only for a brief period into the Mesozoic.

The Coniferae underwent important

parts of their final evolution and segregation in the course of the first part of the Mesozoic: Triassic and Jurassic, and all the families which exist today were found by the middle of the Cretaceous. Most of them belonged to temperate climates, only the Araucariaceae were restricted to the tropics and sub-tropics. Many of the original, now extinct forms: Protopinaceae and archaic Araucariaceae, which then were widely spread throughout Holarctis and which may have played a considerable role in the production of amber, lived under tropical and in part warm boreal conditions. This also appears to have been the case with the archaic pinacean, *Pinites succinifer.*

The young floral elements represented by the angiosperms had in fact just commenced their development when the conifers reached their culmination. Takhtajan (1969, p.123) considers it likely that the angiosperms developed early in the Mesozoic, in the Triassic or latest in the Jurassic, probably by neoteny in pteridosperms. In his opinion this has taken place in tropical mountainous territory which must have lain within the East-Indian region; he considers its possible boundaries to be Burma, Assam and the Fiji Islands, i.e. the south-east corner of Laurasia. Fossil angiosperms are not known with certainty before the Cretaceous, and it is considered that these millions of years have elapsed while the group has become stabilized, spread vertically and become specialized within the possibilities presented by the local mountainous territory. It is considered that even at that early stage, forms were developed which were adapted to many biotopes, including different temperatures; thus even then, a flora adapted to tropical, subtropical and boreal conditions had become established within a relatively small region.

In older Cretaceous, the angiosperms started to appear all over the world, and gradually began to increase in proportion even in the flora of distant parts. The earliest of these plant groups are all extinct today, but from the end of the older Cretaceous we have *"Magnolia" delgadoi*, which can be referred with certainty to the Magnoliaceae, most often considered the most primitive of all known families. The great floristic world-wide change occurred about the middle of the Cretaceous, when the angiosperms spread all over the world and reached both Arctic and Antarctic. By then, a considerable proportion of the known families had already been established.

From the presumed origin in south-east Asia, a number of the angiosperms (still according to Takhtajan) followed a distribution route westwards along the coasts of the Tethys Ocean and the extensive archipelago between the great continental masses. It was predominantly a tropical-subtropical line, which also sent out arms to the Scandinavian and the English south coasts, and which invaded the Afro-American section of Gondwanaland. Other routes of spread followed mountain chains. A northerly stream spread out fanwise over the Eurasiatic continent. In this stream, the original vertical layering which is considered to have arisen in the mountains where the angiosperms evolved, gradually changed to a horizontal distribution, as under the increasingly colder conditions of life northwards the plants sought increasingly lower growth sites, until finally furthest north even the types most resistent to cold spread out over the lowlands. A temperate flora developed rich in species, and it has been possible to trace it in many places in the Holarctis, also in Scandinavia. However, the flora of that time was very uniform from west to east. Thus, according to Takhtajan (p.138), a horizontal geographic distribution of plants corresponding more or less to that of the present was already established from the start in the Cretaceous.

Numerous fossil floras in the Palaearctis are now known, and according to Takhtajan it is possible to trace how the infiltration of the angiosperms into the coniferous and bracken forests gradually increased from south to north, towards

Europe and Nearctis. In the opinion of Takhtajan, the considerable floristic community which developed between Asia and North America was established across the Pacific and the Bering Strait. If, however, the new continental drift theories are accepted (Dietz & Holden, 1970, and others), which were not known to Takhtajan, the route from Palaearctis to Nearctis across the relatively narrow Atlantic gap must appear far more likely an explanation for a considerable part of boreal flora and land invertebrate fauna in particular (see Text-fig. 6). Another distribution has presumably taken place over Africa to South America. Lebanese amber may well turn out of the greatest significance for studies of this route of distribution (Schlee & Dietrich, 1970; Schlee, 1970; Hennig, 1971c, 1972b; Strassen, 1973), as it is found in that very region where Gondwanaland has been closest to the supposed point of origin of the angiosperms in South-East Asia. In addition to the routes of distribution mentioned here, it is also possible (still according to Takhtajan) to demonstrate other tracks in the direction of the southern continents of the globe.

In Middle Cretaceous, hardly any of the modern angiosperms existed in their present genera, but at the end of the older part of the Tertiary era (Palaeocene, Eocene and the commencement of the Oligocene), almost all were established, and only relatively few main genera have arisen during the subsequent geological periods. Somewhat corresponding findings would appear to hold for the insects. McAlpine & Martin (1969b), discussing insects in Canadian amber from the last half of the Cretaceous (at least 73—74 million years old), state that the majority can be referred to living orders, but that many belong to families which have not been described (e.g. *Jascopus notabilis* Hamilton, Homoptera), and all genera and species are new. In Baltic amber, practically all the insects can be referred to existing families, about half to existing genera, and a few even to still existing species. This is

a very considerable change in the course of about 30 million years, and one which far exceeds the changes in the subsequent 40 million years.

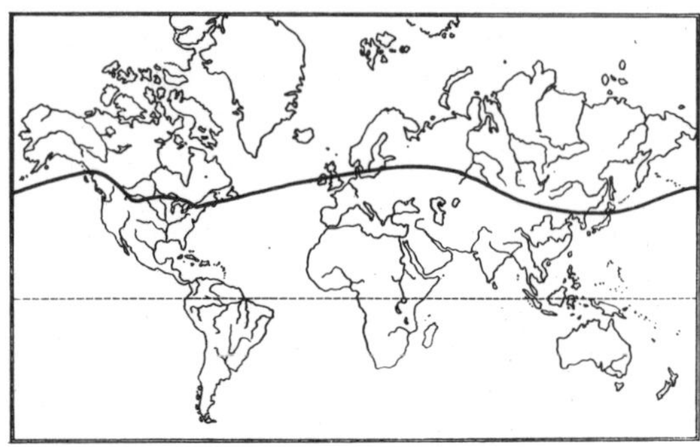

Text-fig.8. Boundary between the boreal and subtropical zones in Eocene. (From Takhtajan, 1969, fig.30).

In the course of the Palaeogenic period, the climatic conditions in our part of Europe changed only slightly. In the Palaeocene there was a minor degree of cooling, but the warm climate was completely restored in the Eocene; the globe is considered to have been wholly without constant polar ice during a period (Schwarzbach, 1950, here from Takhtajan, 1969, p.194). The mild climate is explained above all by the distribution of land and sea and the warm ocean streams resulting from this: the Palaeo-Gulf Stream running along the European west coast and the warm ocean stream of the Tethys Sea running west from the Indian Ocean. East of the Urals, the Sea of Ob joined the Polar Ocean and the Tethys Sea for a long period, and also warmth from this source has possibly exerted an influence on the European climate, at any rate during the presumed ice-free period. The uniform distribution of the plant belts throughout Holarctis (Text-fig. 8) suggests, however, that the earth has been warmer than now. If it was the case that the continental land-mass was not yet completely broken up in the North Atlantic region (Dietz & Holden, 1970), it would be difficult to establish the likelihood of the existence of a Palaeo-Gulf Stream and its course across a possible continental ocean; under all circumstances the dimensions of such an ocean would be much less than those of the present Atlantic. Finally, to judge from the maps published by Dietz & Holden, the Baltic region at the end of the Cretaceous was about 10—12° further south than today, which in itself would involve a considerable climatic improvement. The possible reasons mentioned for the warmer climate which must have dominated in the Baltic at the time, have thus not been fully clarified.

On the basis of the vegetation, Takhtajan (p.184) concludes that in the Baltic region there has been a sub-tropic climate, but at any rate not a tropic one. The boundary separating sub-tropical and boreal climates ran generally speaking from Central England over Southern Scandinavia, North of the Baltic States, through Central Russia, North of China, Korea and Honshu (Text-fig. 8). The boundary between sub-tropical and tropical climates lay correspondingly further north than today, and in the west it reached approximately to England. The sub-tropical vegetation consisted mainly of everygreen trees and bushes, generally with rather slender leaves, while the herbaceous ground vegetation was very scanty. There were a fair amount of lianas and epiphytes. Lauraceae were very dominating, represented for example by the genera *Laurus*, *Persea* and *Cinnamomum*, but also Fagaceae have been particularly numerous, e.g. *Quercus* (numerous species), *Castanopsis* and *Fagus*. Among sub-tropical palms were *Trachycarpus*, *Chamaerops* and *Sabal*. Among the conifers might be mentioned araucarians, but also species of pine, cypress and swamp cypress belong to the sub-tropical flora of the time.

However, from a floristic and faunistic point of view, the sub-tropic belt has not been sharply delimited, either towards the north or the south.

There is a tendency to regard the amber forest as an independent plant society, but it has been far from this. There has been forest all over the globe, wherever the climatic conditions have allowed this, and the amber tree or trees have had their specific, limited placing within this enormous area. The amber forest is thus merely the quite restricted part of the great forest, where it has been possible for these trees to thrive. The plant species have replaced each other to the north and south; in the south the amber trees have grown together with many sub-tropical and also in a number of cases tropical plants, and in the north together with numerous temperate species. Still further north, there have been plant communities which have shown great similarity to the boreal forests of today, where there has been no formation of amber, or it has been based on resin, e.g. from Taxodiaceae. Similar conditions have undoubtedly also held for the vertical distribution of the vegetation, and it appears as if *Pinites succinifer* in the mountains has only grown to moderate heights. It is therefore quite understandable that there has been uncertainty among investigators with regard to the climate which has dominated in the amber territory.

Animals and plants can have been captured in the resin by various routes. It may be a question of organisms endemic to the forest, and it may be objects which have been brought more or less passively by air streams from outside the region of the amber trees, possibly even from very far away, but in this connection especially attention must be given to the organisms which in the amber forest region itself dominated at greater heights than those preferred by *Pinites*. Also these possibilities create uncertainty with regard to the character of the amber forest climate. There is, however, no doubt that the endemic elements are by far the most frequent, and that there are relatively few of the foreign elements. The latter probably include a great proportion of the pollen that has been found in the amber. Particular interest attaches to *Pseudotsuga*, not known to be associated with amber in any other way, and also to the considerable amounts of pollen from *Sequoia*, belonging to the boreal boundaries of the amber tree. Pine pollen was mentioned and pictured already by Goeppert (1883, fig.84), and Conwentz (1890, p.74) writes concerning this topic: "Ich selbst habe öfters ähnlichen Blütenstaub gesehen und bin davon überzeugt, dass man demselben noch viel häufiger begegnen würde, wenn nicht sein auffinden durch die mikroskopische Kleinheit wesentlich erschwert würde." Also small, weakly flying insects such as thrips, aphids and coccids may originate from low-level air plankton.

In the Baltic region, the tropical intermingling exerted very little influence. In the more favourable situated South England, however, the conditions were different. Here, the London Clay has a rich flora, the remains of which are washed out into an estuary. These plant remains probably originate from a climatically varied land region. The flora contains many boreal elements, which have hardly belonged to the coastal vegetation, but the majority of the plant species belong to the subtropics or to the boundary region between tropics and subtropics. A characteristic feature is that some plants, which must have grown in the actual estuary itself, are tropical; this is the case with the trunkless palm *Nypa burtinii* and mangroves (Rhizophoraceae). Traces of this special *Nypa*-flora are found at several sites on a stretch from London to Kiev and to the Odessa region (Takhtajan, 1969), throughout old coastal stretches. The genus *Nypa* is restricted today to saline estuaries in Malaya, Indonesia and adjacent regions of Australia (Darrah, 1939, p.189), the most easterly section of its once wide extent. Also the few tropical elements in the Baltic region seem to be natives of the coastal zone and adjacent regions.

Northwards and at higher level, the subtropics of the Baltic region had a boundary with the warmest districts of the temperate zone, and this neighbourhood has to a high degree set its mark on the plant life of the amber forest. In this connection maple *(Acer)* might be mentioned as having been a common tree, at any rate in certain parts of the East Baltic amber forest, where it is known in several species. This is a tree whose occurrence in the amber forest may appear surprising at first sight, as in the present it is so typically restricted to temperate zones, mainly in mountain forests. Nor can there hardly be any doubt that in the amber territory it has grown in the highlands, and it is most probable that its favourite sites of growth have been at a greater height than those of the amber tree. That maple can easily grow and thrive in a subtropical rain forest, we know for example from Takhtajan's report (1969, p.165), from a personal visit to such a mountain forest in Yunnan, where maple was common among the highest trees in the forest. Maple, however, is only a specially characteristic representative of the temperate element in the forest, other examples may be mentioned including pine and oak.

In due course, the temperate flora underwent a division into a northern, polar flora and a southern flora with a somewhat greater warmth requirement. The more warmth-demanding zone comprised many of those trees which at any rate can grow in present-day Central Europe. They were mainly broad-leaved plants, mostly deciduous trees and bushes: beech, oak, walnut, maple, magnolia, pine, cypress, sequoia and many other. They spread in a belt throughout the whole of Siberia and up to the Atlantic coast of Northern Europe, and the belt continued over the North American continent. In the Eocene, part of the mild-temperate flora penetrated very far north; for example in Ellesmere Land traces have been found right up to present-day 81½°N (but this corresponds to a latitude at that period of 65—70°, according to the map by Dietz & Holden, 1970).

Flora found in amber

The above gives some impression of the plant world of which the amber flora was a section. However, our precise knowledge as to the details of this is very modest. According to the nature of things, what we find of plant parts in the amber can only be fragments. Already before becoming embedded, much was more or less destroyed by drying-in, and is therefore now undeterminable. But part is of considerably better quality and can be used for taxonomic purposes, for example fragments of branches of moss and many other plants, leaves and leaf impressions, at times even complete flowers. As the amber tree has apparently been a high and erect tree, with horizontally placed branches in the fashion of conifers, with the lowest stages dying and falling off in due course, it must be assumed that the resin has above all consisted of flows down the trunk, epiphytic plants have been included in this, but otherwise it has received airborne material. This material may originate from the amber tree itself, but it may also have come from other plants in the forest. Presumably, however, the actual forest floor vegetation has been under-represented in relation to the higher trees and epiphytes.

Fragments of wood are often found in amber, at any rate apparently more or less decayed; in some amber pieces they lie quite close together, in others only singly. With regard to these wood fragments, Schubert (1939, p.42, as explanation of his fig. 33), writes: "Zum Teil aufgefaserter Holzsplitter in klarem Bernstein. Ähnliche Splitter bildeten die grösste Anzahl der Untersuchungsobjekte, deren wahre Natur erst die Lackfilmmethode auswies. In Anbetracht der meistens allzeitigen Einbettung in Harz besteht die Vermutung, dass es sich hier in erster Linie um Holzspäne handelt, die vielleicht von spechtähnlichen Vögeln (entsprechende Federn sind im Bernstein bereits nachgewiesen) herrühren."

In sections of what are most often healthy fragments of wood, Schubert also considers (p.38) having demonstrated tracks of wood-boring insects, "in denen sich Frassreste in regelloser Lage, wohlgeformte Kotballen und zu Detrituswol-

ken verschieden stark aufgelöste Exkremente finden. Die Einzelteilchen des Detritus lassen keinerlei organisierte Struktur mehr erkennen." - - - "Als unmittelbaren und unerträglichen Beweis der nagendbeissenden Tätigkeit der Mundwerkzeuge der diesen Detritus liefernden Insekten finden sich fast regelmässig auf diesen Filmen auch feine Nagesplitter. Es ist anzunehmen, dass das aus den Wunden der Bohrgänge ausfliessende Harz diese Gänge ausspülte und Nagesplitter wie zerfallende Kotballen als "Mulm" fixierte. Die Häufigkeit dieser Erscheinung ist ein nicht unwesentlicher Hinweis auf die Ökologie der Bernsteinkiefern."

The common little mite *"Acarus" rhombeus* Koch & Berendt often occurs in quite large numbers in the detritus clouds mentioned here, and at times small wingless Psocoptera are found. These small animals have apparently a permanent association with the biotope: larval burrows in the wood and bark of trees (see later).

Not least, the vegetation shows that the climate has been seasonal — to a higher degree, it would appear, than that of the present subtropical regions, but there is nothing remarkable about this, since the subtropics of the Baltic Eocene lay at a slightly higher latitude than at present. An important documentation for this is for example the co-existent fossilized tree trunks from other deposits, e.g. from the North Jutland Mo-clay. These trunks show clearly marked annual rings for sharply delimited growth periods.

A very pronounced spring sign is the large number of stellate hairs from oak which are found in amber (Plate 1). These stellate hairs, which vary considerably in their form, are to a special degree a sign of foliation. At the present time they are found on the young shoots in relatively modest amounts, but they are very dominating on bud-scales and on the male inflorescences (Plate 2,A). The foliation and flowering of the individual oak tree last for only a short period, but the overall flowering period in the whole of the amber forest has undoubtedly been of considerably longer duration. For one thing the flora included a number of different species, for another the oaks were subject to individual variations as judged by present-day observation, and finally the extensive territory covered has led to by no means insignificant deviations. According to recent conditions, it may be accepted that in the region of the amber forest there have been flowering oaks for at least one month, and for some time after, withered male flowers have been common everywhere.

Stellate hairs arc not found in all specimens of amber, but they occur so commonly that it is difficult to avoid the conclusion that the resin secretion of the amber trees has to a great degree coincided with the flowering time of the oaks, in other words the amber production has itself been the result of a growth period. To a high degree, this is evidence that the amber flow has been something normal, a stage in a regularly repeated cycle; that it has been a sign of health and not of disease. Just as also today the violent rise in sap in the spring can cause stresses and sap pressure to increase in the trunks of spruce and other conifers, so that the bark splits and the resin exudes, the same must have taken place with the amber trees of former times, except that their production of resin has been far greater, more or less corresponding to that of the present-day *Agathis* species.

It is hardly possible to delimit the exact period for this particularly violent secretion of resin. It must however be assumed that the rise in sap has started early in the year, and that the pressure has culminated at the time the new shoots have grown out. The resin flow has presumably just started simultaneously with the wakening of the buds, has culminated during the development of the shoots and has then gradually declined in strength during and after the maturation of the "candles." It has been a phenomenon especially connected with spring and early summer.

There can be no doubt that the resin flow could also be produced at any time by mechanical damage to the trunks and branches; the quotations from Schubert (1939) given above are clear proof of this. But it has been the natural flow due to growth that has had the decisive influence.

Such a seasonally limited secretion of resin has to a very high degree influenced the choice of local fauna which has been trapped in the resin, and it must be considered probable that insects which are mainly found in amber with many stellate hairs have been spring species. For example, it might be mentioned that the chironomids, which have their rather brief maximum swarming just in this period, are a quite dominating insect group among the fossils.

During the process of fossilization, and particularly during later vitrification, amber has formed a number of artefacts which often show a surprising resemblance to plant structures such as roots and leaves. Not least these products, as well as the inadequacy of many of the actual plant remains, has created some uncertainty with regard to older determinations; a total of about 750 taxo-

Plate 5.
A. So many dung midges (Scatopsidae) are rarely found together as in the present piece of amber. It must mean that the midges have been caught near to an excellent breeding place: mammalian dung or other decaying organic matter. (G. Brovad phot.).
B. Male of the midge family Sciaridae, a family often found inside present-day windows. Many sciarids breed in the litter of humid dark forests, where their larvae sometimes form "armyworms", which consist of thousands of individuals which march straight ahead, all in the same direction. (G. Brovad phot.).
C. Rhagionid fly, a robber, whose prey mostly consists of other flies. The larvae, also predators, are normally found in the ground and are thus never found as amber fossils. (G. Brovad. phot.).

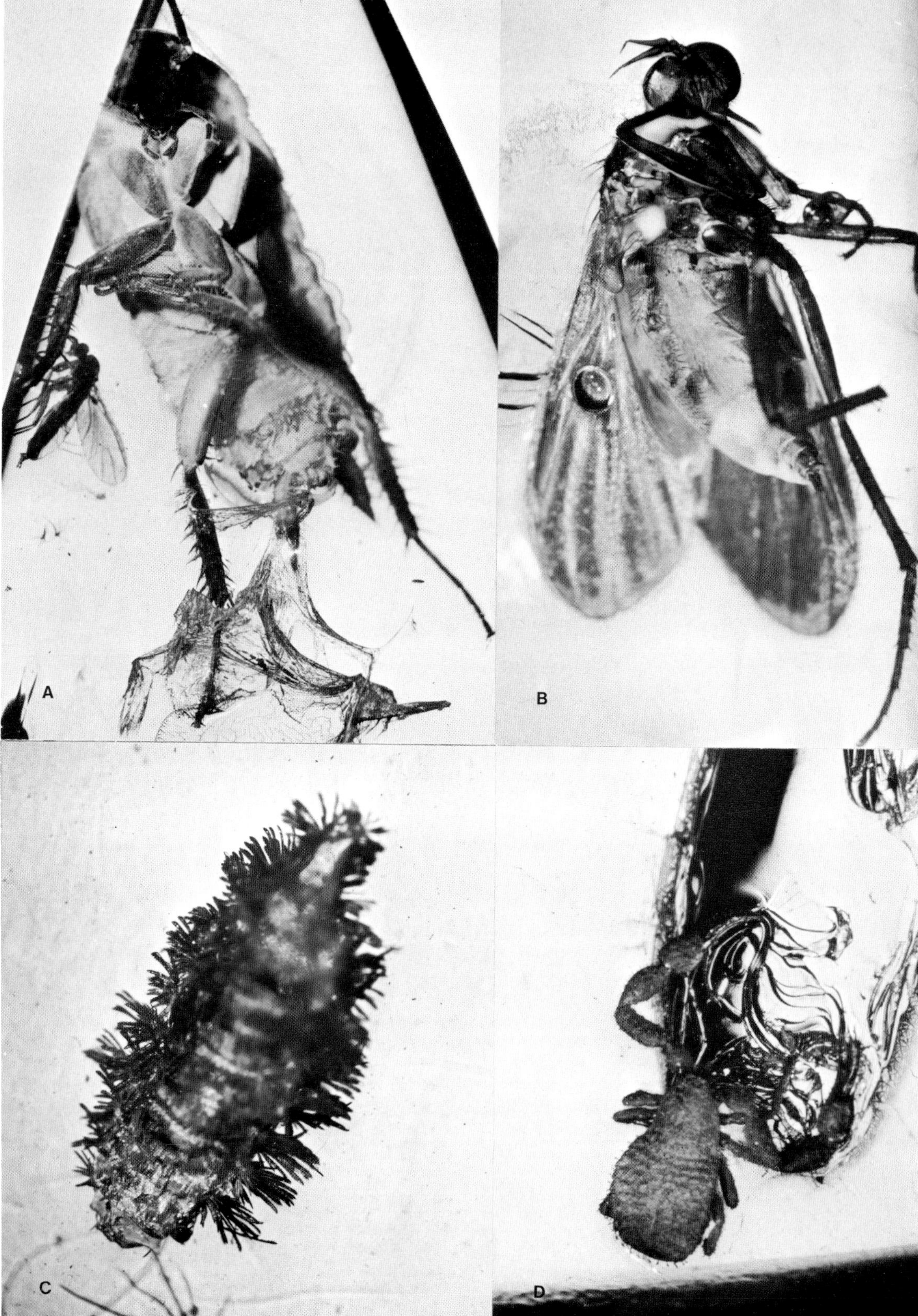

nomic units were described in the course of time. The extensive material from the literature has recently been critically revised by Hanna Czeczott (1961, p.119), who has reduced the number to 216 units by omitting uncertain determinations and synonyms as well as the names of plant parts which were not found in direct contact with the amber. In spite of this revision, there are still a number of uncertain factors in the determination of much of the material.

Bacteria have been found in Baltic amber, 5 genera with one species in each (Blunck, 1929, p.554). Kirchheimer (1937, p.443), however, points to the possibility that these microbes may have penetrated the amber on a later occasion.

Myxomycetes should be mentioned here, by tradition, although they do not belong to the plant kingdom. With reasonable certainty there is only one species, *Stemonites splendens* forma *succinifera* Domke (1952, p.152); accordingly it is thus considered as belonging to a recent species. The majority of Myxomycetes live on dead organic material, particularly tree stumps, windfalls and year-old leaves. They are thus not related directly to the living amber tree.

Fungi are represented by 12 genera with a total of 18 species. Here, too, Kirchheimer expressed doubt with respect to several forms not included in the above, but established by Grüss (1931a, b). The species list comprises among others yeasts (Saccharomycetes), moulds *(Penicillium)*, Pezizaceae and Entomophthoraceae. In the last-mentioned family, Goeppert & Berendt have described and illustrated *Sporotrichites heterospermus* on a fly of the family Dolichopodidae. Among the higher fungi, Caspary & Klebs (1906) described several species under the generic name *Fungites*, but in addition 3 Polyporaceae are known which may have had a quite special significance in the amber forest, since they not merely live in dead and dying trees; many species attack fully viable trees, whose heartwood they destroy. These 3 pore fungi are all classified by Conwentz (1890) as forma *succinea* of a now-living species: *Trametes pini*, living in the trunks and tree-stubs of both conifers and deciduous trees, and which forms red rot in the tree trunks attacked; *Polyporus mollis; Polyporus vaporarius*, which in present-day Denmark is common on timber exposed to continuous moisture, e.g. in greenhouses, but which otherwise belongs to warmer latitudes. The two last-mentioned species were very common, at least in the Baltic amber trees. Typical for these fungi is that the fungal hyphae at the commencement of the attack penetrate into the heartwood, which is gradually broken down. From this point the attack extends in rays following the heartwood of the branches through the sapwood, and it is not until the fungal hyphae reach the surface that the characteristic fruiting bodies are formed. Thus,

Plate 6.

A. A larval cockroach seen from below. The two rafterlike black lines are parts of the bent leg of a limoniid crane fly: tibia to the left, femur to the right. The small midge to the left belongs to the Sciaridae. Note the dichotomous veins at the tip of the wing, however, not reaching the wing-base, characteristic of this family. (G. Brovad phot.).

B. Empididae. Small robber flies. Most of their larvae live as predators in the forest floor, commonly preying on sciarid larvae. (G. Brovad phot.).

C. *Polyxenus* (Diplopoda). The small millipede is seen from below and, therefore, the body appears naked along the whole midline. (G. Brovad phot.).

D. Larval *Oligochernes* (Pseudoscorpionida), 1.8 mm long. The wavy lines in the amber around the chelicers show that the animal was alive for some time before it was drowned by the decisive flow of amber. (G. Brovad phot.).

the fungal attack does not reach the vital cambium until a late stage. Amber fossil insects confirm that *Polyporus* species have been common in the amber forest and also on the amber trees; they have undoubtedly caused the death of many old trees. Also in the case of the fungi, the more detailed determination must in general be regarded as uncertain.

According to Caspary & Klebs (1906), Baltic amber contains remains of **Lichenes** belonging to the genera *Cladonia* and *Cetraria*. Both these genera are found at the present day on moorland and other chalk-poor biotopes. Many species of *Cetraria* grow preferentially on tree-trunks and twigs, so that it is possible that they may also be found in amber. On the other hand, there are good biological reasons to doubt the correctness of the *Cladonia* observations, unless it has been wind-borne. Goeppert (1853) also mentions *Parmelia*, *Sphaerophoron*, *Ramalina* and *Cornicularia*. Caspary & Klebs are however of the opinion that his material was too inadequate.

Bryophyta (Caspary & Klebs, 1906). Among the liverworts (**Hepaticae**), 7 genera are mentioned with a total of 18 species. Almost all these genera have species today which typically grow on tree-trunks or even on damp forest floors. Of the mosses (**Musci**), 17 species are mentioned, distributed into 10 genera, many of which today are known as forest floor types, often forming dense cushions around the foot of the tree-trunks, a number spreading from here to considerable heights. The great exuberance of species of liverwort and moss is striking and suggests that the humidity of the atmosphere has been high in the environment in question, at any rate during periods. It points in fact to conditions observed by Takhtajan (1969, p.165) in the tropical rain forest in southern Yunnan, where the humidity was very great, and where all the trees were closely covered with carpets of moss and numerous other epiphytes. A revision of the mosses found in amber would appear necessary.

Among the ferns (**Felicinae**), Caspary & Klebs (1906) mention the species *(Pecopteris humboldtiana* Goeppert & Berendt and *Alethopteris serrata* Caspary & Klebs, both belonging to the Polypodiaceae. Czeczott (1961, p.122) is however of the opinion that the last mentioned is more probably an *Asplenium*. To judge from the many fern spores found in amber from time to time, there are grounds for assuming that one fern species, at any rate, has been common as an epiphyte in the amber forest, and this may have been one of the known species.

Among cycads (**Cycadales**), Caspary & Klebs mention a single species, but the determination is uncertain. However, cycads existed in Europe at that time. The same is the case with ginkgoes (**Ginkgoales**), which are quite unknown in Baltic amber, but found in the Mo-clay (Koch, 1960, p.283).

Conifers (**Coniferales**). 52 species are identified from materials of varying kinds: needles, scales and small twigs are common, and the same is the case with the male flowers. In some few cases the description is based on cones or wood. Thus, the description of the classic *Pinites succinifer* Conwentz is based on wood. 8 out of the 52 species have been described on very uncertain indications. One of them is *Podocarpites kowalewskii* Caspary & Klebs, the only possible species of Podocarpaceae which is mentioned from amber. *Podocarpus*, however, is known from several Eocene Central European localities (Zalewska, 1961, p.23).

Considering alone those 44 names which designate more or less definitely determined coniferous species, synonyms must be expected among them. Among pines, 8 species have been described on the basis of needles, 6 on flowers and 2 on wood. In the opinion of Czeczott (1961, p.141), it must be considered probable that the last-mentioned 6 + 2 species are also represented by their needles, and are thus synonymous with the 8 species first mentioned, except that it is not possible to know which name is synonymous with which. In a similar manner she reduces the number of species within the other genera. Even if this method of calculation is not unassailable, it nevertheless provides a more acceptable picture of the quantitative composition of the flora than the original list of names, and the number of conifers is reduced to 33. Among these, the pine family (Pinaceae) is represented by 8 species of *Pinus*, including *P. succinifera*, 1 *Picea* and 2 *Abies*. Among Taxodiaceae there are 1 *Glyptostrobus* and 3 *Sequoia*. Cupressaceae are richly represented: 3 *Widdringtonia*, 4 *Thujites* (comprising *Thuja*, *Thujopsis* and *Biota*), 1 *Libocedrus*, 4 *Chamaecyparis*, 4 undetermined species and 2 *Juniperus*. It is remarkable that no trace has been found of Araucariaceae, although they are known to have existed in Eurasia (Darrah, 1939, p.147), and at that time were in fact widespread throughout the tropical and subtropical regions. However, quite recently a number of pieces of petrified tree branches have been found in the contemporary Danish Mo-clay (M. Breiner, Fur Museum, personal communication). The palaeobotanist Fr. J. Mathiesen has determined this material as belonging to the Araucariaceae, genus uncertain. The pieces (Plate 2,C) are very heavy, and the spaces between the annual rings are very narrow. The material lies deep in the Mo-clay horizon (among the striped cement stones, also containing the Mo-clay amber).

Among pines, but especially among Taxodiaceae and Araucariaceae, there are species with high-sitting crowns. At the present day they occur in two- or multi-layered forests in the subtropics, where they produce light conditions which stimulate the development of plentiful undergrowth. Many cypresses prefer such a semi-shady biotope, and can form dense plant communities here. Several of the genera found in amber, for example several of the Taxodiaceae, prefer quite considerable moisture in the earth, while others such as pine can thrive almost everywhere. If *Pinites succinifer* has belonged to Pinaceae with a chemical constitution similar to that of *Agathis* and the recent *Pinus lambertiana*, which is probable, its climatic requirements must have corresponded very closely to those of many now-living *Agathis* species.

The monocotyledonous flowering plants (**Monocotyledones**) in amber comprise only few certain species (Czeczott). They include a number of leaf impressions from subtropical palms (in particular *Sabal*), a palm flower *(Phoenix)* and a few flowers of the liliacean creeper *Smilax* (Bachofen-Echt, 1949, p.29). In addition, a number of indeterminable fragments of grass leaves have been found, although not quite so many as would have been found if the ground had been covered with a continuous herbaceous vegetation. Monocotyledonous plants are far more common in the Mo-clay. In all probability, the monocotyledons commenced their existence as herbaceous swamp plants. Takhtajan and others consider that they have developed from the order of water lilies (Nymphaeales) and consider the families with flowery rushes (Butomaceae) and water plantains (Alismaceae) as particularly primitive. On this basis it is quite natural that in particular monocotyledonous plants should be found in the flora of the Mo-clay and only as secondary adaptations and relatively species-poor islets in the flora of the amber forest, represented in particular by palms.

Of the dicotyledonous flowering plants (**Dicotyledones**), 94 species are known distributed over 57 genera (Czeczott). By comparison with the 33 species of conifers, this is only a modest figure, the probable reason for which is that the relatively young angiospermic flora at this place and time still had not achieved its final equilibrium in relation to the ancient gymnosperms. To some extent, this relative poverty of species is no doubt also due to the densely-leaved undergrowth (with in consequence a poorly developed forest floor) and relatively few meadow-like biotopes with growth of herbaceous plants.

Fagaceae are very richly represented. *Quercus* is an absolutely dominant deciduous tree and known in many species, which no doubt mainly have been evergreens. Both parts of flowers, leaves, buds and bud scales are found, but above all the oaks are known by the innumerable stellate hairs (Plate 1) which are found in most specimens of amber. Other large forest trees are *Fagus succinea*, which is very close to the present genus *Trigonobalanus* found in South-East Asia (Takhtajan, 1969), and *Castanea*, of which the flowers are known. Lauraceae is strongly represented; to this, among others, belong *Cinnamomum* and *Trianthera;* the latter is closely related to the present-day *Eusideroxylon* from Borneo and Sumatra (Czeczott, 1961).

In addition to these dominating families, the following might be mentioned (with notes on their recent occurrence). Aceraceae, woody plants, mostly large trees, from temperate Holarctis. Apocynaceae, which is widespread today, especially in the subtropics and the tropics; mostly trees and bushes, but many lianas. Aquifoliaceae are evergreen bushes or small trees, belonging mainly to the subtropics. Campanulaceae, mostly herbaceous plants from temperate countries. Caprifoliaceae, mostly bushes, small trees and lianas from temperate Holarctis. Celastraceae, today especially small woody plants from warm and temperate zones. Cistaceae, bushes, half-bushes and herbaceous plants from hot-temperate, often arid zones. Clethraceae, subtropic and tropic trees. Dilleniaceae, tropic woody plants, in part lianas. Ericaceae, evergreen bushes, very widely spread. Geraniaceae, the great majority of them herbaceous plants from moderately hot zones, especially South Africa. Hamamelidaceae, woody plants, partly bushes, partly large trees. Hydrangeaceae, bushes, especially from hot-temperate zones in Holarctis. Linaceae, herbaceous plants from temperate and hot countries. Loranthaceae, semi-parasitic bushes, especially in the tropics and subtropics.

Magnoliaceae, from very large trees to small bushes, mainly tropical and subtropical. Myricaceae, trees to small bushes, very wide distribution. Oleaceae, trees and bushes from hot-temperate zones. Oxalidaceae, mainly herbaceous plants, most of them natives of South Africa and South America. Papilionaceae, widespread in multiple life forms, many forest trees in the tropics. Pittosporaceae, especially from Australia. Polygonaceae, especially herbaceous plants from hot-temperate zones, a few lianas. Rhamnaceae, trees and bushes from temperate and tropical regions. Rosaceae, herbaceous plants and bushes, especially from temperate parts of Holarctis. Rubiaceae, from herbaceous annuals to large trees, mainly tropical and subtropical. Salicaceae, bushes and trees from arctic and temperate zones. Santalaceae, semi-parasites, mostly from the tropics. Saxifragaceae, mostly herbaceous plants from temperate zones. Theaceae, trees and bushes from tropics and subtropics. Thymelaeaceae, mostly bushes and small trees from the subtropics. Ulmaceae, trees from tropic and temperate regions. Urticaceae, mainly herbaceous plants, widely distributed.

Both from a systematic and biological point of view, this is a very variegated flora as far as composition is concerned, and our knowledge of it must still be regarded as inadequate. It contains a large element of holarctic flora, represented in particular by dominating genera, whereas the subtropical elements have been less dominant, although they have been represented by far more genera. There is no doubt that a considerable proportion of the flora has been hill types, and that it has had a vertical stratification. Takhtajan (1969, p.193) is of the opinion that both systematic relationships and ecological characteristics indicate that the above flora has been derived from a subtropical flora of the Assam-Burma-Yunnan type.

The European amber forests have had a very great geographical cover from west to east, penetrating deep into the Asiatic continent, even possibly as far as the Pacific coast. It is therefore reasonable that there should be some variation in flora and fauna within the region. Almost all the amber which has been studied botanically, however, originates from the East Baltic region and therefore probably from forests in West and Central Russia. In the case of amber from Denmark and Skåne which must be presumed to originate, at least partly, from a South Scandinavian or Central European, western part of the amber forest, no floral lists are available.

An increased number of pollen analyses, also of western amber, will perhaps provide new information, but will be difficult on account of the often incomplete influence of chemicals on amber, and quite special technical requirements must be satisfied. There can be no question of any real determination of spores and pollen without special preparation. Kirchheimer (1937) released pollen by triturating amber to quite small grains, dissolving it in alcohol and centrifuging the precipitate. Wetzel (1953), working with Holstein amber, was unable to dissolve it completely (using a mixture of alcohol and xylol); he made microscopic preparations of the remainder. Schubert (1961) is of the opinion that it should be possible to find pollen in particularly clear amber by means of the lacquer film method (Voigt, 1936; Schubert, 1939), a technique whereby the pollen is released from the surface of the amber mass by means of a hardened lacquer layer.

The majority of the spores which have been classified are fern spores. Large amounts of pollen are known from *Pinus*, *Sequoia* and a close relative of oak. Pollen has also been found belonging to *Pseudotsuga* and *Tsuga*, trees which have otherwise left no record in amber, as well as one species belonging to Ericaceae and one to Compositae (Langenheim, 1964, p.248). It is remarkable that the pollen originates in particular from other high trees which are wind pollinated, but only exceptionally from plants of lower height; this confirms that

also the amber tree has been among the real giants of the forest. As appears from condition at Green River, however, pollen is a floristic element which should be used with great caution. This flora from Rocky Mountains province is one of the known Eocene floras which is most rich in species, and more than 40 of its genera have been determined by pollen study. While the other vegetation must be characterized as subtropical, most pollen originates from a temperate plant community. We thus have a subtropical lowland flora which is combined with wind-borne pollen grains from more temperate highlands (Knowlton, 1899, here quoted from Darrah, 1939, p.189). Finally, this temperate landscape has perhaps been even further away than has been assumed.

Biological decline from Eocene optimum

As already mentioned, the flora in the "amber age" has been grouped into plant belts which have corresponded more or less to those of the present, merely more displaced polewards. During the period from the beginning of the Cretaceous until the Eocene, considerable floristic changes took place. During this period, the individual plant genera achieved a very considerable spread, especially in an east-west direction within Holarctis, although there were also many common features between Holarctis and the continents to the south. One of the main reasons for this great spread was that the young angiosperms, at any rate, had so far never experienced a climatically critical period, during which their power of adaptation under unusual living conditions would be put to the test. If the recently published theories on continental drift (Dietz & Holden, 1970, among others) should turn out to be realistic, as would appear to be the case, so that even as late as in the Mesozoic the continents have been considerably nearer than they are today, most particularly in the North Atlantic, the resulting palaeogeographic conditions must be ascribed the greatest significance with respect to the ease and rapidity with which the genera and families were distributed.

At the end of the Eocene, however, that climatic deterioration commenced which was to be the basis for the present distribution of plants and animals. Perhaps it would be more correct to talk of a series of varying climatic changes moving towards an exacerbation, which is actually what happened, especially during the last millions of years of the period, with alternating glacial ages and interglacial ages. The unfavourable events had actually commenced during the Mesozoic, with underground disturbances along the northern coasts of the Tethys Sea, where deep trenches were formed corresponding to the new plate tectonic theories. As the African and Indian plates gradually became submerged under the Eurasiatic, the coastal regions of the latter became raised above sea level, and during the course of the Tertiary that great system of mountain chains was formed which extends from the Atlantic to the Pacific, the most prominent sections of which are the Alps and the Himalayas. The decisive phase occurred when the evolving mountains reached such a magnitude that the connection between the Mediterranean and the Indian Ocean was definitely interrupted. The Baltic became a closed inland sea; connections with the south became interrupted, so that for a long period it was only open to the Atlantic.

Under the influence of steady fall in temperature throughout the Oligocene and the entire Neogenic period, a gradual change took place in the composition of the flora and fauna of the countries. There was a slow southward displacement of the areas of dissemination of the species. This movement was general for the entire Holarctis, most pronounced near the pole and decreasing towards the equator, and was not uniform in the various regions. Towards the end of the period, the southwards pressure strengthened in the western regions of the two continents, but especially of the Palaearctis, where the commencing inland ice

was to leave its mark.

All the subtropical elements of the flora and fauna rapidly disappeared. The boreal plant world, hitherto a belt north of the amber forests, pushed forward and filled in the niches which became available throughout the continually changing situation. Both in the Palaearctic and Nearctic regions the territory was now dominated by a uniform boreal flora and fauna. All major secretion of resin stopped, and now took place solely in countries south of the alpine chain and in corresponding regions in the American continent, in spite of the fact that these boreal regions were actually very rich in pine and spruce, and to begin with also in *Sequoia*.

The floral development which was a consequence of the Ice Age tells us much about the varying fate of the various regions of Holarctis. In the Palaearctic region the young Tertiary mountains formed an almost continuous barrier from the Pyrenees in the west to the spurs of the Himalayas in the east. There were only few points at which openings were available which allowed the hemmed-in life in the north to spread southwards. The climatic threats were strongest in the west, where the consequences of the advancing ice were of great significance, while the threats were considerably less serious in the east, where the ice formation was inhibited by a drier climate. The passage southwards was easiest in the beginning, but as the glaciers spread from the Alps out over the adjacent territories, this passage closed almost completely, and the flora and fauna which now existed in Central Europe was as good as doomed to extinction. It must be assumed that all life which characterized the amber forests of North and Central Europe was wipped out without having left any descendants, perhaps with the exception of those organisms which early in the period got through to the Mediterranean region, namely in the west, and here became included in the surviving Mediterranean fauna and flora. The threat to Eocene life has been much less further eastwards, and in South China, Further India and Indonesia there have been considerable possibilities of survival.

In America, the climatic conditions have been about the same as in the Old World; here too, the environmental conditions have been severe in the west and probably more moderate towards the east. The conditions of terrain, however, have been more favourable, as the mountain chains here lie parallel to the lines of longitude and not the parallels of latitude, as in the Old World. The flora and fauna have therefore not been locked-in, but have been able to move in a north-south direction. This has been a factor contributing in a high degree to the maintenance of the animal and plant life.

During the climatic deterioration, all forms of life were subject to the great trials associated with this, and for the first time in the case of the angiosperms and all the newly developed insect fauna in connection with and depending on them. This development was destructive, and many species and genera were wipped out completely, while many survived in those conditions which were most favourable, but became extinct in other circumstances, and hardly any form managed to escape without some effect. The fauna, originally so rich in species and widely distributed, shrank everywhere, many genera persisted only in isolated regions, where they became relics, often called "endemic" or "bipolar". The greatest wealth of species was maintained in the tropics (Text-fig. 9), and under the relatively cool conditions which must have held also here at that time, many boreal species have undoubtedly been among those "surviving the winter" here.

In the course of the great dispersal which terminated during the Cretaceous and Palaeogenic, the flora and fauna met barriers of various kinds, but they were generally of a modest character by comparison with those which the survivors experienced after the Ice Age (and in the Interglacial Ages). In the first place, the distance between the continents had become considerably greater, and extensive mountain systems had arisen in the course of the Tertiary. In addition, the

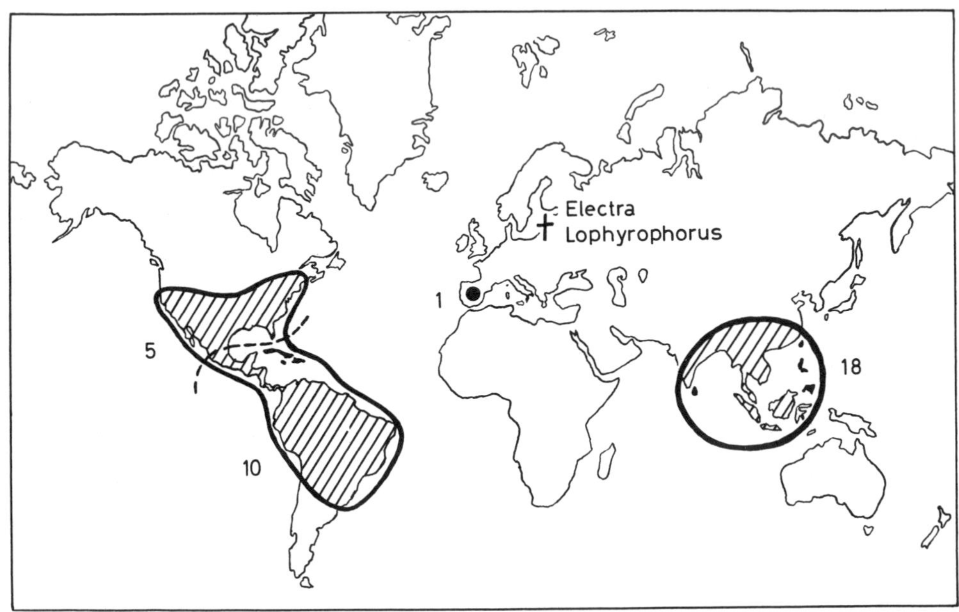

Text-fig.9. Recent distribution of the family Rachiceridae (Diptera), which was represented in the Baltic amber forest by two genera. The number of recent species is indicated. (From Hennig, 1967a, fig.11).

temperature conditions were unfavourable for overcoming great barriers; temperature was low, and the ability to overcome barriers rises and falls in step with this — at any rate in the case of terrestrial arthropods. In the Baltic region, an exchange across the Atlantic has in practice been excluded, before anthropochorous dispersal became effective. A number of species entered the region from the south-west, these species being favoured by an oceanic climate. Considerably more came from Asiatic and South Russian regions, in other words an eastern and south-eastern islet, whose species have mainly had continental features. The whole of this immigration comprises boreal species and genera, and present conditions have not furthered immigration of any subtropical species. This gives a composition of flora and fauna which more or less corresponds to the boreal elements in amber, and to the life which must have dominated in the territory north of the amber forests during the Baltic amber period. It has many points of similarity with the life of the amber forest, but nevertheless does not consist of descendants of the Early Tertiary Baltic flora and fauna.

Recent "amber floras"

On the other hand, a flora whose main features are related to those found in the amber forest is still encountered in various places in the tropics of today. Such a flora, with more oceanic features, is found in parts of Indonesia or even better examples on Formosa or in Southern Japan, and a more continental type in the interior of Further India. Drawing comparisons with the amber flora of the Eastern Baltic, since this is the one we know something about, particular interest attaches to the flora of Further India or South China, its vertical and horizontal classification and the associated climatic features. These forests, characterized by primaeval features, might well be regarded as the remains of the once widespread amber forest belt.

The forests of the Burmese coastlands are populated in particular with deciduous trees, but up in the mountains, where the precipitation is even greater than on the coast and where the dry season is quite short, the rain forest is extensive. Here quite a number of trees are found which rise above the level of the general leaf canopy (to a height of 50 to 60 m). The rain forest consists of a large number of tree species, including many palms. Most of the deciduous trees lose their leaves during the relatively short dry season, while others are evergreens. Among the numerous plant families might be mentioned in particular trees of Dipterocarpaceae, Guttiferae, Sterculiaceae, Meliaceae, Caesalpinaceae and Moraceae, not least of all here *Ficus*. In the undergrowth there are palms and other lower trees, and bushes below these, but only few herbaceous plants; lianas and epiphytes are plentiful. The rain forest here has a purely tropical characteristic, some of the families are found in the amber flora, but as a whole the flora is a somewhat more southern one.

Up in the mountains in the subtropical region, the wealth of species decreases with increasing height. Evergreen oak species are also found in the rain forests of the lowlands, but they become even more plentiful in the mountain forests, just as Magnoliaceae, Juglandaceae and Theaceae are common. Palms are likewise plentiful in the mountain forest. Finally, in many places the forest is transformed into a subtropical rain forest of the same kind as that found on the mountains in the interior of Southern China, e.g. Yunnan. Here, species of pine also play a part, and in certain places they form exclusively pine forests. A spectroscopic study should be made of the resin from these *Pinus* species. It is also possible to find numerous maples (as stated by Takhtajan, 1969, p.165). Such scenery might be almost described as typical amber territory.

In the interior of Further India, the dry season in the districts not including the higher mountains is so prolonged that the majority of evergreen tree species cannot thrive, but are replaced by deciduous species. On some mountain ranges, especially in Burma and Western Siam, the precipitation is so high during the rainy season, however, that a deciduous mountain forest can thrive, the exuberance of which is not much less than that of the rain forest. Only few bushes are found but several herbaceous plants, although these do not constitute a continuous cover. Epiphytes are mostly found in the tree tops. Also these forest types agree to some extent with the features known from the amber territory, but approach very closely its more temperate zone. (This vegetation scenery is cited mainly from the plant geographer Martin Vahl, in Vahl & Hatt, 1925, III, p.265).

On the Philippines, in many places in Indonesia, on New Guinea and on the North Island of New Zealand, *Agathis* species grow which produce large amounts of resin, collecting in the angles of branches and on the tree-trunks. This resin is a trap for insects and other small animals, while at the same time it collects and encloses fragments of leaves, pollen and other plant parts. In *Agathis*, under natural conditions, it is possible to follow the initial stage of the genesis of all ambers. There are quantities of resin in the ground and on the forest floor, and it can also be found partly fossilized to copals in sites where the forest has long since ceased to exist. The fossil resin can be found from the Pliocene and the Miocene. It is found torn out of the forest floor by streams which have varied over the course of time, and have redeposited it in the estuaries, mangrove swamps, bays and river bends of former ages.

This too seems to be a picture of the earliest history of Baltic amber.

References

Bachofen-Echt, A.v., 1949: Der Bernstein und seine Einschlüsse. — Wien. 204 pp.

Beck, C.W., 1972: Aus der Bernstein-Forschung. — Naturwissenschaften 59, p.294-298.

Bengtson, P., 1973: Fossil - en bit i kontinentpusslet. — Forskning och Framsteg 1973 (5), p.26-31.

Bergström, S.M., 1973: När Nordatlanten försvann och nybildades. — Forskning och Framsteg 1973 (5), p.10-16.

Blunck, G., 1929: Bacterieneinschlüsse im Bernstein. — Centralbl. f. Miner, Abt. B, p.554-555.

Brundin, L., 1973: Nutidens fauna speglar kontinenternas delning. — Forskning och Framsteg 1973 (5), p.17-19.

Caspary, R. & R. Klebs, 1906: Die Flora des Bernsteins und anderer fossilen Harze des ostpressischen Tertiärs. — Abh. Kön. Preuss. Geol. Landesanst. (N.F.) 4, p.1-182.

Chandler, M.E., 1961: The Lower Tertiary floras of southern England. I. Paleocene floras, London Clay Flora. — Brit. Mus. (Nat. Hist), London. 354 pp.

Conwentz, H., 1886: Die Flora des Bernsteins. 2. Die Angiospermen des Bernsteins. — Danzig. 140 pp.

— 1890: Monographie der baltischen Bernsteinbäume. — Danzig. 151 pp.

Czeczott, Hanna, 1961: The flora of the Baltic amber and its age. — Prace Mus. Ziemi 4, p.119-145.

Daley, B., 1972: Some problems concerning the Early Tertiary climate of Southern Britain. — Palaeogeogr. Palaeoclimat. Palaeoecol. 11, p.177-190.

Darrah, W.C., 1939: Principles of Paleobotany. — Leiden. 239 pp.

Dietz, R. & J.C. Holden, 1970a: The breakup of Pangaea. — Scient. Amer. 223 (5), p.30-41.

— 1970b: Reconstruction of Pangaea: Break-up and dispersion of continents, Permian to Present. — J. Geophys. Research, 75 (26), p.4939-4956.

Domke, W., 1952: Der erste sichere Fund eines Myxomyceten im baltischen Bernstein. (*Stemonites splendens* Rost. fa. *succinifera* nov. foss). — Mitt. Geol. Staatsinst. Hamburg 21, p.152-161.

Dorf, E., 1960: Climatic changes of the past and present. — Scient. Amer. 48 (3), p.341-364.

Duisburg, H.v., 1860: Urweltlicher Blütenstaub. — N. Preuss. Prov. B1.5, p.294-298.

Goeppert, H.R., 1836: Fossile Pflanzenreste des Eisensandes von Aachen. — N. Acta Acad. C.L.C. Nat. Cur. 19 (2).

— 1853: Über die Bernsteinflora. — Monatsber. Kön. Acad. Wiss. Berlin, p.1-28.

Goeppert, H.R. & G.C. Berendt, 1845: Die Bernstein und die in ihm befindlichen

Pflanzenreste der Vorwelt. — Berendt: Organ. Reste I. p.61-126.

Goeppert, H.R. & A. Menge, 1883: Die Flora des Bernsteins und ihre Beziehungen zur Flora der Tertiärformation und der Gegenwart 1. Danzig. 63 pp.

Gough, L.J. & J.S. Mills, 1972: The composition of succinite (Baltic amber). — Nature 239, p.527-528.

Grüss, J., 1931a: Die Urform des *Anthomyces Reukaufii* und andere Einschlüsse in den Bernstein durch Insekten verschleppt. — Wochenschr. f. Brauerei 14 II (7), 6 pp.

— 1931b: Die Urform der Anthomyceten nebst Zucker und Stärke im Tertiär. — Forsch. Fortschritte 7, p.175-177.

Harris, T.M., 1935: The fossil flora of Scoresby Sound, East Greenland, pt.4, Ginkgoales, Coniferales, Lycopodiales, and associated fructifications. — Medd. Grønland 112.

Heer, O., 1869: Miocäne baltische Flora. — Königsberg.

Helm, O., 1884b; Mitteilungen über Bernstein, IX. Über die Holzreste im Bernstein und unter Bernstein. — Schr. Nat. Ges. Danzig (N.F.) 6, p.127-132.

Hennig, W., 1971c: Insektenfossilien aus der unteren Kreide III. Empidiformia ("Microphorinae") aus der unteren Kreide und aus dem Baltischen Bernstein; ein Vertreter der Cyclorrhapha aus der unteren Kreide. — Stuttg. Beitr. Naturkd. 232, 28 pp.

— 1972b: Insektenfossilien aus der unteren Kreide IV. Psychodidae (Phlebotominae), mit einer kritischen Übersicht über das phylogenetische System der Familie und die bisher beschriebenen Fossilien (Diptera). — Stutt. Beitr. Naturkd. 241, 69 pp.

Hollick, A. & E.C. Jeffrey, 1909: Studies of Cretaceous coniferous remains from Kreischerville, N.Y. — New York Bot. Gard. 3, p.1-136.

Hurley, P.M., 1968: The confirmation of continental drift. — Scient. Amer., April 1968.

Irmscher, E., 1923: Pflanzenverbreitung und die Entwicklung der Kontinente. — Mitt. Inst. Allg. Bot. 1922.

Kirchheimer, F., 1936: Beiträge zur Kenntnis der Tertiärflora: Früchte und Samen aus dem deutschen Tertiär. — Paläontographica B 82, p.73-141.

— 1937: Beiträge zur Kenntnis der Flora des baltischen Bernsteins I. — Beih. Bot. Centralbl. B, 57, p.441-482.

— 1947: Die Laubgewächse der Braunkohlenzeit. — Halle (Saale). 783 pp.

Knowlton, F.H., 1896: American amber-producing tree. — Science 3, p.582-584.

— 1899: The fossil flora of the Yellowstone National Park. — U.S. Geol. Surv. 32 (2), Washington.

Koch, E., 1960: (Ref. from a lecture in Palæontologisk Klub, 9 November 1959). — Medd. Dansk Geol. Foren. 14, p.283.

Köppen, W. & A. Wegener, 1924: Die Klimate der geologischen Vorzeit. — Berlin. 256 pp.

Kräusel, R., 1919: Die fossilen Koniferenhölzer (unter Ausschluss von *Araucarioxylon* Kraus). — Paläontographica 62, p.185-275.

— 1949: Die fossilen Koniferen-Hölzer (unter Ausschluss von *Araucarioxylon* Kraus). II. Teil. — Paläontographica 89 B, p.83-203.

Kurtén, B., 1969: Continental drift and evolution. — Scient. Amer. March 1969.

Langenheim, J.H., 1964: Present status of botanical studies of ambers. — Bot. Mus. Leaflets, Harv. Univ. 20, p.225-287.

— 1966: Botanical source of amber from Chiapas (Fuente botánica del ámbar de Chiapas, México). — Ciencia, Méx. 24, p.201-210.

— 1967: Preliminary investigations of *Hymenaea curbaril* as a resin producer. — J. Arnold Arbor. 48 (3), p.203-227.

— 1969: Amber: A botanical inquiry. — Science 163, p.1157-1169.

Langenheim, J.H., B.L. Hacker & A. Barlett, 1967: Mangrove pollen at the depositional site of Oligo-Miocene amber from Chiapas, Mexico. — Bot. Mus. Leaflets, Harv. Univ. 21 (10), p.289-324.

Liljequist, G.H., 1973: Vandrande klimatområden. — Forskning och Framsteg 1973 (5), p.20-25.

Maack, R., 1969: Kontinentaldrift und Geologie des südatlantischen Ozeans. — Berlin.

Magdefrau, K., 1957: Flechten und Moose im baltischen Bernstein. — Ber. Dtsch. Bot. Ges. 70, p.433-435.

McAlpine, J.F. & J.E.H. Martin, 1969b: Canadian amber — a paleontological treasure-chest. — Canad. Entom. 101, p.819-838.

Menge, A., 1858: Beitrag zur Bernsteinflora. — Schr. Nat. Ges. Danzig 6 (1).

Reid, E.M. & M.E. Chandler, 1933: The flora of the London Clay. — Brit. Mus. (Nat. Hist.) London. 513 pp.

Sarauw, G.F.L., 1897: Cromer-Skovlaget i Frihavnen og Trælevninger i de ravførende Sandlag ved København. — Medd. Dansk Geol. Foren. 4.

Schlee, D., 1970: Insektenfossilien aus der unteren Kreide — 1. Verwandtschaftsforschung an fossilen und rezenten Aleyrodina (Insecta, Hemiptera). — Stuttg. Beitr. Naturkd. 213, 72 pp.

Schlee, D. & H.G. Dietrich, 1970: Insektenführender Bernstein aus der Unterkreide des Libanon. — N. Jahrb. Geol. Paläont. Mh. 1970, p.40-50.

Schubert, K., 1939: Mikroskopische Untersuchungen pflanzlicher Einschlüsse des Bernsteins. I. Holz. — Bernstein-Forsch. 4, p.23-44.

— 1953: Mikroskopische Untersuchungen pflanzlicher Einschlüsse des Bernsteins. II. Rinden und Borken. — Paläontographica B, 93, p.103-119.

— 1961: Neue Untersuchungen über Bau und Leben der Bernsteinkiefern *(Pinus succinifera* (Conw.) emend.). — Beih. Geol. Jahrb. 45, Hannover, 143 pp.

Schwarzbach, M., 1950: Das Klima der Vorzeit. — Stuttgart.

— 1955: Allgemeiner Überblick der Klimageschichte Islands. — Neues Jahrb. Geol. Paläont. Mh.3, p.97-130.

Seward, A.C. & S.O. Ford, 1906: The Araucariae, recent and extinct. — Phil. Trans. R. Soc. London (B) 198.

Stephansson, O., 1973: Kontinentaldrift och den nya globala geologin. — Forskning och Framsteg 1973 (5), p.3-9.

Strassen, R.zur, 1973: Fossile Franzenflügler aus mesozoischem Bernstein des Libanon. — Stuttg. Beitr. Naturkd. 256, 51 pp.

Takhtajan, A., 1969: Flowering plants. Origin and dispersal. — Edinburgh. 284 pp.

Thompson, P.W., 1951: Grundsätzliches zur tertiären Pollen- und Sporenmikrostratigraphie auf Grund einer Untersuchung des Hauptflozes der rheinischen Braunkohlen in Liblar, Neurath, Fortuna und Brühl. — Geol. Jahrb. 65, p.113-126.

— 1958: Die fossilen Früchte und Samen in der niederrheinischen Braunkohlenformation. — Fortschr. Geol. Rheinld. Westf. 2.

Vahl, M. & G. Hatt, 1925: Jorden og Menneskelivet. Geografisk Haandbog, 3. — København. 708 pp.

Voigt, E., 1936: Die Lackfilmmethode, ihre Bedeutung und Anwendung in der Paläontologie, Sedimentpetrographie und Bodenkunde. — Zschr. Dtsch. Geol. Ges. 88, p.272-292.

— 1938: Eine neue Methode zur mikroskopischen Untersuchungen von Bernstein-Einschlüssen. — Forsch. Fortschr. 14, p.55-56.

Wetzel, W., 1953: Mikropaläontologische Untersuchungen des schleswigholsteinischen Bernsteins. — N. Jahrb. Geol. Paläont. Mitt. H.7, p.311-321.

Weyland, H., 1934: Beiträge zur Kenntnis der rheinischen Tertiärflora. I. — Abh. Preuss. Geol. L.A., 161, 122 pp.

Wilson, J.F., 1963: Continental drift. — Scient. Amer. April 1963.

Zalewska, Z., 1961: The fossil flora of Turow near Bogatynia. Second part. Coniferae. — Prace Muz. Siemi 4, p.19-49, 93-102.

The Fauna of the Amber Territory

A study of the amber fossils shows that in its time, the amber tree, *Pinites succinifer*, was inhabited by a whole series of animal species, only a few of which were specific, mainly phytophagous insects, while others were indifferent with regard to the taxonomy of their host plant, or were completely random guests. A comparison with present-day fauna shows that other animal species have been associated with other common plants in the amber forest. The fauna which was specially connected with the living amber tree then, has disappeared today, and on the whole its genera are only distantly related to present forms. Apparently it was a fauna which perished at the same time as its host plant. This characteristic of the specific fauna of the amber tree must be interpreted as strong evidence that on one or other important point, the palaeogenic *Pinites* has differed essentially from now-living *Pinus* species. When it is considered how little there is in common today between the species of animals found on the various conifer families, for example between Taxodiaceae, Araucariaceae and Pinaceae, and recall at the same time the great morphological similarity between *Pinites* and *Pinus*, it would appear reasonable to assume that the varying nature of the resinous acids both is and has been of the greatest significance. The evolution which the Pinaceae have apparently undergone since the commencement of the Tertiary, when the resin, which originally seemed to belong to the diterpene type I, became altered to the abietic acid of diterpene type III, may have been responsible for the change in fauna.

The possibility must also be considered that the resinous acids were primarily, or perhaps exclusively, altered among those Pinaceae which grew in a region outside the subtropical-tropical belt, a flora whose resin composition in the Eocene is unknown to us, and it may be this change in constitution alone which has been the phylogenetic point of origin of the present day Pinaceae. This would explain the strongly boreal features of this group today, and the difference in fauna between them and the extinct *Pinites* of the amber forest. As the Araucariaceae and *Pinites* originally possessed closely related resinous acids, it would not be surprising, should a hitherto unknown East-Asiatic *Agathis* fauna be found, if this turned out to be more closely related to the *Pinites* fauna than to any now living fauna.

The fauna associated specifically with the living amber tree has been above all plant-sucking insects, for example aphids and coccids. Moss and other epiphytes on the tree-trunks and branches, and the cracks and crannies of the bark, have provided accommodation for a rich fauna of small animals, which only exceptionally have had their home here alone, but which under similar conditions would have been found on any other tree-trunk, living or dead. In the humid climate of the sub-tropical forests, all the tree-trunks were covered with thick carpets of moss of various kinds. Takhtajan (1969, p.165) has mentioned this growth in the rain forest from present-day Yunnan, and as already mentioned, moss is common in Baltic amber. This moss has had its own fauna consisting especially of very small animals such as mites, Collembola and various small worms, and these in turn have had their own diminutive predators, for example other mites and pseudoscorpions. This is a fauna which in its details corresponds to life in the same biotope in the present-day forest. In addition, a more typical forest-floor fauna has been found from cracks and crannies in the bark, especially near the root collar, where the cushions of moss were thickest; these were larger animals such as wood-lice, millipedes and beetles from species very closely related to those species which live under quite the same conditions today.

Many ichneumon flies, especially from the family Proctotrupidae, but also from other groups, have sought hosts for their egg-laying on the tree-trunks: insect larvae, spiders, insects' and spiders' eggs, and much more. Especially during the day, innumerable ants have had their fixed routes up and down the tree-trunks, mainly seeking aphid honey, but many small insects have undoubtedly also been brought back to the nests as booty. After nightfall much of what the ants have not captured during the day has become a victim to harvest spiders and the crowd of spiders which have had their haunts in this biotope.

Many of the arthropods in amber have lived in tree fungi. In the first instance these have been bracket fungi (Polyporaceae), a number of which in their mode of life do much damage to present-day trees, and the situation has hardly changed since 50 million years ago. Many species attack living trunks which to begin with are healthy, and their mycelia often penetrate large parts of the heart-wood, which gradually decays and becomes "red rot" or "white rot." This rotted wood, where the cellulose is more or less broken down, is an excellent medium for many insect larvae, and for example many of the Anobiidae which have been found in amber have undoubtedly undergone their developmental period here. The fruit bodies of the fungi break through the surface of the tree-trunks later in their development, some near the ground, several higher up. These fruit bodies are often perennial, and provide accommodation and nourishment for their own characteristic insect fauna, primarily larvae. Also this fauna is richly represented in amber. Many of the windfalls of the amber forest can probably thank the bracket fungi for their harsh fate.

An important part of this fauna of small animals lived in dead tree-trunks, under the bark or in the wood, but not actually associated with fungi or fungal hyphae. In their case the species of plant often played only a minor role. Here in particular it has been a case of the physico-chemical state of the nutrient material and the degree of decay. None of these species are particularly numerous in amber, but a relatively wide variety of species is represented. They have probably been trapped when, during their search after suitable places for egg deposition, they have used the tree-trunk as a chance resting place. To judge from the total number of specimens, many of these species must have been very common, and the forest must have been rich in dead and dying trees of many different species, not least fallen tree-trunks in rapid decay. A rich termite fauna has undoubtedly played a quite considerable part in the further destruction of these masses of wood.

By far the greater proportion of the insects found in the amber are imagines, which actually are not native to the locality, but which have regularly sought in the vegetation for a shadowy and moist hiding place from the sun during its hours of greatest effect. They are predominantly species which deposit their eggs in moist earth or in nearby watercourses and dams. There are midges, especially Chironomidae, Sciaridae and Mycetophilidae, as well as Trichoptera. There are by no means few Lepidoptera of different families, although in many cases these are species whose larvae live on and in epiphytes on tree-trunks. Finally, there are larvae whose occurrence in the amber is completely by chance and often quite inexplicable, for example a few larvae of may-flies and other amphibian insects.

With regard to its fauna, the amber forest as we know it from the amber itself did not differ very much from today's forest under corresponding conditions. However, there are certain aspects of its biology which we only learn very little about. We only obtain indirect knowledge about the life on the forest floor. It would appear that when the resin fell from the trees down to the ground, in most cases it already has such a hardened surface that material on the forest floor hardly became attached firmly enough to be protected against rubbing-off subsequently. Almost all the forest floor animals enclosed in amber seem to have been trapped by the resin from those cushions of moss which have surrounded the lowest part of the tree-trunks, the root-collar. The viscous resin has apparently flowed slowly down the moss-covered trunk, and its pressure has driven out the concealed animal life, so that some of the animals have been trapped by the sticky material. This is only an attempt at a reconstruction, but if the animals had not already emerged from the moss, it is difficult to explain why they were usually not captured together with particles of the moss they undoubtedly lived in.

Insect groups, particularly diurnal animals, which mainly are found in low vegetation or on steppe- and meadow-like regions, are weakly represented, including for example Heteroptera. The same holds for flower-seeking insects (bees and many flies), and gall-forming types (Cynipidae and Cecidomyiidae), also those genera which deposit their eggs on oak, and which might be expected to appear frequently. The infrequent occurrence in amber of dragon-flies and other large and powerful animals is due in particular to their strength, but often they have had to pay for their freedom by the loss of one or more legs or the tip of a wing. Parasites on vertebrates occur very rarely in amber, but a situation where an animal with such a mode of existence could have the opportunity of coming into contact with amber resin would also be quite exceptional. Nevertheless, this has in fact happened a few times in the case of Ixodidae and fleas.

The literature which has been employed in preparing the following pages is of an extremely mixed quality. Many of the studies, also some of the older ones, are of a very high quality from a taxonomic point of view, and have been composed by specialists with an intimate knowledge of the present-day world fauna, but the opposite must be said of a number of others. Further, the literature is handicapped with many nomina nuda and an unknown number of synonyms. The present author does not have the comprehensive special knowledge which is a prerequisite for correctly evaluating this aspect of the material available, but must in each separate case leave criticism to the specialists.

It is also to be regretted that not all that is reported as fossils from Baltic amber is in fact of this origin. This is particularly the case with Giebel's publication from 1862, where a gecko and numerous insects are described. This fossil resin is an East-Indian agatho-copal (Klebs, 1910; Hennig, 1966d). The species are almost exclusively of East-Indian origin (including the gecko). In other cases they are African, most of them belonging to recent genera. Hennig's study gives a list of the species in question, and these are not included here.

Vertebrata

The palaeogenic vertebrate fauna has not left many traces in the Baltic amber: some feathers and hairs, which can hardly be identified with any certainty, and a single footprint of a small mammal. However, the vertebrate fauna has undoubtedly played a considerable role in the natural history of that age. The terrestrial vertebrates undoubtedly differed considerably from those of today, but nevertheless they had undergone a very extensive development since the oldest amber forest of the Mesozoic spread over the continental mass of Laurasia. Giant reptiles, the dinosaurs, had definitely become extinct everywhere during the course of this period, and the warm-blooded birds and mammals had undergone a development which in extent could well be compared to that undergone by the angiosperms. Among the "ancient" animals, the snakes in particular appear to have been favoured by the new development, but these were also animals which to a very special degree benefited from the new myriads of small mammals, which became their chosen prey.

All the finds to date of amphibians (**Amphibia**) in Baltic amber have been shown to be forgeries, as is the case with fish (**Pisces**) as "amber fossils". However, salamanders and pelobatids are known from contemporary European deposits of other kinds. The amphibians have probably been at least as common as in present day regions in the tropics and sub-tropics with rich vegetation.

Reptilia. Apart from numerous forgeries, only a single lizard is known from Baltic amber, but on the other hand it was in a very well-preserved state (Klebs, 1910). It appears to be closely related to the South-African genus *Nucreas.* We know from the Central European deposits, however, that in the region of the amber forests there were both tortoises (e.g. *Eosphargis breineri* Nielsen (1959) from the Danish Mo-clay), lizards (including iguana, skink and monitor), snakes (especially relatively primitive forms) and crocodiles in considerable numbers. The last-mentioned have undoubtedly been favoured by the great abundance on tropical and sub-tropical ocean coasts of Central Europe.

It is difficult to evaluate how prominent a part has been played by birds (**Aves**) in the amber forests; for one thing the recent types of birds had just started their development, and for another, bird fossils have always been relatively rare, on account of the very delicate and fragile skeletons. In the Eocene, however, various European deposits have yielded skeletal parts of a number of sea and shore birds as well as of fresh-water birds such as heron, stork, duck, crane and rail, but in other genera than those now living. In addition, remains have been found of Diatrymidae, stump-winged and long-legged, very powerful giant birds, whose mode of life is unknown.

Among those birds which must to a higher degree be described as woodland birds, a few birds of prey and pheasants are known, as well as a shrike *(Laurillardia)* and the titmouse genus *Palaegithalus.* The cement stone of the Danish Mo-clay also contains well-preserved skeletal remains and feathers (Hoch, 1976). Specimens of down have been found in amber, which according to their microscopic structure must be titmouse down (Bachofen-Echt, 1949, p.183, 1944). These down specimens have important constructional features in common with the recent genera *Parus* and *Sitta* and may possibly originate from *Palaegithalus*, mentioned above. Other specimens of down are ascribed by Bachofen-Echt to species of small woodpeckers. Bachofen-Echt (1944) considers that in a few feathers it is possible to recognize the very characteristic construction found today in the South American genus *Momotus.*

There can hardly be any doubt that birds have been far more common in the amber forests than the few specimens of feathers and down found would seem to suggest. For example, innumerable insects captured in the resin have been partly eaten by birds before being completely enclosed in the protective mass. It is easy for belly and breast feathers to come into contact with the resin during this procedure, and they easily remained in contact, especially during moulting. This is also evidence for the presence of birds, able to clamber around the tree-trunks seeking for food, like the titmouse and woodpecker.

Mammals (**Mammalia**) have represented a constant element in the fauna of the amber forest, but they have deviated greatly from our present mammalian fauna, probably with greater deviations than any other animal group. In the amber itself, only hairs have been found so far, lost by the animal when roving around the resin-secreting plants. Comparative studies of these hairs led Eckstein (1890) to the view that in the great majority of cases, at any rate, they represented species closely related to the squirrel and dormouse. Bachofen-Echt (1944, 1949) has also studied hairs which, from their structure, can only have come from bats. The Copenhagen collection of amber fossils contains a few specimens which at any rate do not originate from bats.

In one piece of amber, impressions have been found of the forefoot of a small mammal (Bachofen-Echt, 1949, fig.188, p.188). This consists of four closely placed balls of the toes (no.2-5), behind and between which there is an elongated ball of the foot. The fact that there are no impressions of claws makes it likely that this is the front paw of a small climbing carnivore landing after a spring.

Two amber fossil fleas of the genus *Palaeopsylla* show with certainty that Insectivora have been represented (see later).

It is thus impossible by means of these random finds in the amber itself to construct any complete picture of the mammalian fauna in the amber forests. However, other European deposits from the same period have provided somewhat more information on this (quoted here from Romer, 1946). Marsupials were found, but at that time were almost extinct. Insectivores and bats, on the other hand, were common throughout the entire Tertiary. Lemurs and tarsiers were common during the Eocene, both in Europe and North America, but only in primitive forms, which presented clear evidence of their relationship with the insectivores. As early as the end of the Eocene, these primates disappeared from the European forests; already by then, it had become too cold here for these animals. No trace has been found of true monkeys, which at that time had just commenced their development, and which are known at the earliest from the lower Oligocene (in Egypt).

Those carnivores which roamed through the amber forests were generally small animals, belonging in the majority of cases to the now completely extinct group Creodontia, but towards the end of the Eocene the first representatives appeared of several now-living groups, although no now-living genus. The creodonts varied very considerably in their mode of existence, their biological range corresponding more or less to that of recent carnivores. Strikingly many of them were omnivorous or insectivorous. Also *Arctocyon*, which was as large as a bear, was presumably omnivorous.

The various groups of ungulates, many of which are now completely extinct, varied considerably in size. Most of those known, however, have been small, about the size of a present-day hare. This is also the case, for example, with the earliest European horse, *Hyracotherium*. Other forms were giants resembling the present-day pachyderms, and a number of them appear to have lived on roots, which they dug out themselves. Deer and Cavicornia did not exist as yet; the ruminants were in fact represented by forerunners of the camel stock: the hare-sized *Caenotheria* and the deer-sized *Anoplotheria*. In addition, some very primitive tragulids were found, which were the same size as the now-living forms.

Plate 7.
A. Silverfish (Lepismatidae) seen from above. This is the same specimen pictured on Plate 2 B.
B. Bristletail (Machilidae), seen from below. The rough background is due to fissures in the amber, the animal having been caught ventrally by the most dominant fissure. (G. Brovad phot.).
C. Polydesmid millipede (Diplopoda) from above. The anterior limits of the animal are at the base of the right club-shaped antenna; the dark spot in front of this is not a part of the animal.

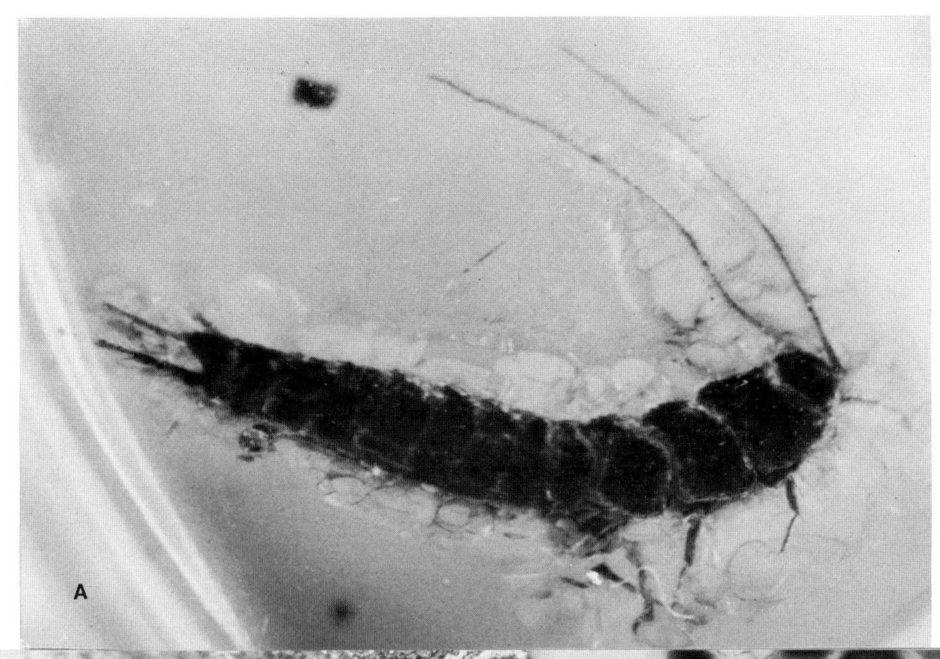

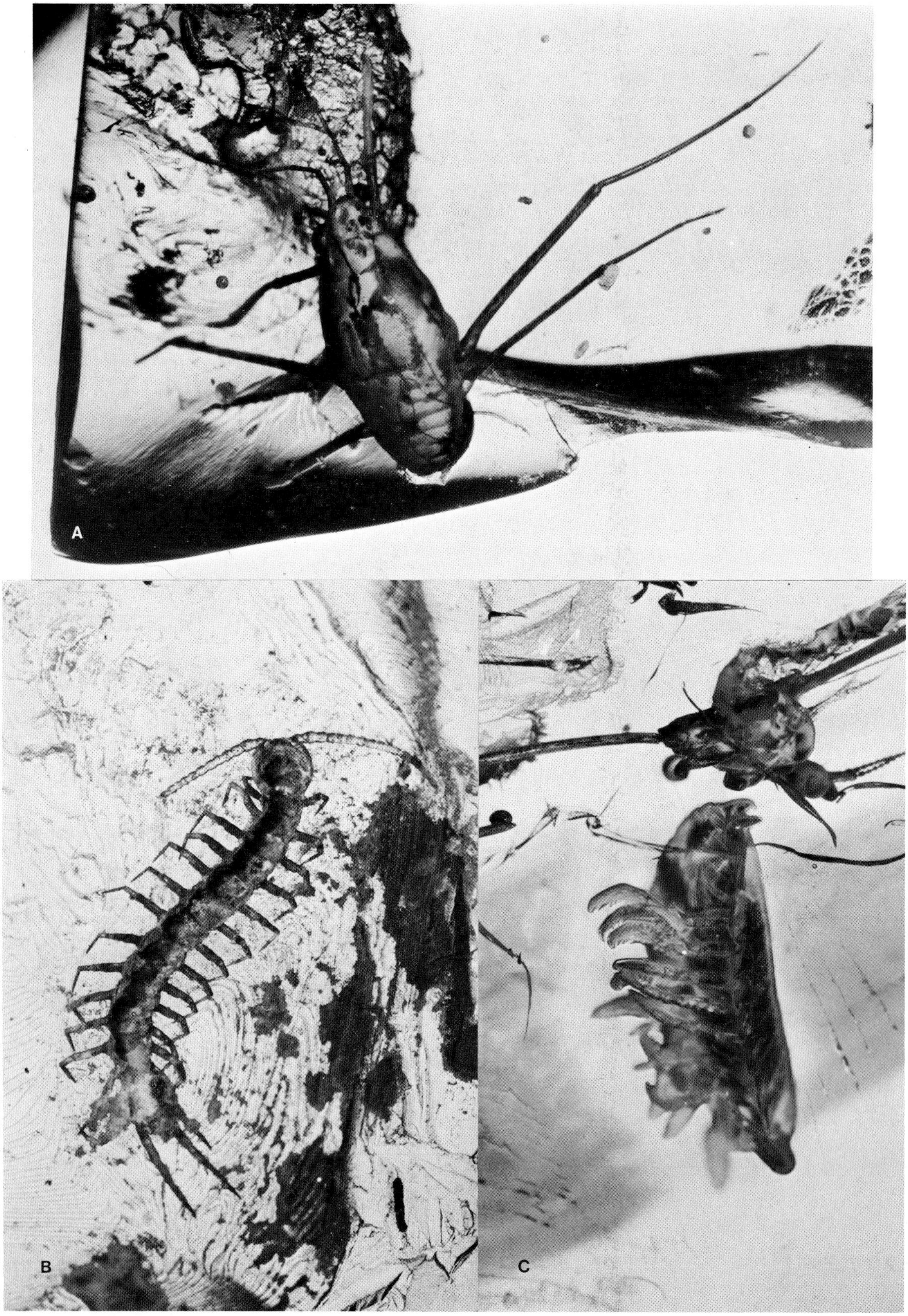

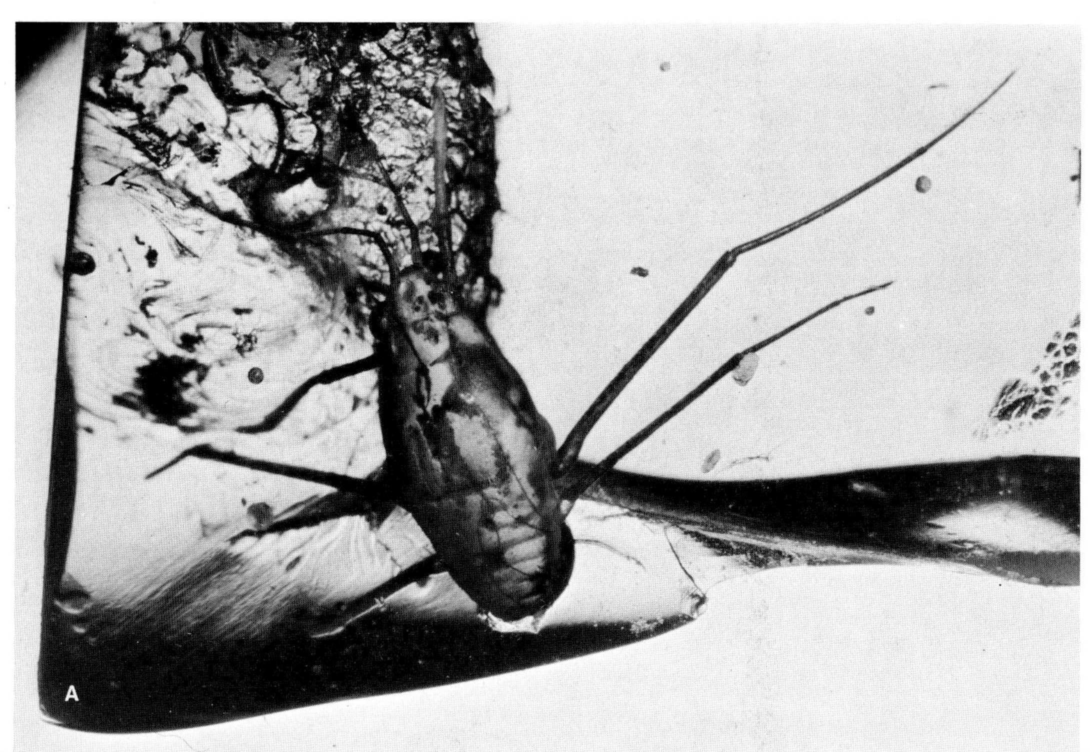

Nor had the rodents completed their development during the Eocene period. They were represented above all by the very primitive group, Ischyromyinae, which had its prime in Eocene and Oligocene, but which very soon became almost extinct. This group has only one now-living genus, the North-American *Aplodontia*, which are burrowing animals; it may be regarded as the primaeval group for all now-living true rodents. The rodent hairs found in the amber must be regarded as having come from the skin of one or other ischyromyid. In addition, the amber forests may have contained specially primitive forms of rodent hedgehog (Hystrichomorpha).

It is not possible to say precisely how far north these various vertebrates have lived, a number of them have probably inhabited only the most tropically featured parts of Central Europe.

Chapter A

Plant-sucking Insects

This section comprises fossil insects in amber, which obtained their food direct by sucking nourishment from amber forest plants belonging to highly developed groups as gymnosperms and angiosperms. In most cases it concerns insects bound to definite plant genera (at times plant species) to a more or less pronounced degree. By comparison with recent fauna and flora it will be possible, with relatively great certainty, to refer several of them to quite specific host plants. These insects, namely aphids and scale insects, which are rather immobile in their wingless phases, represent a very significant element in the fauna of the amber forest.

Aphids **(Aphidoidea)**. Originally, each single species was bound to a narrow circle of host plants, closely related to each other, or they lived exclusively on a single plant species. They lived here in a regular alternation of generations of different morphology and biology. In general, they were also bound to a single plant part: trunk, young shoot + foliage, or roots, at any rate in the case of each single generation. However, several aphid families have had a regular host change, most often between a bush or tree (primary host) and a herbaceous plant (secondary host), and these constantly alternating host plants have no mutual family relationships. Host changes of this nature are secondary and developed polyphyletically, and have not been anything like so common in the amber forest as in the boreal regions of the present. The aphid obtains plant juice from the vascular bundles of the plant, which implies a sedentary mode of life except for the migratory and dispersal periods characteristic for each species. Thus, this is a group of insects which can provide information on the flora of the amber forest. In addition, the group has recently been studied from a palaeontological point of view (Heie, 1967, 1968, 1969a, 1969b, 1970, 1971, 1972 and 1976), and this adds to our knowledge of the group to a high degree. The Copenhagen collection comprises more than 215 specimens, discussed by Heie in previous or coming papers. However, as it will appear from the above mentioned literature (Heie, 1967-76) the taxonomy of these amber aphids is characterized by some uncertainty.

Interesting among these Eocene aphids is the genus *Germaraphis* Heie (Text-fig.10) which constitutes about 60% of

Plate 8.
A. Young larva belonging to the Gerroidea, the pond skaters. (G. Brovad phot.).
B. Centipede (Lithobiidae). The rough background indicates that the animal was lively for a while after being trapped. (G. Brovad phot.).
C. A woodlouse probably belonging to the Trichoniscidae. Above the woodlouse is seen the head, thorax and most of the abdomen of a limoniid crane fly. (G. Brovad phot.).

the known material. It is not only its abundance which attracts attention, but also the striking fact that it is only known wingless, as parthenogenetic females (somewhat problematically, Heie) and especially as larvae of various ages. Not only winged parthenogenetic females, but also sexuales, are unknown. As the material contains several species, the genus has quite possibly been still in its prime at that period. Of 94 individuals, 63 belong to the species *G. (G.) dryoides* (Germar & Berendt) and 13 to *G. (Balticorostrum) oblonga* Heie, while the remainder is distributed with 1-8 individuals in each of the remaining species. The material has later been increased considerably.

Heie (1967, p.47) discusses in detail the systematic position of the genus and concludes that among present-day aphids it is most probably related to *Phloeomyzus*. He even refers it to the same tribe as this genus: Phloeomyzini. Nevertheless, the available biological data are evidence enough to suggest that *Germaraphis* has a rather isolated position. The genus is (together with *Phloeomyzus* and others) referred to the family Thelaxidae, an archaic family which as a whole has not developed any form of host-shift. The possibility can, however, hardly be denied that the genus will prove to be a rather heterogenous complex, should more complete material be found. Thus, Heie (1968, p.5-6) discusses the possibility proposed by Baker (1922), that Germar & Berendt's old species *dryoides* may have been wingless individuals of *Mindarus*.

Features of the morphology of the animal, namely the very long rostrum and the well developed claws of the feet, suggest that the animals have lived on rough bark with cracks and crannies, and the long stylet proboscis has been able to penetrate thick layers of bark into the fluid-conducting cambium. We also know that often, at any rate, they have been sitting in groups, as 2 or 3 individuals in the same piece of amber are not unusual, in a few cases up to ten larvae of almost the same age have been found together, while in one piece of amber even more larvae of different sizes were found, with at least 2 wingless adults (or last-stage larvae?) and half a score of ants of the genus *Iridomyrmex*. This is an example which also shows that already in the Palaeogenic period there was a symbiotic relationship between aphids and ants. Similar examples, although without any generic determination, are known from the amber museum in the former Königsberg (Bachofen-Echt, 1949, p.125). All this suggests that the genus, or at any rate its main species, has formed colonies of a wingless generation on tree-trunks and possibly on larger branches of the actual amber-producing tree.

It is remarkable that *Germaraphis*, which constituted a particularly flourishing group of aphids, apparently has no near relatives in the present-day fauna. It was a very characteristic representative

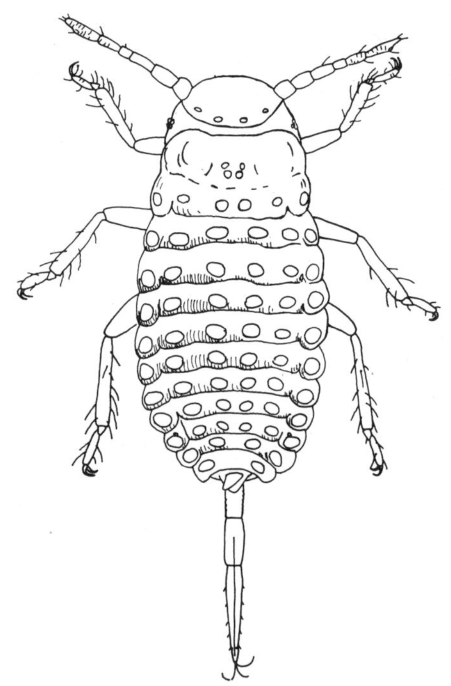

Text-fig.10. *Germaraphis dryoides* (Germar & Berendt), reconstruction of apterous specimen. Characteristic subcircular wax gland plates are seen on the dorsum, the wax having been dissolved by the fresh resin. Natural size about 1.3 mm. (From Heie, 1967, fig.8).

of the amber forest, apparently even of the fauna specially associated with the amber tree itself. The fact that it is extinct today seems incompatible with the idea that the amber tree could have been a pine of ordinary type. Pines have been common ever since, both in the subtropic and, especially, the boreal regions, and *Germaraphis* has hardly been associated solely with a single or some few, now extinct, *Pinus* species, unless such species have had characteristics which deviated significantly, chemically or physically, from that of the common recent type. The investigations by Beck and Langenheim and other authors on the infra-red spectra associated with the different resin-producers are of great interest in this connection. It seems certain that *Pinites succinifer* was the original resin tree, possibly a collective name, covering several mutually related species. It also seems to be certain that amber, developed from the resin from this tree, contains acids which, when dehydrogenated develop diterpenes of type I, in contrast to the majority of recent Pinaceae, where the resinous acids are of the abietic type (diterpenes of type III). This is a considerable deviation from the norm of the family, and one which the amber tree shares with only a few now-living *Pinus* species, whereas it is a feature in common with the Araucariaceae. It is probable that the extinction of *Germaraphis* was connected with a special relationship to the above chemical structure of the resinous acids. It is uncertain, however, whether Baltic amber and related types of North American amber originate exclusively from *Pinites*. There may have been larger or smaller quantities in between, originating from "Protopinaceae" or primitive araucarians, plants with a very great holarctic spread, and among whose aphids there have quite likely been included close relatives of *Germaraphis*, and perhaps even this genus itself. From these considerations, a study of the fauna of aphids on the recent *Agathis* species in the East would be of great interest; the trunk of these trees is the biotope where there is the greatest theoretical possibility of finding close relatives of *Germaraphis*. Or possibly they might be found as fossils or subfossils in *Agathis* copals.

There can be no doubt that *Germaraphis* has had at least one winged generation, but since this has never been found in amber in spite of a probably high incidence, it must have been present at a season when the amber production was down to a minimum, when in other words the host plants have been preparing their resting period. Biologically, therefore, it is likely that it was the amphigenetic generation which was winged. This would be a very archaic characteristic, also in agreement with the conditions in *Phloeomyzus* and other aphids with the same generic relationships.

In the present-day forest, the genus *Mindarus* is very common under boreal climatic conditions on different conifers of the pine-tree family, *Abies* being specially preferred by the most common present-day species. The eggs are deposited on the young twigs, and the larvae suck the juice from the needles on the fresh shoots. The wingless fundatrix, hatched from an overwintering egg, produces a generation which becomes the winged parthenogenetic migratory generation, which produces the wingless amphigenetic generation. The eggs are deposited in mid-sommer, just when the host plant's annual shoots have terminated their growth, so that the genus obtains a very long resting period at this stage.

Conifers like *Abies* and *Picea* are the most likely hosts for this genus, and these trees are known to have grown in the amber forests, although not common. The length of the time the aphids have been on the wing has been short, as in recent species, and has been during the seasonally-determined strongest growth period of the amber trees. The winged forms have therefore had every posssible opportunity of being trapped by the resin. Three winged specimens are known from Danish amber and three from the East Baltic amber, all migratory specimens which by chance have come into contact with the fresh resin. If the genus had been of merely common occurrence in the amber forest, one would have expected its winged generation to be more numerous in amber. Flying aphids, however, are very easily carried up into the air plancton. It is therefore not likely that the genus has lived and laid its eggs on *Pinites* or other conifers in the amber forest, as a regular occurrence. It is much more likely that the winged aphids are brought with the air stream from more boreally characterized regions (more northerly or higher-lying forests) where the amber tree has not grown.

Palaeothelaxes Heie, to some degree related to above mentioned aphids, occurs in the Copenhagen amber in a winged and a wingless specimen. Like its recent near relative *Thelaxes* it has probably lived on shoot-ends and leaves of Fagaceae (oak, chestnut and others) without host shift. It has presumably lived in the amber forest, but its appearance in the amber seems to be due to chance.

Callaphididae (Text-fig.11) have been a widespread aphid family in the amber forest, and probably far more common north of this boundary. Also nowadays it is common and very widespread mainly in boreal regions. It is found on woody plants, at least particularly deciduous trees. Some species are mentioned from conifers (Patch, 1938, p.41), but none from Araucariaceae. There is no change of host.

The most interesting of these Callaphididae is *Mengeaphis glandulosa* (Menge), 9 specimens of which are found in the Copenhagen amber, all wingless and presumably larvae (Heie, 1967, p.115). Morphologically it shows that it has lived on rough tree bark, and on the same grounds as in *Germaraphis*, it must be presumed to have lived on the trunk of the amber tree at any rate during this one stage of its life cycle, which is otherwise unknown.

Most other callaphidid species are known in a few specimens only, and the amber forest is unlikely to have been their actual home. They are known in very different stages, winged and wingless females with almost equal frequency, there are small and large larvae and several empty exuvia from late larval stages. They have presumably lived mainly on oak or other Fagaceae, but hardly on the amber tree. Some genera belong to the group Drepanosiphonini, which today live exclusively on the maple. In all probability they also lived so in the Eocene, and maple has been quite common at any rate locally in the amber forest. The systematic placing of several genera is very doubtful.

Drepanosiphonini are also represented in the Mo-clay. A few species of the genus *Siphonophoroides* Buckton have been found here, and the genus has also been found in Oligocene layers in North America (Florissant, Heie, 1967, p.208; 1970a, p.166).

In addition to the aphid groups already mentioned, a single winged specimen of *Succinaphis flauensgaardi* Heie (1967), belonging to the subfamily Pemphiginae, is known from Danish amber. In all probability it lived on the poplar. As all recent Pemphiginae are associated with poplar, this find makes it likely that

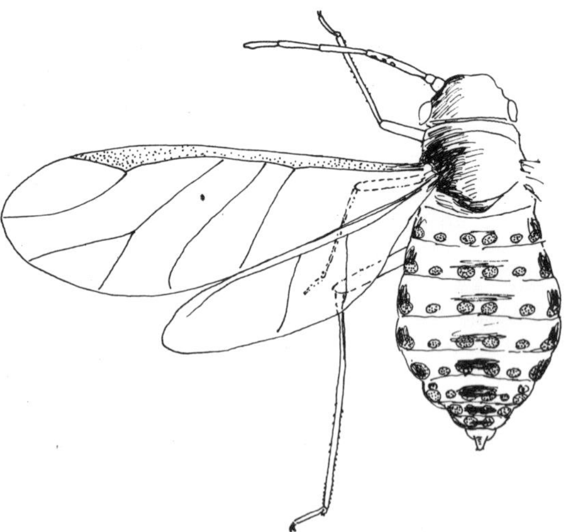

Text-fig.11. *Palaeophyllaphis longirostris* Heie (Callaphididae), reconstruction of alate morph. Natural size 1.2 mm. (From Heie, 1967, fig.30).

poplar, which is known from other plant communities at that time, has had a place in the neighbourhood of the flora of the amber territory, but undoubtedly a very modest place. The specimen may also have been carried over a great distance as air plankton, and this is perhaps the most likely explanation.

Aphididae. A very significant find is *Diatomyzus eocaenicus* Heie (1970, p.163) from the Danish Mo-clay (oldest Eocene). In many ways this is a primitive form which cannot be placed within the known recent subfamilies. This aphid, unknown from amber, has at any rate made its last journey across the North Jutland sea as air plankton. It is only possible to guess as to how far this last journey has stretched, but as Aphididae in general are boreal insects, it seems reasonable to imagine that this specimen has come from a country north of the amber forest belt.

Elektraphididae (Cockerell, 1915; Heie, 1967 (under the name *Antiquaphis)*, 1970, 1976; Steffan, 1968) are represented by seven fossil species, in all by less than a dozen specimens, most of them known from Baltic amber (Text-fig.12). They belong most probably all to one and the same genus: *Schizoneurites* Cockerell. They seem to be intimately related to the Adelgidae of the oviparous aphids. Only alate specimens are known.

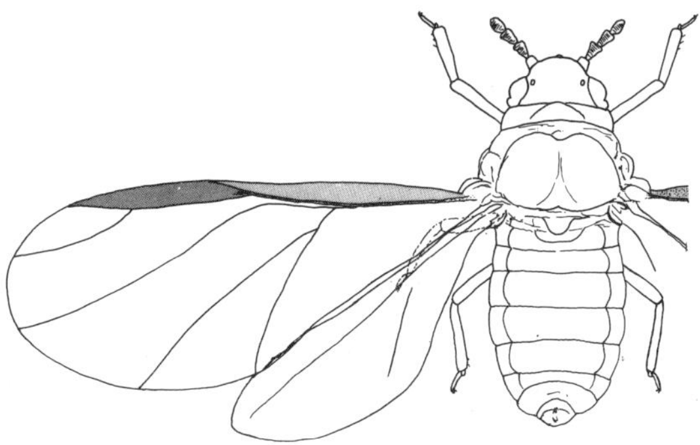

Text-fig.12. *Schizoneurites obliquus* Heie (Electraphididae), reconstruction of alate morph. Length of forewing 2.0 mm. (From Heie, 1976, fig.1).

It is not possible with certainty to know anything on the biology of the family. Adelgidae are stricktly bound to conifers, possessing a quite special type of host-shift, and the more distant relatives, the Phylloxeridae, are exclusively living on deciduous trees but without change of host. Most probably the Elektraphididae possess a single alate generation in the summer, developed on a host plant which now is extinct. This host can possibly have been the amber pine, and the intake of food can in most of the life-cycle have taken place on young shoots and needles of this tree. The problem can, however, hardly be solved.

If we ignore *Germaraphis* (and probably also *Mengeaphis)* as being quite obviously the specific aphid of the amber tree, and one whose biological preference we for that matter know nothing of, as it is extinct like its host plant, then actually only the Elektraphididae seem to be aphids really native to the amber forest. This latter family is a group of aphids which possibly may have been more intimately connected with the tropics than other fossil aphids found in amber. The other fossil aphids found here are today more or less pronouncedly boreal groups, and they are present in amber in such surprisingly small numbers (even the callaphidids), that they could hardly indicate more clearly that here they had been out of their natural climatic region — or at any rate in a boundary region. As a whole, the aphids are a pronouncedly boreal insect group, whose forms of host-shift almost all, in various ways, indicate that they must have developed under boreal conditions.

It is therefore not surprising that there are a number of families which are still quite unknown from amber. These include Adelgidae, Phylloxeridae, Lachnidae and Eriosomatinae. As already mentioned, among the Pemphiginae only a single winged specimen is known from amber, and among the Aphididae one from the Mo-clay. Even though a number of these families have undoubtedly undergone an important part of their evolution in the course of the Tertiary, this being the case above all for the Aphididae, which today include more than 2000 species, they must all have been in existence during the Eocene, and already at that time have undergone a significant part of their evolution and specialization. We know, however, that north of the real amber forest there have been forest regions where the climate has been temperate, and here there has been a flora deviating from that of the amber forest. As a result, there must also have been a corresponding boreal fauna. Within these northern forest regions it would have been natural to find Adelgidae on spruce and other conifers; Phylloxeridae on walnut and oak; a far greater wealth of Callaphididae; Lachnidae partly on conifers, partly on various deciduous trees; primitive Aphididae on poplar, in due course extending over the entire flora; Eriosomatinae on elm (primary host) and Pemphiginae on poplar (original primary host?). The two last subfamilies have probably had roots of spruce (or perhaps some herbaceous plants) as original secondary host. However, definite knowledge as to such a fauna is not available.

Among scale insects (**Coccoidea**) only a few species are known from older literature (Germar & Berendt, 1856; Menge, 1856), and from this it seems that the genus *Monophlebus* (Margarodidae) (Text-fig.13) dominates. This is highly confirmed by the Copenhagen material, which at the moment is being studied by Dr. Barbara Ogaza, Poland. This collection comprises 91 pieces of amber with scale insects, containing 63 specimens of Margarodidae. According to Dr. Ogaza (personal communication) the families Ortheziidae, Eriococcidae, Pseudococcidae, Coccidae and Diaspidae are also represented, although some of them by only a very few specimens. The collection contains a number of wingless females, almost all of them Margarodidae.

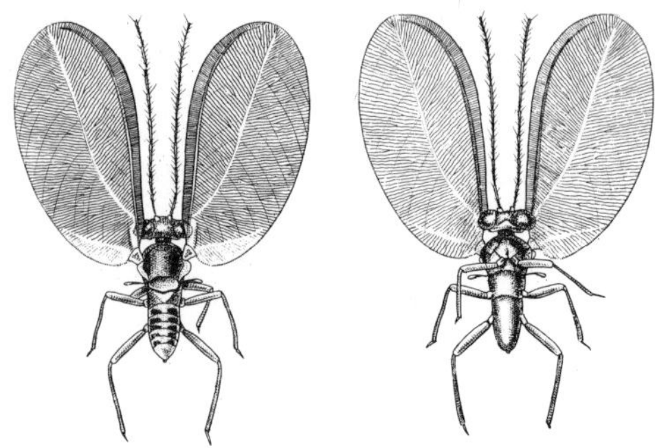

Text-fig.13. *Monophlebus pinnatus* Germar & Berendt (Coccoidea), reconstruction of winged male in dorsal and ventral views. (From Berendt, 1856, pl.I, fig.1).

All these females possess a pronouncedly articulate body and well developed legs, and they have all been freely movable. First-stage larvae are found in most of the families, although pseudococcids are far more common than any others. Males are present in 52 amber pieces; most individuals of this sex are winged.

The scale insects constitute a mainly tropical and subtropical insect order, although some of them are boreal, but only very few species live in really cold areas. Thus, they must be said to be true inhabitants of the amber forest zone. It looks as if the coccids are evolved concurrently with the angiospermous flora, and are therefore a relatively young group. Only few taxonomic units depend on coniferous trees, these units in addition being mutually unrelated. Among recent species may be mentioned *Physokermes abietis* (Coccidae) on spruce and the diaspid *Aonidiella taxus* on *Taxus* and *Podocarpus*. Regarding the comparatively great number of female margarodids it seems likely that at least one species of this family has been an ordinary member of the fauna living on the amber tree trunk. On the other hand, the many first-stage larvae tell nothing about their host plants, since these young larvae represent the dispersal stage of the coccids; they are very active and easily become part of the air plankton.

White flies (**Aleuroidea**) in the Baltic amber have so far only been known from one species, *Aleurodes aculeatus* Menge (1856, p.18). In the Copenhagen collection there is a material of 15 specimens, partly studied by Schlee (1970). However, as only imagines are known, and no larvae or pupae, it is not possible to distinguish reliable species in this material, but on the same grounds it may be said with relatively great certainty that the group has not bred on the amber tree, at any rate not on the trunks and large branches.

Aleurodicus burmiticus Cockerell(1916) is known from Burmese amber. Schlee (1970, p.32), however, is of the opinion that its relationship to the Aleurodidae is doubtful.

Of much deeper interest is the discovery in Lebanesian amber of the genera *Heidea* and *Bernaea*, each in one specimen (Schlee, 1970). These animals, 2 to 3 times as ancient as the insects in Baltic amber, deviate significantly on many points from the white flies of the Tertiary and Quaternary. They represent apparently an early stage during the specialization of the Aleuroidea. Schlee writes (p.39): "Diese mesozoischen Fossilien weisen einen Körperbau auf, der in mehreren Punkten vom Bauplan rezenter Mottenläuse abweicht. So könnte es — bei relativ grober Betrachtung — scheinen, dass die Tiere den Kopf einer Zikade, das Hinterende eines Blattflohs und den Rüssel einer Blattlaus aufweisen, während die übrigen Merkmale den Mottenläusen entsprechen."

The Lebanesian species do not appear to have lived on smooth, soft plant surfaces like the recent species, but much more probably on uneven bark (the long proboscis). This could be conifer bark, as Schlee thinks, but it need not be the case. It is probable that already by the time of the transition to the Cretaceous period, the angiosperms had achieved a quite considerable development, and in fact the region where the Lebanon amber occurs was quite near to the region of the theoretical development of the angiosperms in South-Eastern Laurasia (according to the ideas of Takhtajan, 1969).

Among jumping plant lice (**Psylloidea**), only a very few specimens are known from amber, among which Bachofen-Echt (1949, p.173) mentions a species of *Strophingia* (Psyllidae). The Copenhagen collection contains 3 undetermined specimens.

Cicadas (**Auchenorrhyncha**) are relatively common in the amber forest fauna, but in the meadow and steppe fauna of the Mo-clay they are together with bugs a completely dominating insect group.

As early as the middle of the 19th century, many species were described in amber, above all by Germar & Berendt (1856). These species belonged to the superfamilies Fulgoroidea, Cercopoidea and Jassoidea, and in addition with some doubt, Cicadoidea. It is, however, uncertain whether the species described actually belong to the genera to which they have been ascribed. It is interesting that Fulgoroidea was a group rich in species, as this corresponds very well with the findings from the rich Mo-clay fauna of the same age. The Fulgoroidea (Plate 3,A) apparently played a considerable role in tropical and subtropical faunas already at that time. Those genera mentioned in the older literature are *Cixius* (9 species), *Flata* (2 species), *Ricania* (1 species), *Poeocera* (3 species). Bachofen-Echt (1949, p.173) adds the genera *Oliarius*, *Pseudophana* and *Issus*. Of Cercopoidea, 3 species of *Arthrophora* and 2 species of *Cercopis* are known. Of Jassoidea, 1 species of *Bythoscopus*, 2 species of *Tettigonia* Geoffr., 2 species of *Typhlocyba* and 2 species of *Jassus* are known, and the number of genera is increased by Bachofen-Echt with *Thamnotettix*, *Cicadula* and *Deltocephalus*. Bervoets (1910, p.125) and Jakobi (1938, p.188) have described a few cicadas in amber, belonging to the above genera.

The Copenhagen amber collection contains 102 specimens, among which both Tettigometridae (not previously mentioned from Eocene), Cixiidae, Issidae, Cercopidae and Jassoidea are represented. Danish amber appears to be relatively poor in species living on herbaceous plants. About 50% of Jassoidea are larvae of different ages. Among Fulgoroidea there are also a number of larvae of various ages, the great majority of which appear to belong to one and the same genus. An adult issid and a larva of the same species have been found in one piece of amber (Plate 3,B). Among cercopids, there are only few larvae (*Arthrophora?*). These observations suggest that the larval stage of cicadas coincides with the most active period of resin secretion by the amber tree, that is, in the local spring and early summer, so that it is likely that the amber trees were not everywhere part of a massive and dark forest thicket.

The possibilities cannot be excluded that some cicadas have been among the special fauna of the amber tree. Not least on these grounds, a closer evaluation of the material, especially of the larvae, would be of interest. Biologically, however, cicadas are such an actively moving insect group that the larvae found in the amber could easily have originated from nearby deciduous trees. At the present day, not least of all oak houses many species. On the other hand, they have hardly spread with air plankton.

Most of the Danish amber cicadas are well preserved. In some, namely cercopids, parts of the body have however been lost, as is so often the case with large strongly sclerotized insects in amber (e.g. among cockroaches and click beetles).

A remarkable cicada is known from the Canadian Cretaceous amber, *Jascopus notabilis* Hamilton (1971). This apparently occupies an intermediate stage between Cercopidae and Cicadellidae. Hamilton has therefore established a new family for it: Jascopidae.

Bugs (**Heteroptera**, pars). Water bugs (Hydrocorisae and Gerridae) are mentioned later, and the myrmecophilous *Prophilocerus* after the ants. The terrestrial bugs are relatively rare in amber, but very common (indeed) in the cementstone of the Mo-clay, a difference which undoubtedly depends to a great degree on the mode of life of these insects. Bugs live almost exclusively free-moving on vegetation, especially herbaceous plants and low bushes, so that it is quite natural to find more of them on the open coastland (the Mo-clay fossils) than in the actual, more or less wooded, amber region. Most of them are plant suckers (dominating among Miridae, Lygaeidae, Berytidae, Tingidae, Pentatomidae, Scutellidae and Cydnidae, to mention some of the families found in Baltic Eocene deposits). Other families are more or less pronounced predators (among those occurring: Saldidae, Anthocoridae, Nabidae and Reduviidae). Only Aradidae live exclusively in forest, where they are found under loose bark on both deciduous and coniferous trees or in tree fungi, where they suck the juice of other small animals. They are only known from amber, where they occur in three species, and in addition an otherwise non-specified larva is known (Germar & Berendt, 1856, p.22).

The Miridae dominate in amber. Germar & Berendt (1856, p.24-29) describe no less than 13 species. They are all grouped in the genus *Phytocoris*, but undoubtedly belong to a greater number of genera. Among the plant sucking families, Germar & Berendt also describe 2 *Pachymerus* (Lygaeidae) and *Tingis quinquecarinata* (Tingidae), and Menge mentions *Berytus* (Berytidae). Scudder (1885) mentions the following genera: *Aetorhinus, Harpocera, Homodemus, Hoplomachus, Lopus, Lygus, Oncotylus, Orthops* and *Systellonotus*. The insect-sucking bugs are found in the same biotopes as the plant-sucking bugs. Also in their case we are indebted to Germar & Berendt for the most of our knowledge, as they describe *Salda exigua* (Saldidae), *Limnacis succini* and *Platymeris insignis*

(Reduviidae) and *Nabis lucida* (Nabidae). In addition they give an illustration of a *Reduvius* larva (pl.2, fig.9). Menge mentions in his publication a similar specimen under the name *Nabis prototypa* (1856, p.20).

The Copenhagen collection of amber includes 53 undetermined specimens, and also here Miridae dominate. More than one third of the Danish amber bugs are larvae, which shows just as in the case of the cicadas that larval development has at any rate mainly coincided with a period of lively resin secretion in spring and early summer.

Fringe-wings (**Thysanoptera**) are represented in the Copenhagen material of Baltic amber by 73 specimens, 9 of which are larvae. These have so far not undergone any scientific study. What follows is especially based on studies by Priesner (1924 and 1929) and Bagnall (1914 and 1924) on Prussian material. Possibly with the exception of a single larva, it refers to recent families. Of the genera present, about 2/3 are extinct. This is a large percentage by comparison with conditions in other insect orders, although this may be due to Priesner's personal evaluation of the extent of the concept genus. In addition there is a minor material of three imagines and two larvae, published by Stannard (1956); this gives no information on the locality of origin of the material.

In the published descriptions, Terebrantia are represented by Aeolothripidae, Heterothripidae, Thripidae and Merothripidae, Tubulifera by Phloeothripidae.

According to its morphology, Aeolothripidae is considered the most primitive family; it has relatively broad fore wings with 3 longitudinal veins and also a few transverse veins, and the antennae are 9-jointed. Priesner even considers that in *Archankothrips* the end joint can be recognized as 5-jointed, which might suggest that the Thysanoptera originated among insects with at least 13-jointed antennae. The recent species are rapid in their movements, also at the larval stage. Many species are predators, found in herbaceous vegetation, often in flowers. Priesner's amber material contains 5 species as well as larvae of the genera *Archankothrips*, *Promelanthrips* (2 specimens) and *Melanthrips*. *Stenurothrips bagnalli* Stannard (1956, p.453) must presumably be included in this family.

Of the Heterothripidae, only *Protothrips speratus* Priesner is known.

The Thripidae are also at the present day the family most rich in species. In the amber material, Priesner recognizes 22 species, distributed among 7 genera, the most dominating of which are *Anaphothrips*, *Oxythrips* and *Taeniothrips*. In the case of *Anaphothrips*, which at the present day appears to be associated with monocotyledonous plants, especially on swampy ground, the Eocene material includes no less than 4 larvae, just as many as there are imagines. In the case of *Oxythrips*, whose recent species live on conifers, no larvae are known. In the case of *Taeniothrips*, today found everywhere in herbaceous vegetation, especially in flowers, but which can also be found on conifers, only a single larva is known. The explanation for this remarkable distribution is probably that only the larvae of *Anaphothrips* have lived free on the vegetation, while the majority of the others have developed while hidden in fruits and flowers, as is also general today within this family. In addition, Stannard (1956, p.456) describes *Hercinothrips extinctus*.

Only 2 species of Merothripidae are known: *Praemerothrips hoodi* Priesner and *Merothrips fritschi* Priesner.

Tubulifera are regarded as the most highly specialized fringe-wings. They are plant suckers, mainly living under dead bark, where their food is fungi and algae. They move slowly. Many species, however, are found in herbaceous vegetation. So far 6 genera have been found in Baltic amber, each with its own species. In addition one unspecified larva is known. Only one of the genera, *Haplothrips*, is known also in the present-day fauna.

It is emphasized in the literature

(Priesner, 1924; Ander, 1942; Bachofen-Echt, 1949), that the Aeolothripidae are relatively more common in amber than in the recent fauna, an observation which should indicate a relatively earlier stage in the phylogenetic development of the fringe-wings. The difference can hardly be said to be strikingly great, however, and is rather due to the relatively great activity and more roving tendency of the family, a necessary characteristic in most predators. It might seem surprising that the bark-living Tubulifera were not found in greater numbers. As however they are associated in particular with dead trees without any resin secretion, and are also characterized by their slowness, they have had no greater opportunity of being captured by the fresh resin than many other species, and not at all as concealed larvae. Everything suggests that in the Eocene, the fringe-wings had long been split up into the known families.

Among recent fringe-wings, both short-winged and completely wingless individuals occur in greater or lesser numbers in many species which are normally macropterous. No short-winged forms are known among the amber Thysanoptera. This might be due to a lesser tendency to be captured by the resin, but as the amber contains numerous larvae, which have shared the same conditions of capture as far as the short-winged forms are concerned, this can hardly be the reason for the absence of the short-winged forms. It is possible that brachypterism had not yet developed within this presumably relatively young order of insects, or the warm climate has hindered its development. The most likely explanation is that the generally warm climate of the earth during this period has hindered brachypterism, at any rate in Eocene Central Europe.

Apparently, no fringe-wings have sucked the sap in the actual living, resin-producing amber tree, so their presence in amber must be due to chance. The relatively many larvae, which would have difficulty in coming into contact with the resin by their own efforts, suggests that wind-spreading in the form of low-level air plankton has played a considerable role. This is not least striking in the case of the relatively many *Anaphothrips*, which mainly live on various monocotyledonous marsh plants, and whose larvae live free on the plants.

Zur Strassen (1973, p.1-51) describes Thysanoptera from Lebanesian amber.

Chapter B
Leaf- and Seed-Consumers

These form a heterogeneous group, both taxonomically and biologically. Their sole common feature is that in the course of at least one developmental stage they have obtained their nourishment by gnawing living seed-plants.

Leaf beetles (**Chrysomelidae**). The main genera of all the now-living sub-families are familiar from the older amber literature, from Germar in 1813 to Klebs in 1910. However, only very few species have been described: *Crioceris pristina* Germar (1813, p.4), *Electrolema baltica* Camillo Schaufuss (1891, p.63), *Sucinagonia javetana* Uhmann (1939, p.18) and *Oposispa scheelei* Uhmann (1939, p.21), the last two belonging to a now extinct group of Hispinae. With regard to *Electrolema baltica* Bachofen-Echt (1949, p.113) writes that it is very common and apparently lived on the needles of the amber tree. However, the description of the species is based on a single specimen from Helm's collection, and Schaufuss does not refer to any larger material which can result in biological conclusions. In Klebs' material of 30 specimens of leaf beetles, the genus is not mentioned, not even merely as *Lema*. As the Criocerinae as a whole are associated to a special degree with the monocotyledonous flora, there are grounds for doubting that this species has lived particularly on the needles of the amber tree. Apart from those mentioned here, Giebel (1856, p.120) has described *Chrysomela succini*.

Both as larvae and imagines, leaf beetles are in the great majority of cases associated with herbaceous plants, bushes and small trees. In general, therefore, it is not very likely that any of the species which are fossils in amber should have been associated with the amber tree, even though Menge also mentions a larva which he refers to the genus *Chrysomela* (1856, p.23).

The Copenhagen collection contains 7 undertermined Chrysomelidae.

Weevils (**Curculionidae, pars**). The classic amber literature mentions numerous genera which are common at the present day: Burmeister (1832, p.635): *Polydrosus, Thylacites* and *Phyllobius;* Keferstein (1834, p.330): *Dorytomus;* Berendt (1845, p.56): *Sitona, Phytonomus, Hylobius, Pissodes, Apion* and *Rhynchites;* Helm (1896c, p.228, and 1899, p.37): *Sitona, Phyllobius, Mecinus, Bagous, Apion* and *Ceutorhynchus.* Only a single species, *Erirhinoides cariniger,* has been characterized in detail (Motschulsky, 1856, p.27). In 1910, Klebs published Reitter's determinations of 47 specimens from his collection: 1 *Acalles,* 1 near *Acalles,* 2 *Anthonomus,* 1 *Apion,* 1 *Balanobius,* 2 Calandrini, 2 Cossini, 1 *Choerorrhinus,* 1 *Cryptorrhynchus,* 1 *Dryophthorus,* 1 *Dorytomus,* 1 *Erirrhinus,* 1 *Sitona,* 1 *Lixus,* 1 *Magdalis,* 1 *Mesites,* 1 *Nanophyes,* 2 *Notaris,* 1 near *Phytonomus,* 13 *Phyllobius,* 1 *Pseudostyphlus,* 2 *Rhinoncus,* 1 *Rhynchites,* 1 *Rhyncolus,* 1 near *Rhyncolus,* 1 *Trachyphloeus,* 1 Tychinini and 3 undetermined. Since then, Wagner (1924, p.134) has described an *Apion* species, Voss (1953, p.119) has described species from the collection in Hamburg's Geological State Institute *(Anchorthorrhinus incertus, Apion subdiscedens, Archimetrioxena electrica, Car succinicus, Isalcidodes macellus, Necrodryophthorus inquilinus, Paonaupactus sitonitoides, Polydrosus scheelei, Synommatus patruelis),* and (1972) the following from the Zoological Museum, Copenhagen: *Involvolus liquidus, Apion anderseni, Otiorrhynchus pellucidipes, Phyllobius sobrinus,* *Phyllobius cephalotes, Ampharthropelma decipiens* and *Pissodes henningseni.*

As already appears from Klebs' collection, it is characteristic that the majority of the genera only occur as one or a few individuals. Only *Phyllobius* occurs in a comparatively large number. In this genus, the well-known green weevils from the spring-time deciduous forest, the imagines of almost all the present-day species live on the young, newly unfolded leaves of many different deciduous trees, and it may be presumed that they have been well represented on the many oak trees of the amber forest. The occurrence of these insects on the amber tree is at any rate only accidental, and is apparently the result of the frequency with which these insects occur in the spring and of their tendency to go on the wing in full daylight. The plentiful occurrence of *Phyllobius,* the recent members of which have such a relatively short annual season at the imago stage, confirms that in the amber territory there have been annual climatic fluctuations which have been expressed among other things by a fixed cycle in the life of the vegetation. There has been a definite time of the year, a spring, when the general resting period of the vegetation has been interrupted, and when the oak, presumably mainly evergreen, the main food of *Phyllobius,* has sprung out. At that time, the flow of resin from the amber trees must also have been fully established.

Other genera which were particularly associated with the flowers and fruits of the deciduous trees are *Balanobius* (oak, willow), *Dorytomus* (willow, poplar), *Anthonomus* (especially Pomaceae and Amygdalaceae) and *Rhynchites* (leaves of deciduous trees).

Eighteen specimens of the Klebs collection belong to genera living in dead wood, in the great majority of cases of deciduous trees. They will be discussed later on.

A further 13 specimens in this collection represent genera which for the great majority live on herbaceous plants, in part on their roots. No less than 4 of these genera *(Lixus, Notaris, Erirrhinus* and *Nanophyes)* are today more or less firmly associated with the rich vegetation of the banks of various fresh waters. This suggests that the amber tree has been able to grow on fairly moist soil. The remaining genera are *Trachyphloeus* (rather low plants and their roots), *Sitona* (Papilionaceae), *Phytonomus* (various herbaceous plants, c.g. *Rumex)*, *Pseudostyphlus (Matricaria), Rhinoncus (Polygonum* and *Rumex)*, Tychinini (specially Papilionaceae) and *Apion*, the larvae of which live in seeds, especially Papilionaceae: herbs and small bushes. They are all genera whose main biotopes lie outside the actual forest regions. Apart from *Trachyphloeus* and *Otiorrhynchus* they are willing flyers.

None of the weevils appears to have had any constant association with the amber tree, and most of them must have lived in the edge of the wood or in the vegetation along the water courses.

Bean beetles **(Bruchidae)** as amber fossils are mentioned only by Helm (1896c, p.228), without further determination. As the larvae of this family live almost exclusively in the seeds of angiosperms, almost always of Leguminales, they cannot be considered as directly dependent on the amber tree.

Sawflies **(Tenthredinoidea)** are heavy-flying insects, the often free-living larvae of which gnaw the vegetation in the same way as the larvae of leaf beetles and many Lepidoptera. They are surprisingly uncommon in amber, and no species have been described. Brischke (1886, p.279), from a material comprising Menge's and Helm's collections, mentions *Lophyrus, Selandria* and 3 imagines and 2 larvae of the genus *Tenthredo*. Menge (1856) mentions in addition to these an *Emphytus* and larvae of *Lophyrus, Pamphilus* and *Cimbex*. However, these larvae are hardly all correctly determined. The Copenhagen collection includes 2 imagines: 1 belonging to Lophyridae and 1 to Tenthredinidae.

The larvae of sawflies live almost exclusively on angiosperm plants, and of the genera found in amber only *Lophyrus* is found exclusively on conifers. Ignoring the quality of the determinations, the material contains a very large percentage of larvae, which observation is related to the difference in the mode of life of the two stages of development. Oak is a very frequent host plant for recent species of these genera, and most probably the fossil larvae in the amber have been captured in the resin during their lowering themselves down from the crown of the trees to the forest floor, where their pupation will have taken place. The imagines, on the other hand, because of their relatively slight interest in flying, have been less exposed than most other wasps.

With regard to *Lophyrus*, it could be imagined that this genus had lived on the actual amber tree, as it is an insect with a pronounced coniferous habitat. If this had really been the case, a far greater number of larvae would have been expected in amber than the single specimen known, as the danger while the larvae were hanging in their self-spun thread would have been very great, and wherever they occur, the *Lophyrus* larvae are always in great numbers. *Lophyrus*, however, is a pronounced boreal animal.

Of the **Cephidae**, whose occurrence in amber must be quite accidental, Menge (1856, p.24) mentions a *Cephus* "mit sehr deutlichen zähnen der legeröhre", and Konow has since described *Electrocephus stralendorffi* (1897, p.17).

Chapter C
Gall Producers

Gall wasps **(Cynipinae)**. It might be expected that the amber forest, being the luxuriant mixed oak forest it undoubtedly was, would have been an excellent locality for gall wasps and quite specially for the gall-making sub-family Cynipinae, but this appears to have been far from the case. The literature mentions only few, Bachofen-Echt (1949, p.131) reckons with 5-6 species of the genus *Cynips*. *Cynips* from amber is mentioned for the first time by Schlotheim (1820, p.43), and Presl describes (1822, p.195) the species *Cynips succinea*. Menge (1856) mentions the group as Gallicolae and has a total of 5 specimens in his collection, one of the specimens being wingless, apparently of the parthenogenetic generation, another is a male. The Copenhagen collection contains only a single specimen.

It is exceedingly likely that the gall wasps have been active during the special season for resin production by the amber tree (spring and early summer), both individuals of parthenogenetic and especially of amphigenetic generation. None, however, have lived on conifers, and the few specimens known from amber have apparently landed on the resin by accident. This is however not enough to explain their rarity when the frequency of oak trees is taken into consideration. Menge's material appears to show that the generation change characteristic of gall-wasps on oak had already been established, and since in the Palaeogene the oak had long reached its complete evolution, there are grounds for assuming that also Cynipinae had been completely established, as apparently has been the case with almost all insect groups. The gall-wasps of the present are mainly boreal insects, however, the spread of which towards North and up into the highlands follows that of the oak, while towards subtropical regions it retreats considerably. The amber forest has presumably been distributed throughout the climatic boundary regions of the cynipids.

Gall midges (**Cecidomyiidae**, pars). Within this family, a total of 36 species is known from Baltic amber, but the actual number appears to have been greater. Loew (1850, p.32) has named 4 species, Meunier (especially 1904b, p.26-48) has named the remaining 32. All three now-living sub-families are represented in the material, and of these Heteropezinae and Lestremiinae will be treated later.

But even among Cecidomyiinae, comprising about 25 described species from amber, probably at least 10 have been saprophages, namely species of the genera *Bryocrypta*, *Colpodia* and *Palaeocolpodia*(?) as well as an undescribed representative for *Colomyia*. The number of true gall-makers within the family thus appears to have been rather modest on a percentage basis.

Conditions in the territory as a whole, however, have not been quite so severe as the amber material might suggest. In the environment today, and that is after all our only basis for comparison, the gall midges which occur in living plant tissue, most often as gall-makers, are very specific in their choice of host plant. The great majority are found on herbaceous vegetation, which has been relatively scantily represented in the amber forest, many are found on deciduous trees, but only rather few on conifers (here might be mentioned *Juniperus* and *Taxus*). It is therefore quite natural that this biological group should be rather scanty with respect to those genera which are saprophages and which at the larval stage have most often been living under dead and loose bark. In addition, as the gall midges are poor flyers, easily driven along by air streams, part of the existing material can have been added from outside and thus be foreign to the biotopes of the amber forest.

The Copenhagen collection has an unanalyzed material of 174 Cecidomyiidae, namely 2 Heteropezinae, 68 Lestremiinae and 104 Cecidomyiinae. Nothing can be said as to the distribution between saprophages and phytophages within the last-mentioned sub-family.

Chapter D
Nectar Seekers

Bees (**Apoidea**). Bees in amber have always attracted attention, and yet the number known is surprisingly small. Earliest, Burmeister (1832, p.636) mentions a bee which he considers is closely related to the Orient's *Trigona* (Meliponinae). Menge, in 1856, gave an account (p.26) of his collection of half a score examples in all. He refers one, with some doubt, to *Anthophora*, and another to *Osmia*. He calls two individuals *Apis*, the one being named *Apis proava*, but according to his description, this can hardly be correct. He names 5 specimens in two pieces of amber, grouped 2 and 3 together, as *Bombus pusillus* (the species is only 3 mm long), while he describes another as *Bombus carbonarius*. Likewise these generic determinations cannot be correct, but the small species is probably a social one. Brischke (1886, p.278) later had the opportunity of reviewing Menge's collection, supplemented by material of Helm's, a total of 11 bees. His result was 1 *Bombus*, 4 *Anthophora* (?), 2 *Chalicodoma*, 2 *Andrena* and 2 Apinae whose wing ribs recalled the South American *Melipona*, but whose legs had another structure. Menge's *Osmia* is thus lacking, and at any rate his *Bombus pusillus*, and presumably it is Menge's *Apis proava* which is referred to the *Melipona*-like genus, possibly the same genus as Burmeister's *Trigona*-like specimen.

In 1906 (p.158), Buttel-Reepen (in his study of the honey bee) describes *Apis meliponoides*, since then referred by Salt (1931, p.144) to the genus *Electrapis* Cockerell (1909a, 1909b). In two publications from 1909, Cockerell describes a number of genera and species. One and the same small piece of amber contains 1 male and 7 females (workers?) of *Electrapis tornquisti*, which in Cockerell's opinion makes it likely that it was a case of a social insect, whose mode of life recalled *Apis*. It may be a case of Brischke's *Melipona*-like genus. Of more bumble-bee-like forms he describes *Protobombus* (2 species), *Chalcobombus* (3 species) and *Sophrobombus* (1 species), possibly social bees, and the more primitive *Glyptapis* (4 species) and *Ctenoplectrella* (1 species). This species list is

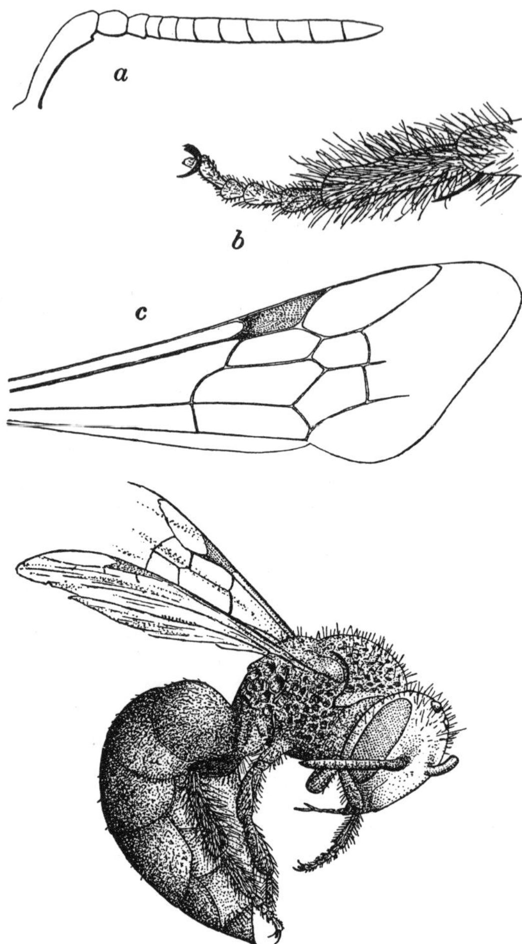

Text-fig.14. *Glyptapis neglecta* Salt (Apoidea), a: antenna, b: right hind tarsus, c: right forewing. (From Salt, 1931).

supplemented by Salt (1931) with 1 *Glyptapis* (Text-fig.14), 1 *Ctenoplectrella* and *Andrena wrisleyi*, and by Manning (1961) with 1 *Electrapis*. Kelner-Pillault describes (1970a) 1 *Ctenoplectrella* and 1 *Electrapis* (3 individuals in the same piece of amber) (Text-fig.15) and (1974) *Electrapis bombusoides*.

It is remarkable that the relatively primitive *Andrena* is the sole example which is not extinct among the definitely determined genera of Eocene bees.

As a supplement to this material, Bachofen-Echt (1949, p.128) mentions a piece of amber in his private collection, in which about 12 bees of the same species lie close together. He concludes that a swarm landed on a tree with flowing resin, and that a section of the swarm was destroyed as a result. This was probably a species of *Electrapis.*

Electrapis appears to have been a dominating genus in the Eocene amber forest. That it has been a social form can hardly be doubted; this is proven by finding more frequently than by chance, in the same piece of amber, several or many individuals which must be ascribed to the same species. When such a relatively great percentage of the bees in amber belong to the genus, it is most likely that

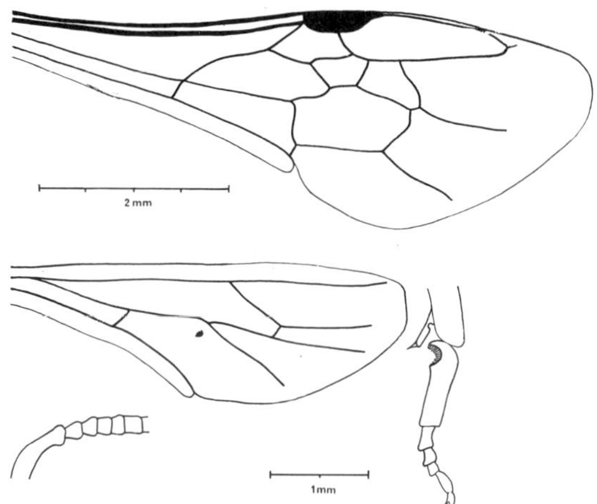

Text-fig.15. *Electrapis apoides* Manning (Apoidea), wings, antennal base and apex of foreleg. (From Kelner-Pillault, 1970).

there has been a definite purpose in the visit to the trunk of the amber tree, and this purpose can hardly have been other than to collect resin to help in building the nest. This is also what happens today in a number of social bees, e.g. the honey bee and the "stingless" bees, *Melipona,* both originally tropical or at any rate subtropical genera. Others among the genera must be regarded as primitive stages in the evolution towards the state of social bees of the present, including the bumble-bee, and some of them have probably themselves been social, although not so differentiated as *Electrapis.* True bumble-bees, on the other hand, appear to have been lacking

in the amber forest, and the same holds for the more primitive families of solitary bees. Of these, only one species is known with certainty in a single specimen, namely *Andrena wrisleyi* Salt. While the family Apidae, with its many social genera, except for the bumble-bees, has more or less tropical features, the other families are boreal to an extensive degree. Many of them are also associated to a special degree with the vegetation of the open territory, with its far greater wealth of flowers. It is therefore likely that there has been a rich fauna of primitive, solitary bees and possibly in addition of true bumble-bees in the boreal districts, which have had a great extent north of the amber forests, a more boreal element which is almost completely missing in the amber forest.

The Copenhagen collection contains five undescribed fossil bees in amber, of which two specimens (the one in contemporary copal) have been studied by J.S. Moure, Brazil.

Chapter E

Insects Trapped while Resting

The animal groups already discussed had their home on the forest vegetation. Most of them were associated with the trees in the forest, and in a few cases it was possible to establish with reasonable certainty that they had been associated with the amber tree itself. In many cases, however, their appearance in amber is not the result of a direct biological connection, but of pure chance; a large number of these animals apparently originated from the innumerable oak trees in the forest. This chance feature is shown to an even higher degree by the insects to be mentioned, which in the amber forest have merely sought a resting place hidden in the folds of the bark, during the part of the day when their activity has been reduced, just in the same way as their now-living relatives in the present-day forest. These insects have usually not developed on the plants of the forest, some pass their larval period in various kinds of fresh water, others belong to the forest floor, while others again live in the ground in the adjacent meadow areas of various degrees of humidity. Finally, some of these animals have apparently finished-up in the resin by purely unexpected events.

Stone flies **(Plecoptera)** are rare in Baltic amber. A few species have been described by Pictet (1854), but our modest knowledge of the group is based primarily on Hagen's publication (1856), including Pictet's material and animals from Menge's collection. This study comprises 3 *Perla* speciés (Text-fig.16) (9 imagines and 3 larval exuvia), 2 *Taeniopteryx* (1 imago of each species), 4 *Nemoura* species (11 imagines, 1 larval exuvium and 1 larva 7 mm long), and 4 *Leuctra* species (Plate 3,C) (22 imagines and no young stages). It is however uncertain whether the views on genera held by these classic authors are in agreement with modern ideas. The Danish collection contains 16 imagines, which are being investigated by Professor J. Illies, Germany.

At the present day, almost all stone flies are restricted to running water, often strong rivers. This is also the case for most of the species in the genera mentioned. The larvae show a definite preference for cold water, and many of them will in fact only grow during the cold season (Brink, 1949). They are therefore most common in the upper regions of the rivers, where the water has still to some degree retained the temperature of the source spring. The larvae have also considerable oxygen requirements, so that they are most numerous where the water runs most strongly or even bubbles over the stones, to whose under-surface they hold firmly. *Perla* larvae are predacious, while the larvae of the other genera are vegetarians, living in particular on deposits of diatoms and green algae, which they brush into their pharynx by means of their mouth-parts.

When the larvae are fully grown, they move to the dry land, often at night. They can wander quite a long way before settling down on rocks or tree-trunks, where the last moult takes place. The life of the imagines is brief, and they do not fly very much. Egg deposition may take place at any time during the year, but in the case of the individual species, the period extends at most over a couple of months. Stone flies prefer not to go far from the vicinity of the river in which they have developed.

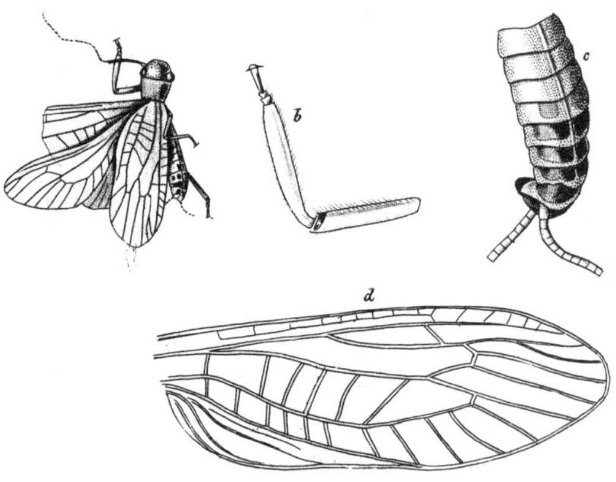

Text-fig.16. *Perla prisca* Pictet (Plecoptera), separate: leg, abdomen, forewing. (From Berendt, 1856, pl.VI, fig.7).

As stone flies are a very ancient order of insects, and as their mode of life is very uniform, whatever their location on the earth's surface, it should be possible

Plate 9.
A. Lacewing (Neuroptera) in left lateral view. (G. Brovad phot.).
B. Psocopteran larva with wing-buds, right lateral view. (G. Brovad phot.).
C. Adult Psocoptera in a crack between two amber resin flows. The lines in the amber (the "fingerprint") show that the psocid fly at first was trapped by the fluid resin with its dorsal side where it then rowed with the wings on the surface of the resin some time before it was submerged.

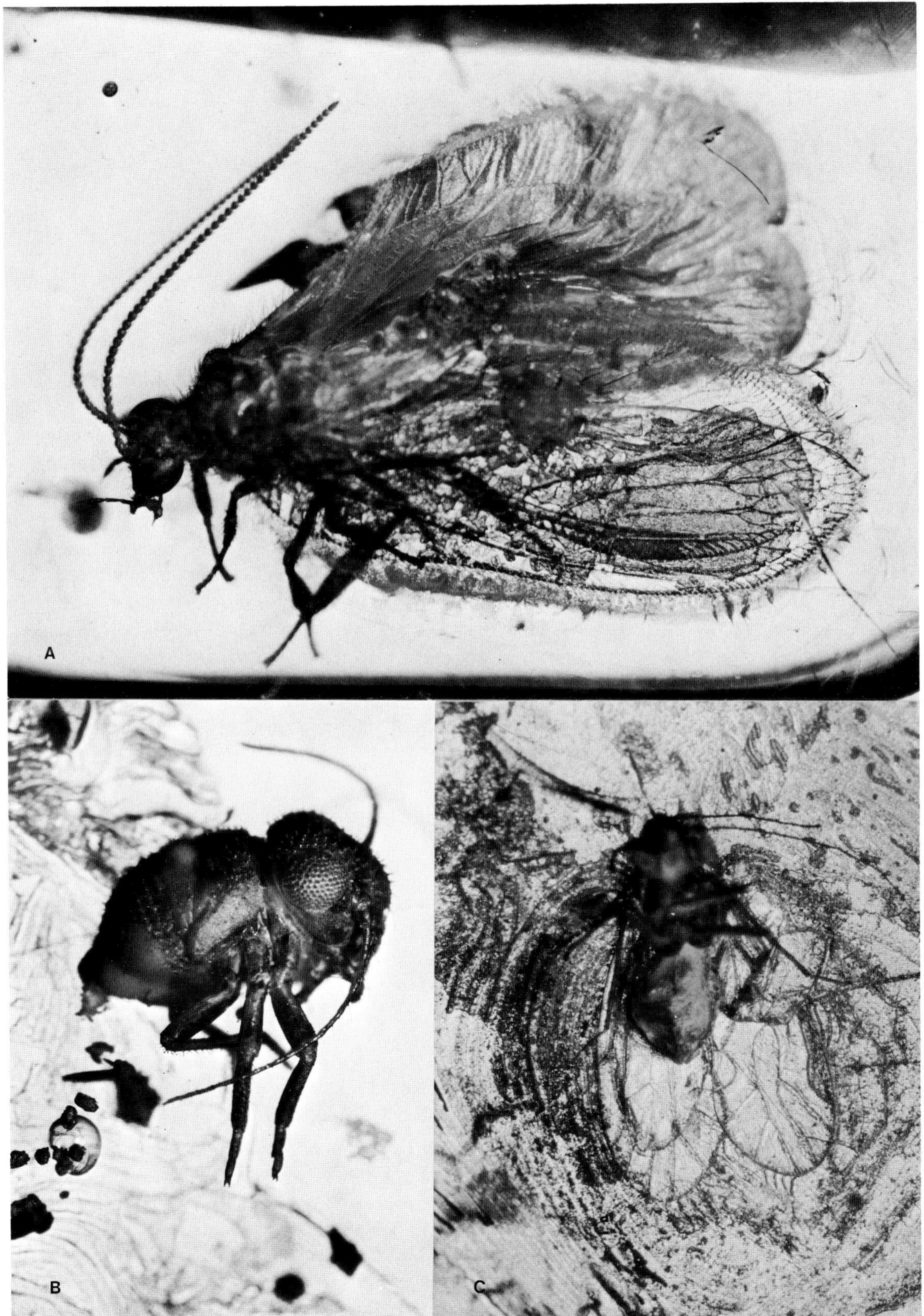

to conclude that also in the European Eocene they have had more or less similar habits. It might seem surprising to find insects which appear so sensitive to heat during their growth, in a landscape which as far as can be determined has had in fact a very mild climate, but it must be assumed that they have in the main lived in the vicinity of springs in wooded districts, where the water still retained an essential part of the low underground temperature, and where shadows from overhanging trees, including the amber tree, have protected the site from too rapid overheating. Stone flies also live today in corresponding climatic conditions. It is also possible that stone flies and stone fly larvae have been carried to amber forest biotopes on driftwood from higher levels with a more alpine climate. Highland districts have undoubtedly been preferred. It is therefore quite natural that only a single specimen of this insect order is known among the many insect fossils from the Mo-clay cement stone.

To judge from the relatively many exuvia, the larvae have achieved their final moult on land, just as at the present day, and the trunks of the amber tree have apparently been suitable for the purpose. It seems that the amber tree has been able to grow and thrive relatively close to streams, even though this tells us nothing as to the requirements for ground drainage.

May-flies **(Ephemeroptera)**. The three earliest descriptions are from Pictet (1854). Hagen (1856), who also had Pictet's material available, supplemented it with a further 3 species. Hagen's material includes almost 50 specimens,

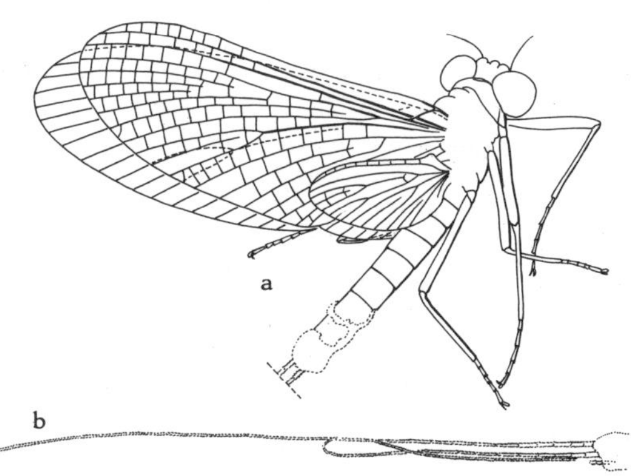

Text-fig.17. *Heptagenia ?bachofeni* Demoulin (Ephemeroptera, Heptageniidae), a: subimago (female), b: cerci of same. (From Demoulin, 1970, fig.5).

referred to the genera *Palingenia, Baetis* and *Potamanthus*, but these generic names have all been changed since. Eaton (1871), on the basis of the literature alone, referred *Palingenia macrops* Pictet to *Polymitarcys, Baetis anomala* Pictet to *Cronicus* and *Potamanthus priscus* Pictet to *Leptophlebia*. Demoulin (1955, p.1), on the basis of the original material, confirms the correctness of *Cronicus anomalus* (Pictet), and justifies the placing of this extinct genus in Isonychiidae. In addition, he alters (1968) *Polymitarcys macrops* (Pictet) to *Siphloplecton macrops* (Pictet) and *Leptophlebia priscus* (Pictet) to *Paraleptophlebia priscus* (Pictet). According to Demoulin (1965, corrected and supplemented 1968), the known amber fauna today comprises 1 Ephemeridae (which however does not provide sufficient basis for description), 3 Siphlonuridae, 2 Isonychiidae, 5 Ametropodidae, 9 Hepta-

Plate 10.
A. Opilionida in amber containing many fissures and bubbles. (G. Brovad phot.).
B. *Scraptia*, a member of the heteromerous beetle family Scraptiidae. Both adults and larvae are fungivorous. (G. Brovad phot.).
C. Male of *Lasius schiefferdeckeri* Mayr. Some authors maintain that this species is synonymous with *Lasius niger* L., the little black ant so common in our recent European gardens. (G. Brovad phot.).

geniidae (Text-fig.17), 5 Leptophlebiidae, and 1 Ephemerellidae. In addition must be added *Succinogenia larssoni* (1965, Heptageniidae), described from a larva (Text-fig.18), and which may belong to one of the above-mentioned 9 species of Heptageniidae.

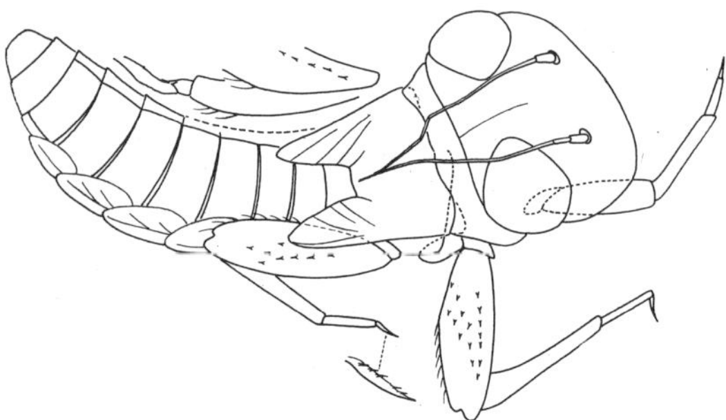

Text-fig.18. *Succinogenia larssoni* Demoulin (Ephemeroptera, Heptageniidae), larva in dorsal view, body-length 6-7 mm. (From Demoulin, 1965, fig.5).

May-flies are amphibian and have been so throughout all their phylogenetic development. Present-day species show such a great biological variation that it is possible to find may-fly larvae in almost all kinds of fresh water, both running and stagnant. This does not appear to have been the case for the amber forest; here all the species have apparently belonged to the running water fauna (Demoulin, 1968, p.272), and for example the larvae of Caenidae from stagnant water appear to be missing, at any rate no imago is known from Baltic amber as yet. It is strange that Baetidae are not represented at all in the amber fauna. "Cette famille est pratiquement cosmopolite et ses représentants offrent des tailles, des formes et des exigences écologiques extrémement diversifiées." (Demoulin, 1968, p.273).

May-fly larvae are almost exclusively algae eaters. When metamorphosis approaches, they move in stagnant water to the surface during the evening, and the change into the sub-imago takes place with almost explosive speed. In flowing or disturbed water, they seek the surface at any object above water, and it has presumably been in this situation that the larva of *Succinogenia* (Text-fig. 18) has met its fate. This finding thus confirms that the amber tree was able to grow quite close to water, under all circumstances where this has had a considerable fall in level. The sex organs in may-flies are mature already by the last larval stage, and copulation and egg-deposition take place very soon after the metamorphosis, in some species even without the final ecdysis of the female from sub-imago to imago. The imagines then die. However, in some species the imago has a lifetime of a few days. This brief existence at the imago stage is one of the main reasons for the relative rarity of the may-flies in amber. A relatively large number of fossils are sub-imagines, and a few of the individuals are just at the stage of moulting to imago at the moment of capture. This suggests that capture has occurred mainly just around this last ecdysis, in other words during a very brief interval of the entire life of the may-fly. There can therefore be no doubt that they have been far more common in the amber forest than the fossil material would suggest.

In the recent species, the time of the year for flight varies very much from one species to another, but in Denmark it is in the early summer in most cases. In southern latitudes, one and the same species often flies both during spring and autumn, and it may be presumed that this was the case for at any rate some of the species in the amber fauna. May-flies undergo a very large number of ecdyses.

Only very few may-flies are known from the Mo-clay, including 2 imagines in the same slate, apparently a small part of a swarm which had been driven out to sea by the wind. The may-flies in the Mo-clay are very poorly preserved.

Dragonflies (**Odonata**) do not fly into the dense forest. Occasionally they rest on trees and bushes in the edge of woods, but if they come into contact with fresh resin as a result, they will almost always be able to release themselves because of their own robust strength. It is therefore very rare to find dragonflies or even portions of dragonflies in amber. A few Zeugoptera are known from Pictet & Hagen (1956), these being described under the name *Agrion antiquum* (p.78). According to Hagen (p.79), however, the species, also some small wing fragments in other amber pieces, must be referred to *Platycnemis* (Coenagriidae). In addition, a few larval exuvia are known. Bachofen-Echt (1949, p.73, fig.63) gives an illustration of a very fine but undescribed specimen of a Zeugoptera. Among the Anisoptera, Hagen (1956, p.81) has quite briefly described *Gomphoides occulta* on the basis of a wing fragment from Menge's collection. The Copenhagen collection contains no dragonfly specimens.

Dragonflies are amphibious, and the larvae are found in fresh water of very different characteristics. Most frequently they are found in dams and other still or slowly-flowing waters, but a small number are found in rapidly flowing streams or in the surf of large lakes. Just before metamorphosis, the fully grown larvae crawl out of the water on to straws, stones, etc., and the abandoned exuvia are often found in large numbers on the vegetation of the shore. It must have been as a result of such circumstances that the few known larval exuvia have found their way into the amber (Hagen, 1856, p.80; Weidner, 1958, p.52). Not only the larvae but also the imagines are greedy predators. Propagation takes place at any time within a longer period of the year, depending on each individual and species. Egg deposition begins early in the summer and continues until late in the autumn.

Helodidae. Among the beetles found in amber, only Helodidae are really amphibian, and just for this reason are very common (Text-fig.19). The incidence is very striking as far as their individual count is concerned, constituting between 10 and 20% of all beetles. On the other hand, they constitute only a very small

percentage of the number of beetle species found. For example, in Klebs' collection (1910), the individual count is 19.3%, the species count 2.2%. The distribution in the Copenhagen collection seems to point in the same direction, although not so striking; here the individuals amount to 10.9% of all beetles.

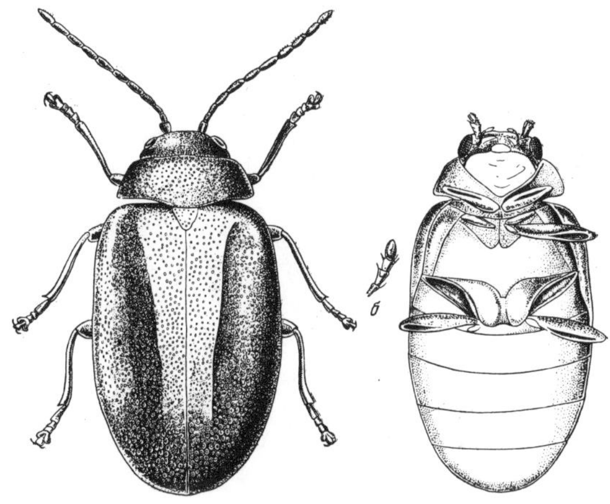

Text-fig.19. *Microcara dokhturovi* Yabl. (Helodidae), dorsal and ventral view. (From Yablokov-Khnzorian, 1960, fig.5).

In Klebs' collection of Helodidae, 5 genera are known which have been determined by Edm.Reitter. They are all common in present-day Europe: 1 *Hydrocyphon*, whose larvae live in rapidly flowing water, 1 *Helodes*, whose larvae are found in running water of more moderate force, 3 near *Helodes*, 3 *Scirtes* from stagnant water, 282 *Cyphon* from dams, where the larva is found in particular among the duckweed on the surface, 21 near *Cyphon* and 39 *Microcara*, where the larva can be found between rotting leaves at the bottom of waterholes and ditches in the woods. 25 were not determined. *Cyphon* has a very strong dominance over the other genera. Among the 68 Danish specimens are 53 *Cyphon*, 4 *Helodes*, 6 *Microcara* and 5 *Prionocyphon*, the larvae of which live in puddles in the forest or in collections of water in holes in trees. *Hydrocyphon* and *Scirtes* do not seem to be present in this collection, which is now being studied by Dr. Bernhard Klausnitzer, Germany. In the main, the distribution here appears to be as in the Klebs collection.

It is however hardly likely that the Helodidae have had such a relatively great incidence in the amber forest as the numbers quoted above might suggest. We know from the present-day that it is only unwillingly that imagines leave their hatching biotope, but crawl up on to the surrounding vegetation, also tree trunks. It has thus been possible to find amber trees in the immediate vicinity of all the various kinds of hatching waters in which the Helodidae have lived.

The greatest taxonomic significance must be attached to two studies by Yablokov-Khnzorian (1960 and 1961a), in which he described several new species from amber. The 1960 study mentions 3 *Microcara* species (p.96-101). The 1961a material comprises in addition new descriptions of the genera: *Brachelodes*, belonging to Microcarini, with the species *motschulskyi;* the new tribus Helodopsini with *Helodopsis solskyi* and *Cyphonogenius zakhvatkini*. In addition, within Cyphonini, *Plagiocyphon plavilschikovi* and 3 new *Cyphon* species. There is a considerable likelihood that these and possibly also still undescribed forms are included in the above-mentioned material from the Klebs collection and from the Copenhagen collection.

However, Klausnitser quite recently published a paper on *Helodes* species in Baltic amber, partly based on the Copenhagen material (1976, p.53). The publication discusses morphological details necessarily for reliable descriptions of species belonging to the genus *Helodes*. Klausnitzer states: ''Taxonomische Untersuchungen an rezenten Arten der Gattung *Helodes* Latreille zeigten, dass eine sichere Artdiagnose nur unter Einbeziehung des Aedeagus und der Tergite und Sternite des 8. und 9. Abdominalsegments möglich ist. Für die Diagnose besonders bedeutsam ist der Bau des Penis, des Tegmens und des 8. Tergits. Für die Bestimmung von Bernsteininklusen muss man grundsätzlich die gleichen Kriterien heranziehen wie für rezentes Material. Das bedeutet, dass nur solche Inklusen bestimmbar und damit auch als Arten beschreibbar sind, die eine Beurteilung wenigstens von Teilen des Genitalapparates und der Terminalia zulassen.'' This statement may possibly also be of some significance for taxonomic evaluations in other genera. Five new species are described.

In connection with Helodidae, some small families should be mentioned which are hardly truly amphibious, but where the larval development in many respects nevertheless has points of similarity with several helodid species. What is named **Dascillidae** in the older literature appear to be helodids. Klebs' collection contained 5 specimens of true dascillids: 2 *Dascillus*, 1 *Pseudodactylus* and 2 undetermined. The Copenhagen collection contained no specimens. This is a family in which the larvae in particular live on luxuriant meadows, and where the larvae, which live in the ground, gnaw the plant roots in the same way as the *Melolontha* larvae.

The **Ptilodactylidae** include *Ptilodactyloides stipulicornis* described by Motschulsky (1856, p.26), the imago of which has most probably sought a hiding place in the moss and cracks on the lower section of the tree-trunks. Its mode of life as larva is uncertain. Among recent forms, some genera live in water (*Eulichas, Anchytarsus?*), others in the moist tropical forest floor or rotting wood with a high water content (*Ptilodactyla*). In the Danish Noona-Dan expedition to the Indo-Australian Archipelago, a large number of *Ptilodactyla* larvae were found in Berlese samples from the forest floor. The family is not represented in the Copenhagen collection of amber.

Otherwise unknown is **Heteroceridae**, of which the Copenhagen collection contains a single specimen of the very characteristic larvae which live in self-constructed tunnels in the ground. The reason for its presence in amber is obscure. The fact that a few **Byrrhidae** are found in Baltic amber is not in itself particular strange, what is remarkable is that Klebs (1910) only mentions one specimen which is taxonomically close to *Syncalypta*, a genus which is mainly found on the sandy shores of fresh water. The same is the case for *Limnichus* (**Limnichidae**), of which Klebs has likewise a single example. The Copenhagen collection contains a single *Limnichus* and 4 small Byrrhidae (probably belonging to several species), which are apparently close to *Syncalypta*. Berendt (1845) has mentioned *Byrrhus* and *Limnichus*. The few finds of Heteroceridae, Byrrhidae and Limnichidae appear to show that the amber tree could also grow on sandy shores or at any rate in their immediate vicinity.

Alder flies (**Megaloptera**). Both families, Sialidae and Corydalidae, are represented in amber, but only a very few specimens are known, and none are represented in the Copenhagen collection. Nor is the group known with certainty from the Mo-clay in either of the two Danish collections. *Chauliodes prisca* Pictet (1856, p.82) undoubtedly belongs to the family Corydalidae (Chaulioninae). What has been described in the older authors (Burmeister, 1832, p.637; Gravenhorst, 1834, p.92) under the name *Semblis* (= *Sialis*) is probably not Megaloptera, but true lacewings (Hagen, 1956, p.81); at any rate the determinations are uncertain. Weidner, however (1958, p.60), has described the abdomen of a larva which is quite definitely a *Sialis* larva. At the present day, hardly more than 30 species of alder flies exist, but they are extensively distributed throughout the hot and temperate zones.
They are amphibian insects. The larvae are found mainly in running water; the Sialidae on quite soft, vegetation-free river beds, where they dig themselves into the bottom, in other words in slowly flowing water or in dams, while the Corydalidae, on the other hand, are found in streams with greater rate of fall, on gravelly and stony beds, where they are found most often under stones and driftwood. They are predators. In present-day Denmark, pupation occurs in the spring in an excavation on land, and it must be supposed that Weidner's *Sialis* larva has met its fate during a similar wandering. Just before the final metamorphosis, the pupa, which is very mobile towards the end of the period, works its way to the surface. The imagines take only slight nourishment, and die in the course of a few weeks. The eggs are deposited in sheets on the leaves of bushes or, in the case of Corydalidae, also on stones and twigs. The newly hatched larvae let themselves fall from here into the water, but are able to undertake quite long travels on the ground, if this should be necessary because of unsatisfactory placing of the eggs.
As adult alder flies take to the wing to only a limited extent, they are mainly encountered in the neighbourhood of their breeding biotopes, where they are accustomed to creep around in the vegetation.

Lacewings (**Neuroptera**, pars). Most lacewings are exclusively terrestrial, and both as adults and larvae live on scale-insects, aphids and other small arthropods. However, two families exist which each in its characteristic way live an amphibian existence, and which have a deviant diet: Osmylidae and Sisyridae. At the present day both these families are widely spread in tropical and temperate zones, but are limited in species (with about 50 and 20 species, respectively). Both families are known from Baltic amber, but only in a few specimens. It is unknown whether the amphibian lacewings are represented in the rather comprehensive, but unprepared Mo-clay material; it seems, however, most likely. The small Copenhagen amber collection contains only a single specimen of Osmylidae and none of Sisyridae. The Copenhagen amber lacewings are at present being worked up by Professor E.G. MacLeod, U.S.A.
Both larvae and imagines are predators.

The season for flight is late spring and summer.

Osmylus (at any rate the present European species) is found as larva in quite shallow water between moss, leaves, and under stones, where it feeds on other insect larvae, mainly chironomids. In the spring, pupation occurs between damp moss, in a cocoon which it has spun itself. After a couple of weeks the pupa bites through the cocoon and seeks a suitable spot for the final metamorphosis. Hagen (1856, p.86), has described *Osmylus pictus*, referred by Krüger (1923) to *Protosmylus.*

In the Sisyridae, the larvae live on colonies of fresh water sponges and fresh water bryozoans, which they eat. In the spring the larvae crawl above the water surface and spin their pupal cocoon. Hagen (1956, p.87) has described *Sisyra (Rophalis) relicta* and *Sisyra (Rophalis) amissa* as belonging to this family, while Krüger (1923) has referred them to one species, *Rophalis relicta.*

The two lacewing families thus tell of flat swampy areas with diffuse springs, or conditions in which the water-table has at any rate been high, as well as of quiet dams with a profuse growth of animal sponges and bryozoans, just that form of mud-covered biotope which can have housed *Sialis* larvae.

Caddis flies (**Trichoptera**) have been very common in the amber forest territory. Following Ulmer's large study (1912), there are known from the Baltic amber 152 species distributed over 56 genera, 30 of which are also known from the present day. An analysis of the about 150 amber fossils included in the Copenhagen collection will undoubtedly increase the number of species. In addition the Mo-clay contains quite a large number of undetermined caddis flies, so far about 150 pieces in all in the two Danish collections of Mo-clay fossils, but in general these insects are poorly preserved (Larsson, 1975). Among the Copenhagen amber fossils there is a single larva (Plate 3,D) belonging to the case-building family Leptoceridae (A.

Nielsen det.). It is obscure how this has come to be enclosed in amber. It has managed to release itself from its case before death, and the case is not preserved.

Caddies flies have always been amphibian insects. The larvae are found in fresh water of very varying characteristics, many genera are bound to stagnant water, most often dams with thick growth of vegetation (Limnophilidae, Phryganeidae), while others live in the surf zone of large lakes (*Molanna*, some Leptoceridae), and a large number are found in rivers with very varying strength of flow (many Polycentropidae, Hydropsychidae, Philopotamidae and Rhyacophilidae).

Adult caddis flies can be found throughout almost the entire year, but their period of propagation is most often in the course of the summer. They only take liquid nourishment, e.g. nectar from flowers, and the individual insect usually lives only a couple of weeks at this stage. The technique of egg-deposition is very varied, but for example those species whose larvae live in fast-flowing water, creep down into the water and fasten their eggs on stones and the like. The larvae of many species spin net snares, and are carnivorous. The case-bearing larvae, which are mainly vegetarian, spin a tube to which in various special ways they fasten different kinds of foreign bodies (sand grains, small snail shells, plant parts). Some tubes are fast-spun to stones from the start (some Rhyacophilidae), but most of them are not fastened to fixed objects until pupation. The pupa seeks up above the surface of the water or just to the surface itself, immediately before the final metamorphosis, which takes at most only a few minutes.

Ulmer's material, which among other items includes the classic collections in Königsberg, amounted to more than 2000 specimens, almost exclusively of East Baltic origin, while the small Danish collection consists mainly of West Baltic amber. Ulmer's material comprised Rhyacophilidae (6 specimens in

3 species), Hydroptilidae (21 specimens in 3 species), Philopotamidae (82 specimens in 6 species), Polycentropidae (1398 specimens in 64 species), Psychomyiidae (340 specimens in 16 species), Hydropsychidae (20 specimens in 4 species), Phryganeidae (71 specimens in 7 species), Calamoceratidae (2 specimens in 2 species), Odontoceridae (9 specimens in 1 species), Molannidae (5 specimens in 3 species), Leptoceridae (25 specimens in 5 species), Limnophilidae are completely missing, Sericostomatidae (85 specimens in 6 species) and 6 species which could not be referred to a family. In the undetermined Danish material there are 2 Rhyacophilidae, 4 Hydroptilidae, 6 Philopotamidae, 82 Polycentropidae, 17 Psychomyiidae, 2 Hydropsychidae, 0 Phryganeidae, 0 Calamoceratidae, 0 Odontoceratidae, 0 Molannidae, 2 Leptoceridae (including one larva), 0 Limnophilidae, 0 Sericostomatidae, while the remainder cannot be referred to any family. Generally speaking, therefore, the composition appears to be the same in the two collections.

It appears as if the number of genera from running water was relatively greater in the Eocene than at the present day, but that the relationship between the numbers of species was about the same then as now (Ulmer, 1912, p.366).

As stated by Ulmer, a quite striking difference between the Eocene and the recent caddis flies is that the most common caddis fly family of to-day, Limnophilidae, to which 25% of present day caddis fly species belong, is not known at all from amber, although it is quite unreasonable to assume that it did not exist at that time, when the order of caddis flies was otherwise fully differentiated. Hagen (1856, p.102) admittedly describes *Hallesus retusus*, but neither that species nor any of the species referred by Hagen to *Limnophilus* are recognized by Ulmer as Limnophilidae. Ulmer draws attention to the fact that also today the family avoids warm climates, and can therefore have lived in other parts of Holarctis, where it has been less warm. Ander (1942, p.17) is not quite satisfied with this explanation, as in his opinion the family is found today so near the subtropics that sensitivity to warmth should not have been a hindrance to finding it among the amber insects. He considers that if the family existed at all in Europe at the time of the amber forest, then it can have been only very scantily represented, and not yet have reached its possible distribution. He stresses that it is the youngest of the known families of caddis flies. Even though Ander is also correct in stating that some species occur quite close to the tropic zones, this does not alter the truth of Ulmer's argument, since the family as a whole is a pronouncedly boreal one. The family could quite well have been very common in the boreal region of that time (no one can know this), but nevertheless so uncommon in the amber territory (in agreement with its present distribution) that it was not represented among those species, admittedly numerically scanty, that we know from the amber fossils.

Conversely, Polycentropidae were represented in amber by 44% of all species (67% of the total of individuals), while at the present day they constitute only 6%. Polycentropidae are found under varying climatic conditions, and the reduction in the family in recent times is in Ulmer's opinion associated with the fact that it itself is in rapid decline.

It is also remarkable that Ulmer's figures (ibid, p.347) show that species with typical net-spinning larvae constituted about 62.5% in the amber fauna (only), while at the present day, net-spinning larvae constitute 27.6% (but of the total world fauna of caddis fly species). At the same time, the number of species with larvae living in shelters has increased from 28.3% to 59.5%. This suggests that the actual amber territory has been relatively rich in running water. Ulmer's calculations did not include Rhyacophilidae and species whose systematic placing was uncertain.

Those caddis flies found in Mo-clay are generally a little larger than the forms in amber, and the species must be regarded as different in a number of cases. However, the material has not been sorted in detail, and many of the fossils are unsuited to scientific study. The sometimes heavily grained surface of the cement stone often hinders the interpretation of the delicate and hairy surface of the caddis flies, especially the wing structures.

It is probable that Ulmer was right, at any rate to some degree, in his view on the systematic differences between the caddis flies of the fossil period and today. However, he appears to have overlooked the possible explanation that the amber forest was associated with highland territory, and in all probability this was to some degree the case. The composition of the flora and of the many insect groups suggests this. It is quite natural to find relatively many caddis flies with net-spinning larvae in highlands, namely species associated with rapid streams, and it is just as natural that there should only be relatively few shelter-building species, with the opportunity of suitable biotopes, where the reduced rate of flow of the water provided possibilities of existence for these larvae, which in most cases were able to creep about freely. The firmly fixed larval cases of Rhyacophilidae are not considered in this connection. It is therefore not quite correct to compare the rather one-sided subtropical highland fauna of the amber forest with the total fauna of today, from all the world's biotopes. A systematic study of the caddis-fly fauna of the Mo-clay, which as far as can be judged has been a purely lowland fauna in a nearby region, both in time and place, will probably provide a considerably better basis for comparison between then and now. It is probable that in this fauna, a considerable preponderance of case-dwelling species will be found, and the possibility cannot be excluded that Limnophilidae will actually be found in this collection, in spite of the subtropical or perhaps even warmer climate, and also an increased number of Phryganeidae.

Among the **Diptera,** numerous amphibian genera are found, especially among the **Nematocera.** In most of the families belonging here, many purely terrestrial genera are also found, and often the transition between the aquatic and terrestrial larval morphology and biology in one and the same family is quite uniform. In many cases it is impossible to say whether a fossil imago belongs to an amphibian or to a terrestrial species; this is the case to an outstanding degree for the daddy-long-legs, for example. Many larvae live on land or among moss, but are so dependent on the immediate vicinity of water that there can be some doubt as to whether they are aquatic or terrestrial.

Black flies (**Simuliidae**) are only a small family of world-wide distribution, but hardly comprising more than a thousand recent species. According to Bachofen-Echt (1949, p.156), about a dozen species were known from Baltic amber at that date, a figure which appears to be based on Handlirsch (1908, p.981). This estimate, however, is very uncertain, as only 3 species of *Simulium* have been described, all by Meunier (1902a, p.397) and all based on material already published by Loew (1850, p.38), the first investigator who seriously interested himself in the difficult content of Diptera in amber. Apart from this Loew-Meunier material, Helm (1896c, p.223) states that many species were found in his collection; this is probably exaggerated, at any rate it is undocumented. Only 9 Simuliidae specimens have been found in the untreated material in the Copenhagen collection, and none are known from Mo-clay.

Both larvae and imagines of Simuliidae have quite special demands on the environment, so that they clearly show details of nature as it was in the past. In Denmark, imagines can be found from early spring to late autumn, with a culmination in the warmer months, but in the Danish climate they are unable to overwinter at this stage. Some species, however, have two generations annually, and others have even three (Petersen,

1924, p.309), so that it must be presumed that where the conditions permit, these species show a permanent course of short-lived generations, and the picture may have been like this in the amber forest. The feature which is specially characteristic in Simuliidae is that the females are particularly plaguesome blood-suckers on mammals, among which ungulates appear to be especially exposed to this scourge. As black flies also often occur in large swarms, they can be a considerable stress for the victims, which at times pay with their lives as a result of the violent attack. If it was not already confirmed direct by findings of fossil bones in other European Eocene deposits, the existence of Simuliidae in amber would provide a weighty piece of evidence that ungulates (or at any rate mammals) of suitable size and mode of life must have existed at that time.

The eggs are laid just at the water's edge, in clear running water. Only here can the larva be found, which is microphagous, always attached by its posterior end to stones or the like; it appears unwilling to change its site. The pupa is also found here, in a cocoon which the larva spins shortly before pupation. As an introduction to the appearance of the imago, large amounts of air are produced in the pupa between the old and the new cuticles, and on hatching the imago ascends together with this air bubble, so that it is ejected quite dry and ready to fly out into its new element. As a result of this explosive development, torn-off parts of the pupal skin containing air can be brought up to the surface, and such a characteristic part of a pupa is also known from amber, presumably brought by the wind from where it has been washed ashore at the edge of a rivulet or stream.

Black flies are admittedly not woodland insects, but where woodland meadows have spread alongside watercourses and provided grazing to the ungulates of that time, the Simuliidae have had their opportunity. The relative rarity with which this family of gnats occurs, although its members nowadays often appear in great swarms, must however signify a relative poverty of suitable food-stuffs in the immediate vicinity of the amber trees; among the ungulates of that time, there have hardly been species with a tendency to occur in large flocks in the tree-covered terrain. The larval biology confirms the presence of lively flowing, oxygenated water in the amber territory, in accordance with the conclusions which could be drawn from the biology of other amphibian insects.

Non-biting midges (**Chironomidae**) are exceedingly common in amber and during certain periods must have constituted a considerable proportion of the flying insect world of the amber forest (Plate 4, B). The Copenhagen collection contains about 900 specimens, approximately 11% of the entire collection, and the museum has been offered at least ten times as many. It is remarkable that the number of males is considerably smaller than the number of females, about 1/3 in the collection, but in reality much less, although under recent conditions far more males are seen than females. Even though parthenogenesis occurs among some chironomids, it is nevertheless unusual; the most likely cause of the difference in sex proportions is more probably the difference in the life patterns of the two sexes, including their duration of life. Chironomids are so far unknown from the Mo-clay, presumably because of the manner in which the Mo-clay fossils have been selected from the fauna of that period, and not because they have been lacking on the coastal regions of South-East Scandinavia, which must have possessed many favourable biotopes.

A total of 60 species have been described, almost all by Meunier (1904b, p.210). The earliest mention is by Burmeister (1832, p.617) and later likewise collectively by Helm (1896c, p.222). Loew (1850) mentions the genera *Chironomus* and *Jentzschiella*, on which material Meunier (1899b, p.162) describes *Chironomus obscurus* and *Jentz-*

schiella sp. In 1856 (p.251), Giebel described 2 species, and in 1868 (p.23) Duisburg described *Sendelia mirabilis.* The total material of described species is distributed as follows: Chironominae: 19 *Chironomus*, 21 *Cricotopus*, 3 *Tanytarsus*, 6 *Eurycnemus* and 2 *Camptocladius;* Tanypodinae: 8 *Tanypus;* in addition 2 *Sendelia* and 1 *Jentzschiella.* A revision of this comprehensive material in connection with a study of the specimens later included would be highly desirable. Meunier in particular has often been strongly criticized (Handlirsch, 1906; Alexander, 1931). The male genitalia are often directly accessible to study in the amber fossils.

In Chironomidae, the imagines are only slightly affected by the territorial conditions. They have but a brief lifetime and do not take much more nourishment than a little nectar and perhaps other plant juices. Swarming occurs in the course of a few weeks, especially in the early summer, and at that time they can occur in exceedingly great numbers around and over their breeding waters, while later in the year they occur more singly. Swarming thus takes place within the period in which the amber trees probably have their maximum production of resin. The larvae are found on very different kinds of biotope, preferably at the bottom of still or quietly flowing water, and often in lakes at great depths. Most of them manage well in very oxygen-poor water, although *Tanytarsus* prefers clear, clean lakes with a high content of oxygen. It is remarkable, therefore, that exactly this latter genus occur with so few species in amber. The larvae among Chironominae are microphagous, while among Tanypodinae they are predators.

Biting midges (**Ceratopogonidae**). Among the amber fossils belonging to this family, 23 species have been described: 3 by Loew (1850, p.30), 1 by Giebel (1856, p.252), and 18 by Meunier (1904b, p.239), who referred all species to the main genus *Ceratopogon*, as well as a species by Petrunkevitch (1957,

p.208), who established *Eohelea stridulans*, which he referred to its own subfamily. As the recent Ceratopogonidae vary very considerably biologically, it is very regrettable that no effort has been made to group the existing fossil material of *Ceratopogon*, which in the amber fauna must be regarded as a collective genus. As the ceratopogonids are common in the Baltic amber (Plate 1), there must presumably be a considerable number lying in the various museum collections of amber; in Copenhagen, for example, there is a material of more than 270 pieces.

In contrast to the chironomids, the imagines of the Ceratopogonidae take animal nourishment, and their mouth parts are marked characteristically by this mode of life. In recent time some are highly plaguesome blood-suckers on vertebrates, as for example *Culicoides* and at any rate some *Ceratopogon*, while others, as for example *Forcipomyia*, suck the contents of larger insects, while others again capture and suck the contents of insects which are even smaller than themselves. It would be of considerable interest if it were possible by extended taxonomic studies to obtain an estimate of the incidence of the various ways for the uptake of nourishment among the biting midges of the amber forest, for comparison with those of the present day.

Although the imagines are found today mainly in the vicinity of water, or at any rate in moist surroundings, their biology of propagation is nevertheless very varied. Larvae are found in water, mostly stagnant, in moist earth, in rotting vegetation, exuded plant juice and decayed wood, and there are clear morphological differences between aquatic and terrestrial larvae. In most cases the larval biotope is determined by the genus, but among *Culicoides*, for example which are mainly aquatic, also terrestrial species are found, and within the normally terrestrial *Forcipomyia* there are some aquatic species.

Just as in the case of the chironomids, females preponderate among the amber fossil biting midges, less than 40% being males. This is presumably due to sexbound variation in their biology, possibly merely that the females live on the average a few days longer than the males; the proportion is hardly an expression of a true numerical difference.

The description of *Eohelea stridulans* (Text-fig.20) by Petrunkevitch is very detailed and well illustrated. In addition to various characteristics which in Petrunkevitch's opinion in themselves justify establishing the sub-family Eoheleinae, it possesses a very characteristic organ of transverse bars in the wing tips close behind the subcosta, an organ which Petrunkevitch interprets as a stridulation organ. A further 9 specimens showing the same wing organ have since been found in the Copenhagen collection, together with 1 specimen in which the bars have a considerably different structure. A remarkable feature associated with this small material is that hitherto, in spite of efforts to demonstrate the opposite, it has been found to contain only females, and the thought occurs that it is perhaps a case of a secondary sexual characteristic.

Mosquitoes (**Culicidae**) are amphibian, and like the preceding families have a very great geographic distribution. In contrast to the chironomids, they appear to have increased their number of species quite considerably from the Eocene till the present. The mosquitoes described as *Culex* in the older literature (Schlotheim, 1820, p.43; Motschulsky, 1845, p.98; Helm, 1896c, p.222), are uncertain, or belong to the Chaoborinae (= Corethrinae). In the Chaoborinae, specimens named from amber are *Mochlonyx atavus* Loew (1861) and *Mochlonyx sepultus* Meunier (1902) (Text-fig.21), which according to Edwards (1932) and confirmed by Hennig (1966b) are the same species, hereafter carrying the lastmentioned name. In addition, *Corethra ciliata* Meunier (1904b, p.89) has been described, but the identity of this cannot be confirmed, as the sole specimen is now missing (Hennig, 1966b, p.12).

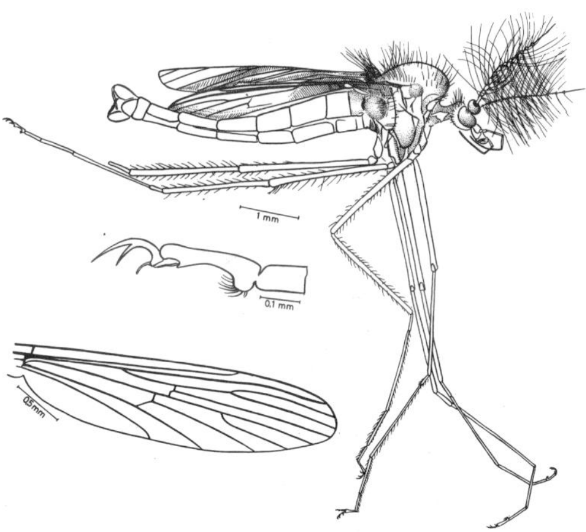

Text-fig.21. *Mochlonyx sepultus* Meunier (Culicidae), male, the enlarged tip of the hind tarsus and the wing are shown below. (From Hennig, 1966b, figs.16-18).

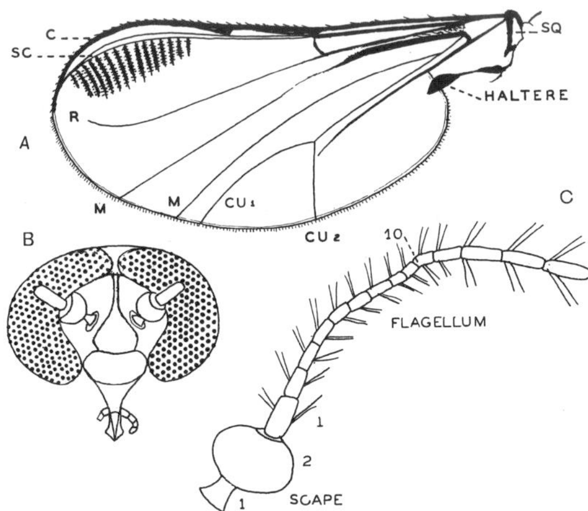

Text-fig.20. *Eohelea stridulans* Petr. (Ceratopogonidae), A: wing with stridulation organ near its tip, B: head in frontal view, C: antenna, all of female. (From Petrunkevitch, 1957).

A few species are known from other deposits from the older Tertiary, and a small number of undescribed specimens (probably Culicinae) appear to occur in the Mo-clay of early Eocene (Larsson, 1975). Only 3 specimens are known from the Danish amber, and they are poorly preserved; one of them is a *Mochlonyx*,

referred by Hennig with some doubt to
M. sepultus Meunier.

The females of the actual biting mosquitoes (Culicinae) must suck blood
from warm-blooded vertebrates if they
are to produce eggs capable of development. In addition they are more or less
specific blood-suckers on definite groups
of hosts, some feeding exclusively on
birds, others on mammals of various
kinds. Both birds and mammals underwent their most important period of
differentiation in the Tertiary, namely
in its older stage, and quite naturally
this must have set its mark on the differentiation and incidence of their
parasites. This may explain the scarcity
of the group in amber. The explanation
may also be lack of suitable water in
which to breed. Finally, the mammals
of the Eocene have on the average been
small; the larger forms have lived on the
plains, while forms living in the forest,
like deer, had not yet evolved. This too
may have contributed to the fact that
the earliest biting mosquitoes have lived
mainly in the plains and have thus avoid-
ed being trapped in the amber resin.

Dixidae are a little family of small
midges (today about 150 species), with
3 definite species known from amber:
Dixa minuta Meunier (1906c, p.395)
(Text-fig.22), *Paradixa succinea* (Meunier, 1906c, p.395) and *Paradixa filifor-
ceps* Hennig (1966b, p.11). There are no
specimens in the Danish amber collection, nor in any of the Mo-clay collections. The larvae of these midges live in
the surf area of stagnant or quietly flowing water, along banks or upon stones
and other objects projecting from the
surface of the water. The family does not
appear to have undergone any remarkable morphological development from
the Eocene till today.

Moth flies **(Psychodidae)** (Plate 4,C) are
a family with several hundreds now-
living species. 25 species have been described in the Baltic amber, the first ones
by Loew (1850, p.31), the majority by
Meunier (1899b, p.175; 1905a; 1905c,

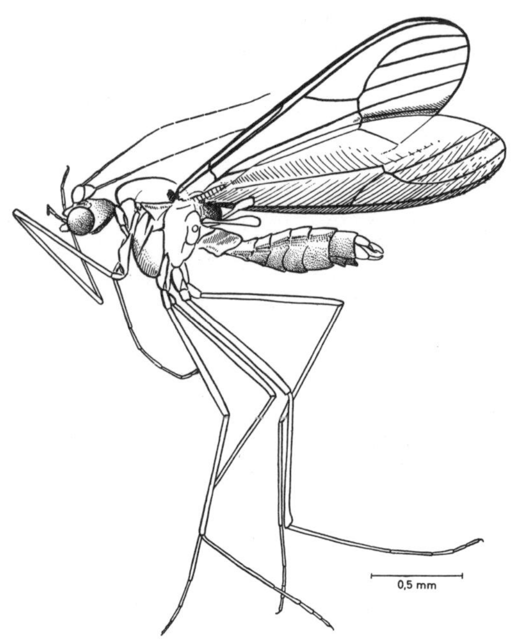

Text-fig.22. *Dixa minuta* Meunier (Dixidae),
male. (From Hennig, 1966b, fig.2).

p.243; 1906a), while Hennig (1972b)
summarized our knowledge of the fossil
forms of these Nematocera. Known from
the Baltic amber are two species of
Nemopalpus (Bruchomyiinae), 1 *Phlebotomus* (Phlebotominae), 1 *Eatonisca*, 2
Posthon, 1 *Syracorax* and 12 *Trichomyia*
(all Trichomyiinae), and 2 *"Pericoma"*, 1
Telmatoscopus and 3 *"Psychoda"* (all
Psychodinae). Later, Stuckenberg (1975,
p.456) additionally described one *Phlebotomus* species. The Danish amber
collection contains about 200 specimens
which have not been examined. The
family is unknown in the Mo-clay.

Psychodidae are photophobic and are
always found in damp localities. The
imagines, which are poor fliers, stay
close to their breeding biotope; it is
therefore almost certain that the larvae
of the species found in amber have lived
in the actual amber forest and almost
probably in the near vicinity of the
amber tree. *Trichomyia* is the genus
which appears to have had the best conditions of life. Today it breeds in
heavily rotted wood and similar plant
products. Its relative frequency in

92

amber is evidence of a damp forest floor with fallen tree-trunks in the process of decomposition. *Pericoma* is amphibian, and today species are found whose larvae live in fast-flowing streams, while others are found in strongly contaminated water. Among the *Psychoda* species, some live in gently flowing water or in stagnant contaminated water, while others live under more terrestrial conditions, although always where there is plentiful liquid organic material as a source of nourishment. Others, again, can be found in rotting fungi. The blood-sucking genus *Phlebotomus* appears to have developed in connection with thin-skinned reptiles, which are the preferred source of feeding for a considerable number of present-day species.

An important section in the publication by Hennig mentioned above includes a description of 2 species of Phlebotominae from Lebanese amber (Lower Cretaceous): *Phlebotomites brevifilis* and *P. longifilis* (p.39). In contrast to the Aleurodidae described by Schlee (1970), the *Phlebotomites* species do not deviate significantly from the present-day forms.

According to Alexander, the primitive family **Tanyderidae** is represented by only a single species, *Macrochile spectrum* Loew (Text-fig.23), of which about 20 specimens are known. The genus has since become extinct. This family, with few species, is today widespread, but found above all in the Australian region. The larva is aquatic.

Of the winter gnats **(Trichoceridae)**, Loew (1850, p.36) mentions 2 species of *Trichocera*, although he gives no detailed descriptions of these, but merely remarks that their wing venation deviates only insignificantly from that of the recent species. Loew's material of these insects has not been covered since, and no further representatives of the family have been observed. The larvae live in damp humus. At the present day, the family is a large one, but mainly of holarctic distribution. It is actually boreal and alpine, and the imagines are mainly found in the winter season. It may therefore be doubted whether they in fact lived in the mainly subtropical amber territory.

Crane flies **(Tipuloidea)**. The amber specimens of this group have been named and more or less completely described by Loew (1850, 1851), Meunier (1895b, 1899, 1906, 1916b, 1917), Cockerell & Clark (1918) and Crampton (1926). However, a collected study of all available East Baltic material was first published by Alexander (1931). In the introduction to this study, Alexander also provides an analysis of the older publications, and is not least critical of the older work by Meunier. He emphasizes convincingly that only entomologists with a comprehensive specialist knowledge of all the recent fauna will be able to provide an exhaustive taxonomic evaluation of the fossil material. The correctness of this attitude has for that matter been confirmed repeatedly by the literature on fossil insects.

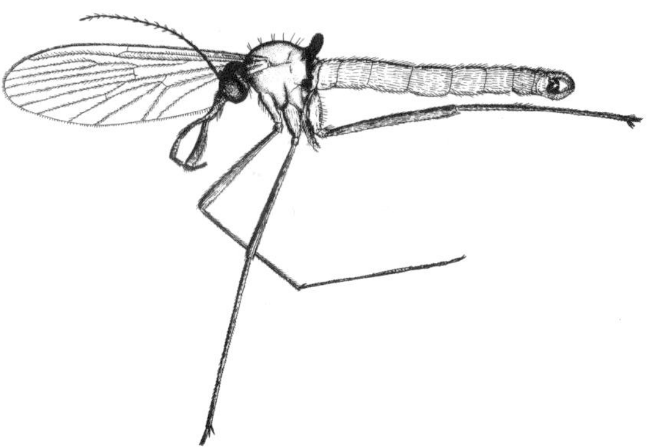

Text-fig.23. *Macrochile spectrum* Loew (Tipuloidea), male. (From Alexander, 1931, fig.1).

Among the actual Tipulidae, there is a single species of *Macromastix*, a genus which today is native to the Southern Hemisphere, and 15 species of *Tipula*, which is now widespread everywhere except in Australia. The larvae are found on many different biotopes, although they always include at least some moisture: fresh water, soil containing humus, living, soft plant parts, and decayed wood.

The family which dominates above all in amber is Limoniidae, represented by not less than 30 genera with a total of 85 species; 5 of the genera are extinct. Today, most of the genera are holarctic, but it is remarkable that 4 *(Limonia, Dicranoptycha, Pseudolimnophila* and *Adelphomyia)* are also Ethiopian. Three genera are cosmopolitan *(Helius, Elephantomyia* and *Rhabdomastix), Eriocera* is found in all the tropics, 3 are palaeotropic *(Styringomyia, Trentopohlia* and *Ceratocheilus),* and only one genus is Nearctic; only the Holarctic *Thaumastoptera* is stated by Alexander to be mainly Palaearctic. This is a geographic distribution which is very common within the amber fauna. Biologically, the Limoniidae in the present are mainly amphibian, although in a number of species the larvae are found in moist, rotten plant remains, e.g. fungi.

In the Danish amber material which has not yet been studied in detail, there are 6 specimens of Tipulidae, 1 of Tanyderidae and 76 of Limoniidae, a proportion which corresponds well to the East Baltic material. However, it is in contrast to the relationship in the partly contemporaneous North Jutland Mo-clay, where the distribution is more or less the inverse, many Tipulidae and fewer Limoniidae. This expresses the influence which the character of a landscape exerts on the choice of items in the composition of a fossil fauna.

Sciaridae (Plate 5,B) is a family which is represented in very great strength. There are often several, at times many of these small midges in the same piece of amber. This fact should in no way be surprising. These insects are weak fliers, and their large surface in relation to their size, as well as their weak muscular power, makes it impossible for them to get free after even a very weak contact with the resin. As is the case with so many other Diptera, namely midges, the fossils include many specimens in which

the male genitalia are almost anatomical preparations for investigations.

The taxonomy of the family presents very great difficulties for the entomologist who is without specialist knowledge. Sciaridae are first mentioned by Burmeister (1832), Berendt (1845) and Helm (1896c). Apart from a very few species, named by Loew (1850), our knowledge is based in particular on Meunier (1895b, 1899, 1901, *1904b,* 1916 and 1917). Meunier has referred by far the majority of the fifty or so known species to the genus *Sciara,* but also *Trichosia, Palaeoheterotricha, Heterotricha, Dianepsis, Zygoneura, Corynoptera* and *Bradysia* are mentioned, the last with a quite considerable number of species. There can be no doubt that this extensive material to a high degree requires revision. In Copenhagen there are about 600 undetermined specimens, by far the majority of Danish origin.

The Sciaridae are exceedingly widespread all over the world, and today are found in a great number of species. They require a damp, dark environment, where fungus grows well, and where organic matter is rapidly broken down. Most of them breed in such biotopes, most often in vegetable materials, for example rotting leaves or in mushrooms, but also in animal products. Some sciarids, however, breed in living plant tissue. The larvae (army-worms) develop very rapidly, and where conditions otherwise permit, many of them can go through several generations in the course of one year. The biology of the family thus demonstrates that there must have been plentiful shadowy localities in the amber forest, most probably as a result of a dense undergrowth of bushes and small trees; this is in good agreement with the scanty herbaceous vegetation of the forest. The Sciaridae, as imagines, have been a dominant faunal element in the deepest layer of the forest, around the root collar of the trees, but their numbers have undoubtedly decreased strongly with increasing height from ground level.

Also fungus gnats (Mycetophilidae) are exceedingly numerous in amber, more than 150 species having been described by Loew (1850), Giebel (1856) and by far the majority by Meunier (1895b, 1899b, 1901, *1904b*, 1916b, 1917, 1922 and 1923). In the material studied by Meunier, fifty or so genera are mentioned, among which *Mycetobia, Macrocera, Platyura, Sciophila, Tetragoneura, Loewiella, Boletina* and *Mycetophila* are the most numerous and also a broad representation of the family. Here, too, a taxonomic revision of the material would be highly desirable. The Danish collection contains about 450 specimens.

Biologically, the Mycetophilidae are closely related to the Sciaridae, particularly in the case of the imagines; they too appear to be most comfortable in the damp shadows near the forest floor. Most of them breed in fungi (Mycetophilinae, Ceroplatinae and in part Sciophilinae). They have a very rapid larval development, and they often manage to completely break down the fruiting body of the fungus. Sciophilinae in particular, also breed in the fruiting bodies of the Polyporaceae, and in the wood rotted and water-saturated by influence of the fungal hyphae. It is probably their larvae and empty pupal skins that are sometimes found in amber. *Mycetobia* breed in particular in the flow of sap from the trees. Although there is hardly likely to be any egg laying, far less development of larvae in the flowing resin, the freshly-flowing viscid sap can nevertheless be thought to have attracted the gnats, so that they swarmed dancing around it, as is their habit in the breeding season. This may explain why they are so common as amber fossils.

Bibionidae are rare in amber, but are nevertheless known right back to Guérin (1825). Two species *(Plecia prisca* Meunier, 1899b, and *Plecia borussica* Meunier, 1904) might well belong to the genus *Amasia* Handlirsch, 1908, p.953), a third species is *Dilophus priscus* Loew. Four specimens are found in the Copenhagen material. This rare incidence may be explained on the basis of the family's biology. Imagines, which have their season in the spring, are heliophilic and seek flowering shrubs, where they drink the nectar, and here mating also takes place. This biotope was not found very often in the amber forest. The larvae live in the ground, probably mainly on fallen leaves which are turning into mould, but young grass roots and other fresh vegetables also play a considerable part in the diet, and there was only little of this in the forest. However, the family could find these requirements satisfied on the extensive plains lying between the amber-producing forests and the Mo-clay sea (Larsson, 1975). The fact that Bibionidae are not known among the fossils of the Mo-clay may be due to their being overlooked. A much more likely explanation, however, is that the interacting factors which were a condition for the insect fossils of the Mo-clay could only have been present in the warmest summer, when the swarming time of the bibionids had long since terminated.

Our knowledge of the **Scatopsidae** (Plate 5,A) of the amber forest is very slight and based exclusively on Meunier, who in 1904 described 4 species of *Scatopse*, mainly on the basis of Loew's old material. It might seem surprising, therefore, that the Copenhagen material contains 26 specimens, primarily of Danish origin. Even though it is not known how many species this collection contains, this would suggest a real difference between the West Baltic and East Baltic fauna. The larvae live in rotting material, and are often found in mammalian dung. It is possible that the Scatopsidae have mostly been present outside the forest area or in other selected regions, where the large herbivores of the time have found their special territory; at any rate they have not been natives of an actual forest biotope.

Brachycera. Most of the flies belonging to this sub-order are exceedingly uncommon in amber, only Empididae and

Dolichopodidae occur in very large numbers.

Xylophagoidea. This includes Rachiceridae and Xylophagidae, the insectivorous larvae of which live in rotting wood or under dead bark, preferably but not exclusively from deciduous trees. Rachiceridae, found at the present day (Text-fig.9) in America from the southern boundary of Canada to the northern boundary of Argentine, as well as the East Indies and Spain, are known from amber by less than 10 specimens, one of them in Copenhagen. Loew (1850) describes *Chrysotemis speciosa* and *Electra formosa*, but the type material of these species has been lost. Meunier described *Lophyrophorus flabellatus* in 1902, this is possibly a synonym for *Chrysotemis speciosa* (Hennig 1967a, p.7).

The Xylophagidae are represented at the present day by 15-20 species, scattered throughout all boreal Holarctis. Only a very few specimens are known from amber. Loew has named the oldest one *(Xylophagus mengeanus)*, but without describing it. Later, Meunier (1909a) described *Xylophagus eridanus* and compared this with a specimen which was presumably Loew's *X.mengeanus*. All these fossils have been lost (Hennig, 1967a, p.11). The Copenhagen collection contains one specimen, which according to Hennig (ibid, p.12) may possibly be *Xylophagus mengeanus*, and in addition an undetermined specimen.

It might appear strange that these primitive fly families have been so uncommon in localities with the characteristics of the amber forest, which might have been considered very suited. They are also spring insects, and at any rate the recent *Xylophagus* likes to feed on the sap oozing out of plants. Even today, however, very little is seen of these flies, even where they might be plentiful considering the larval finds. However, in the amber forest *Xylophagus* has presumably been very near to its southern boundary.

Stratiomyoidea. This group is represented in the amber forest, but only by a very few finds (Hennig, 1967a, p.13—20). Most natural is the occurrence of *Solva nana* (Loew, 1850), sole representative of the family Solvidae. Three specimens of this species are known, the larvae of which have presumably lived in the same way as the xylophagid larvae, and have therefore found their natural requirements for existence in the amber forest.

Only two species of the family Stratiomyidae are known with certainty: *Hermetiella bifurcata* Meunier (1909a) and *Cacosis sexannulata* (Meunier, 1910b); in addition, Helm (1896c) has referred a specimen to the genus *Beris*. The *Beris* larva lives on very moist sites, e.g. near springs; the other species have probably lived as larvae in moist, humus-rich ground or similar places. The reason for the great scarcity of the Stratiomyidae is probably that the imagines are such pronounced sunshine-loving insects and in addition live on the juices of flowers. Neither of these two families have representatives in the Copenhagen amber collection.

Plate 11.

A. Rhipiphoridae, male. Reflecting fissures in the amber around a specimen often makes it difficult to obtain good photographs. (G. Brovad phot.).

B. A member of the beetle family Temnochilidae. Both adults and larvae are predators in galleries of wood-boring insects. Three adults and five larvae are known from the Copenhagen amber collection. Presumably all belong to the species shown here. (G. Brovad phot.).

C. *Mordellistena* (Mordellidae), a heteromerous beetle whose larvae occur in rotten wood which is strongly attacked by fungus. (G. Brovad phot.).

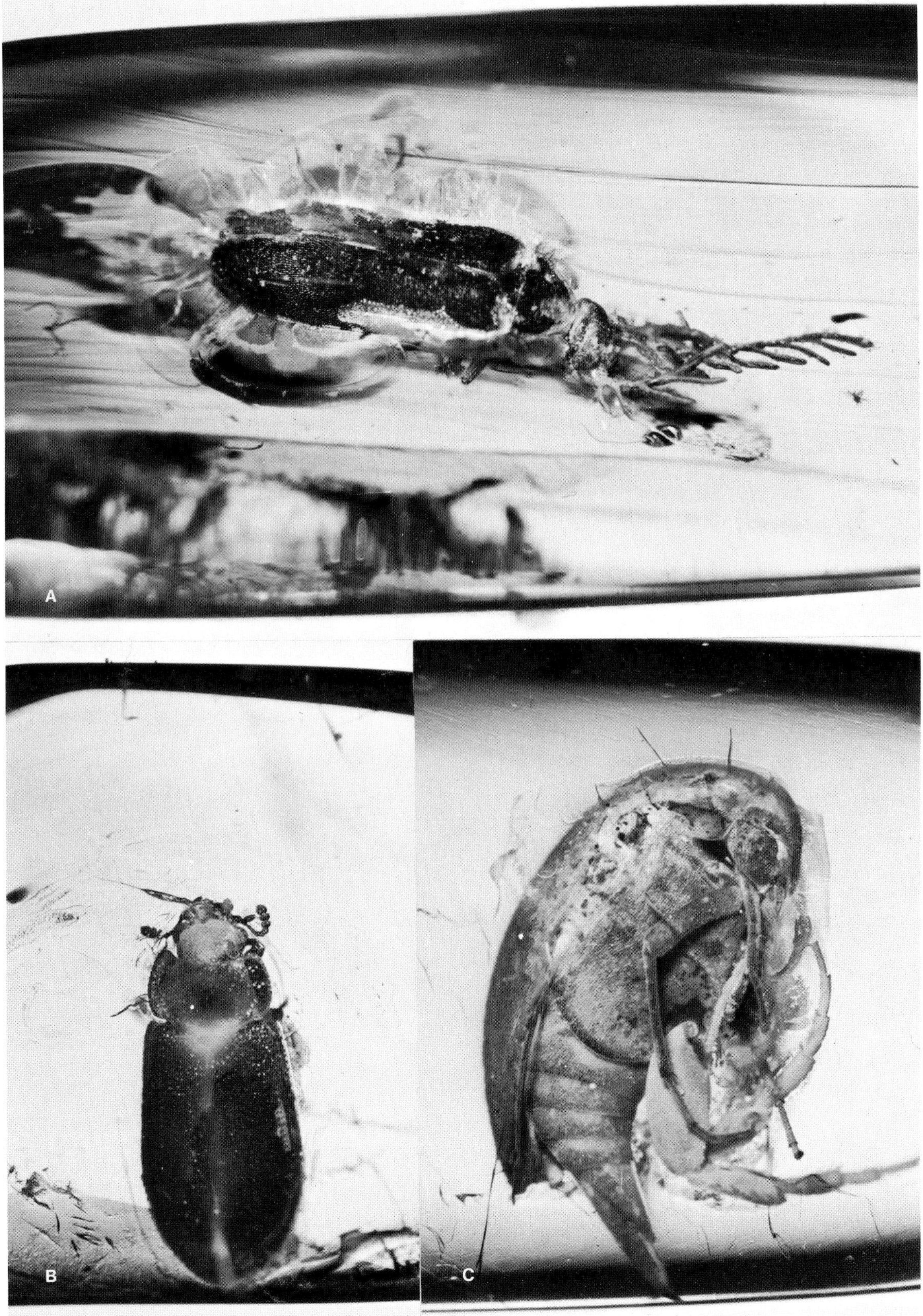

A
B
C

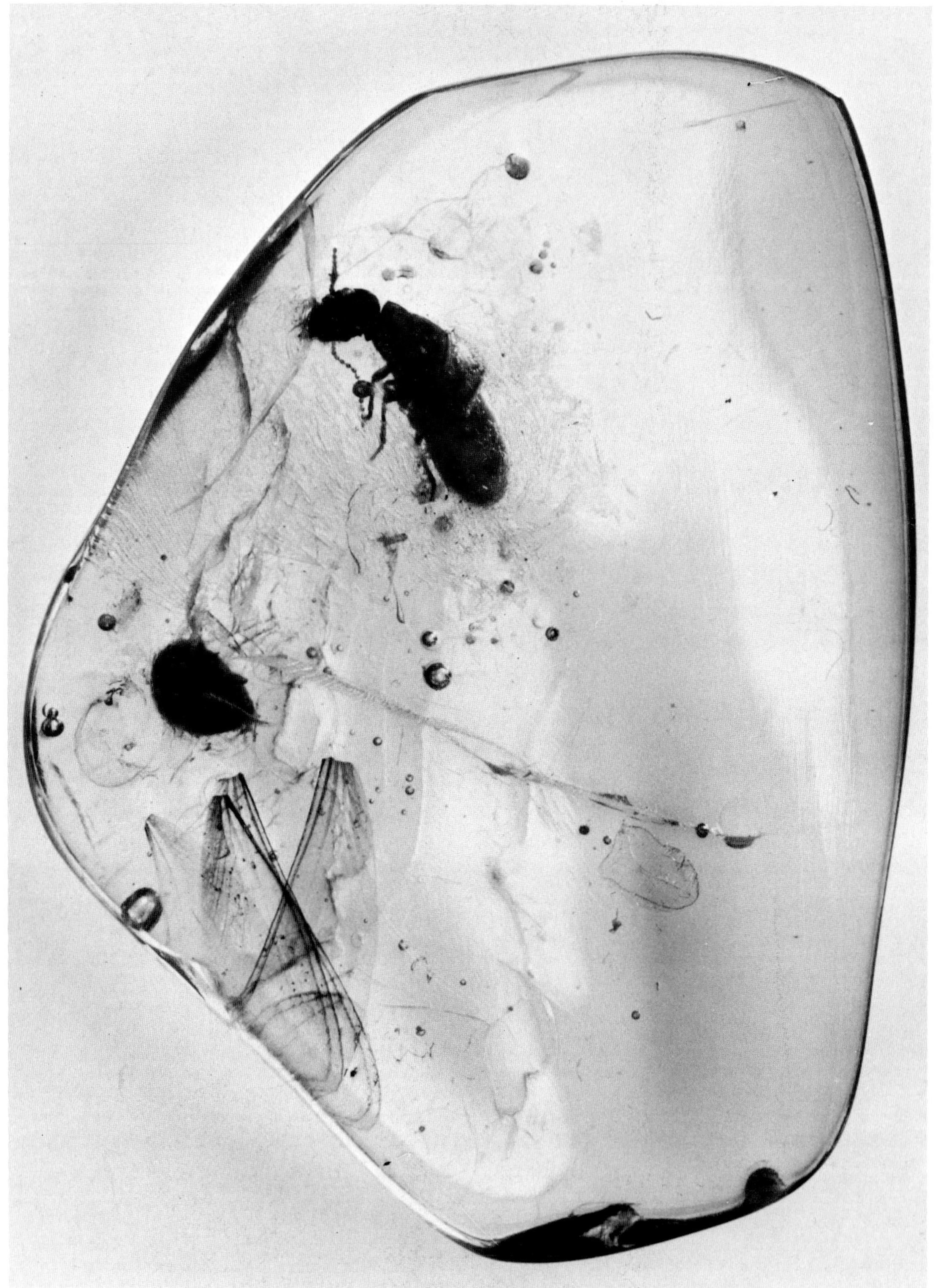

Rhagionidae (Hennig, 1967a, p.21—42) (Plate 5,C) are by far the most common of the primitive brachycerous families in the Baltic amber; a total of about 200 specimens being known. By far the majority belong to *Symphoromyia* and *Rhagio*. Of *Symphoromyia*, 8 species have been named (most of them under the generic name *Atherix;* see Hennig) and of *Rhagio* 14; in addition to these genera, there is in Copenhagen an undetermined material of 32 specimens. These fossils (Loew, 1850; Meunier, 1899, 1909a, 1912 and 1916; Paramonow, 1938) are very much in need of a critical revision. In addition, *Atherix exigua* Meunier (1912, 1916a) has been described; a further 5 specimens near to this genus, the true *Atherix*, are known, including one in Copenhagen, a material which most recently has been studied by Stuckenberg (1974, p.275), who considers **Athericidae** to be a family on its own. Stuckenberg includes here the separate genus *Succinatherix* with 2 species. Of *Bolbomyia loewi* Meunier (1902), which possibly consists of more than one species, 4 specimens are known, one of which is in Copenhagen, while *Protovermileo electricus* Hennig (1967a, p.26) is only known in the type specimen.

When the family is so relatively common in amber, this is not merely because it has been able to find suitable breeding biotopes within the region, but presumably because these flies prefer to settle on tree-trunks where they wait for prey, mainly other flies. This is a characteristic habit at the present-day, and has most probably also been in the past. Most Rhagionidae are forest insects, the larvae of which live as predators in the ground. The larva of *Atherix* may have lived in running water, which was plentiful in the amber forest.

Horse flies **(Tabanidae)** are apparently known in only very few fossil species. One of them is *"Silvius" laticornis* Loew from Baltic amber (Text-fig.24). Hennig (1967a, p.42-49) gives an account of its systematic status and comes to the result that at any rate it is not a *Silvius* species, that it possibly does not even belong to the Chrysopinae. Hennig had. 1 male and 6 females at his disposition, including apparently Loew's specimen. In addition there is an undetermined piece in Copenhagen. Stuckenberg (1975, p.460) describes a new tabanid fly from the Baltic amber, a species of the still existing genus *Haematopota*. He discusses also the problems concerning the *"Silvius."*

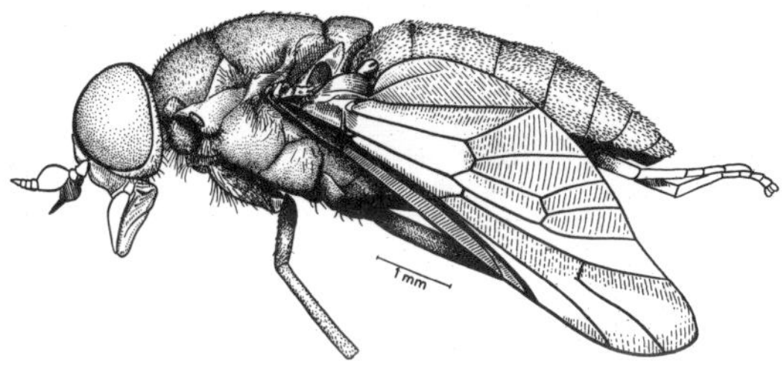

Text-fig.24. *"Silvius" laticornis* Loew (Tabanidae), female. (From Hennig, 1967, fig.72).

Tabanidae like warm, stagnant air and are in general insects of high summer. Many species keep to clearings in the forest and along the edge of the forest, where the females attack in particular larger mammals. The male subsist on the juices of flowers. It is difficult to imagine that the amber territory should only contain these two tabanids, but it is conceivable that the distribution of the somewhat larger mammals may have been such that most of the tabanids have been living in open plains, and that only a few species have been adapted to a

Plate 12.
Termite (Isoptera). Apparently the individual pictured here had just finished its short swarming flight and unfortunately landed upon some fresh resin. After being trapped it immediately shed its four wings at the predestinated basal sutures and succeeded in moving a short distance away from the shed wings before it died. Amber termites are known both before, during, and after wing-shedding.

definite element in the forest territory. In one way or another, the species must have differed in their biology from other tabanids at the time.

Acroceridae. 4 species have been described: *Eulonchiella eocaenica* Meunier (1912; Hennig, 1966e) belongs to a generic group which today is spread over the southern hemisphere, and whose nearest now-living relatives live in South Africa, *Villalites electrica* Hennig (1966e, p.18), and *Glaesoncodes complectinervis* Hennig (1968, p.4), the last-mentioned from the Copenhagen collection. This is also the case with *Prophilopota succinea* Hennig (1966e, p.9), whose recent relations live scattered throughout many countries with warm climate (Neotropics, Mediterranean, South Africa, North India, China). It is this last fossil which has given rise to the remarks on p. 39 concerning Danish amber.

Not only in the Danish amber, but also in the amber from the East Baltic, there are exceedingly few pieces which do not have the same chemical composition as amber from *Pinites succinifer*. As a rule, these pieces are recognized during grinding, because they also differ in their physical reactions, but at times it is possible to have an immediate impression that a difference exists. Professor Hennig was able to do so with the piece in question, and asked Professor Jean Langenheim to test it by infra-red spectroscopy (Hennig, 1969b, p.1). The result was an unknown spectrum. However, as the amber forest was a very mixed forest botanically, where we know that other resin-producing plants than *Pinites* existed, there is nothing remarkable in the fact that also resin from these could occasionally include small fossil animals. There is no obvious reason for doubting the approximate simultaneity of all these amber pieces which have been collected under uniform conditions on the Danish beaches.

Biologically, the Acroceridae deviate from the other Brachycera by their larvae being parasites in spiders or on spiders' eggs in the web cocoon. As these arachnids have lived in the amber forest in uncountable numbers, lack of breeding possibilities has at any rate not been the reason for the relatively modest numbers of Acroceridae among the amber fossils. However, the imagines have been very inactive, when the sun was not at its greatest strength; most often they sit motionless on dry twigs or on the underside of leaves. When they do fly, they can "stand still" in the air for a long time, like syrphids, and do not dash from one object to another like so many other flies. These two characteristics can greatly reduce the possibility of their being caught by the resin.

Bee flies (**Bombyliidae**). These flies are rare in amber, and the majority of the species are only known by single specimens. Considering the life-form of the recent genera, this rarity must be considered natural. Imagines are sun-lovers, nectar seekers, often hovering in the air in front of the flowers, whose nectar chambers they empty; the great majority are found in open territory, where also the hosts for their parasitic larvae live, and only few of them seek biotopes like the edge of the forest. It is in fact more surprising that no less than 6 species have been found with certainty in amber. The species are *Paracorsomyza crassicornis* Loew (1850), whose nearest recent relatives are found in South Africa, *Palaeoamictus spinosus* Meunier (1916a). Here, the type has been lost, and the species must be regarded with some scepsis (Hennig, 1966f) and *Amictites regiomontana* Hennig (1966f); these all originate from East Baltic amber. Further, 3 species are known from Danish amber, the first two at any rate from the coast of the Kattegat: *Glaesamictus hafniensis* Hennig (1966f) (Text-fig.25), *Proglabellula electrica* Hennig (1966f) and *Proplatypygus succineus* Hennig (1969c).

Hennig was surprised that so relatively many of the species came from Danish amber (1969b, p.1), "obwohl die aus Ostpreussen stammenden Bernsteineinschlüsse in den Sammlungen noch immer unvergleichlich viel zahlreicher sind als die dänischen." The position is perhaps explainable when, as here, amber from the West Baltic and the East Baltic is regarded as originating from two widely separated regions, admittedly both within the same giant forest. The Swedish forest regions, from which in many cases the amber from North Jutland must be considered to have originated, were under all circumstances strongly restricted in extent, and not least because of the belt of Swedish lowlands around the great lakes of the present day, the landscape of the forest territory has varied. The Russian forest regions, from which the Samland amber is considered to have originated, have most probably been enormous, rather uniform landscapes. As insects such as the Bombyliidae only skirt the forest edge biotope, and otherwise keep to open territory, it is not especially remarkable that they are relatively well represented in the smaller amber-forest regions of the west.

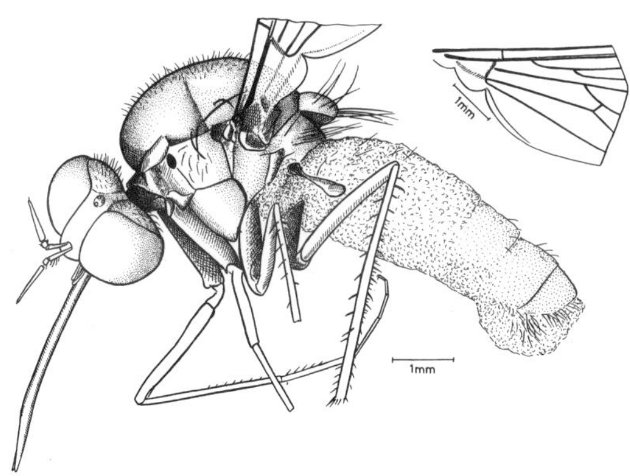

Text-fig.25. *Glaesamictus hafniensis* Hennig (Bombyliidae). (From Hennig, 1966f, figs.17, 18).

Therevidae have been studied most recently by Hennig (1967c), who provides a critical evaluation of pieces described earlier. The material, extremely scanty, is referred by Hennig to 3 species: *Glaesorthactia magnicornis* (Meunier, 1909a), to which he ascribes 5 specimens, *"Psilocephala" agilis* Meunier (1909a, 3 specimens) and *"Psilocephala" pusilla* (1967c, 1 specimen). In addition there are two undetermined specimens in the Copenhagen collection.

Asilidae are at least as rare. Known are *Holopogon pilipes* Loew (1850), *Asilus klebsi* Meunier (1908b), *Asilus angustifrons* Loew (1850) and *Asilus trichurus* Loew (1850), but all these pieces have since disappeared (Hennig, 1967c, p.12). Hennig himself has had the opportunity of seeing 4 specimens, 3 of them from the Danish collection, but the entire material is in such a poor condition that it can hardly be made use of. An excellently preserved asilid has been found in the contemporary Mo-clay from North Jutland (Larsson, 1975, fig.8, p.207).

Empididae (Plate 6,B) are one of the most common insect groups in amber, and in the Copenhagen collection, for example, there are 220 specimens. Furthermore, this is one of the families which to the highest degree needs taxonomic revision, as there has been nothing published on Baltic amber material since Meunier's monograph (1908a). It should be pointed out here that Meunier's material is based all the way through almost exclusively on material from the East Baltic, while the undetermined Danish material is predominantly from the West Baltic. Most of the genera mentioned in the studies by Loew (1850), Giebel (1856) and Meunier (1892, 1895b, 1899, 1902 and *1908a*) are represented by only a single or very few species, the most numerous being apparently *Brachystoma* and *Empis*, but there are also several species of *Hilara* and *Hemerodromia*.

The great incidence of the Empididae can be explained by their biology, even though this covers a wide range of biotopes. A very considerable number of recent species breed in moist, mould-enriched earth among year-old leaves and rotting wood — in other words the natural biotope for Sciaridae. Here, the larvae are predators, just as most other Brachycera larvae, and it is probable that *Sciara* is in fact part of their prey. Thus, the family is to a large extent bound to the humid, shadowy undergrowth. The imagines fly in swarms which typically comprise males and copulation-seeking females. Their flight is very agitated, taking place in a ring dance almost in the same way as that of the chironomids; at one moment the swarm is flying low, at the next they are up in the tree-tops. It only requires a slight disturbance of the air to bring part of such a swarm into contact with the tree-trunks, and in the amber forest these have to some degree been the resin-secreting amber trees. In addition, many of the flies could have become victims when seeking rest on the lower section of the trees.

Dolichopodidae. In the case of these flies, our knowledge is even more scanty than for the Empididae. On the other hand, the family appears to be even more common, since the Copenhagen collection, for example, contains almost 650 specimens, and has been offered several hundred more. The great variation in the material does not suggest that Bachofen-Echt (1949, p.159) exaggerated when he estimated the number of species in the collections as about 100. Here, too, we have the publications by Loew (1850), Giebel (1856) and above all Meunier (1892, 1894, 1895, 1899, 1906, 1908, 1916) to build on. Meunier groups the species into a large number of genera, many with only a single species. Among those most rich in species, recent genera such as *Medeterus, Dolichopus, Chryso-tus* and *Porphyrops* are mentioned. The material is in dire need of revision.

The Dolichopodidae are often encountered grouped in large numbers in low vegetation, e.g. in shrubberies. As the majority of them also like high humidity, it is not remarkable that they are so common in amber, and that it is not unusual to find several together even in rather small pieces of amber. Most of the larvae live in moist, mould-filled earth; they are predators, just like the imagines. The larvae of recent *Medeterus*, however, live under bark, where they consume bark-beetle larvae and other thin-skinned insects. It would therefore be

of interest to determine the fraction represented by this genus within the large material.

Cyclorrhapha Aschiza. This group is represented in Baltic amber by Syrphidae, Pipunculidae, Phoridae, Platypezidae and Sciadoceridae, but only Phoridae occur with greater incidence.

Syrphidae are mentioned for the first time by Loew (1850); he names several species which are closely related to *Ascia* or *Sphegina*, as well as a few *Syrphus* and *Chilosia* species, without describing any, however. Meunier in 1902 and 1904 describes a number of species: *Spheginascia biappendiculata, Palaeoascia uniappendiculata, Palaeosphegina elegantula, Xylota pulchra, Syrphus curvipetiolatus* and *Palaeopipiza xenos.* The Copenhagen collection includes 11 undetermined Syrphidae and a few larvae.

When these decidedly heliophilic, flower-seeking insects occur at all in amber, this must imply that there was something about the tree-trunks which particularly interested them. In the case of the recent species, this is always colony-forming plant lice, representing an assembled food mass large enough for the growth of the larvae. One can picture the Tertiary species floating in the air in front of the tree-trunks, on which at intervals they have attempted to place a couple of eggs among the plant lice, just as at the present day. It is unlikely that any of the species occurring in amber have been associated particularly with the amber tree, they have simply been looking for plant lice wherever they could be found in adequate amounts, but on the amber tree they have undoubtedly been seeking above all the *Germaraphis* colonies. This is confirmed by the larval findings. As the Syrphidae during their egg laying are hardly likely to have gone down into the gloomy Sciaridae stratum of the undergrowth, the conclusion must be drawn that to a certain extent (probably to a major degree) *Germaraphis* has been present in the more airy layers, above the dense undergrowth, where there has been only half-shade.

Among the **Pipunculidae**, the literature reports a total of 8 specimens from the amber in Königsberg, belonging to 5 species: *Verrallia succinia* Meunier, *Verrallia extincta* Meunier, *Nephrocerus oligocenicus* Carpenter & Hull, *Cephalosphaera baltica* Carpenter & Hull and *Pronephrocerus collini* Carpenter & Hull. The latter authors have worked-up the material in 1939. The Copenhagen collection contains a single undetermined piece.

The flies have the same hovering and tireless flight as the syrphids, and are flower-seekers. In particular they prefer rather humid air with shrubs and small trees, where they can lay their eggs on cicadas, which are most common in just these biotopes. The larvae are endoparasitic, leaving their host when fully grown and pupating in the ground.

Phoridae are small flies, most often living hidden on damp sites where there is decomposition of both animal and vegetable matter: leaves, fruit, dead bulbs, dead animals, including both vertebrates, snails and insects. They do not fly much, preferring to run and jump, and their general mode of movement much recalls that of primitive fleas. When they are as frequent in amber as they are, they have most likely commonly been breeding in the forest floor around the amber trees in the *Sciara* zone, at any rate in its lower layer. And in fact they are also often found in the amber pieces together with Sciaridae. Many recent species are myrmecophilous or termitophilous, and there is a considerable possibility that species with a corresponding biology can be encountered in the material of amber fossils. A number of recent Phoridae from the last-mentioned biotope are brachypterous or even apterous; they are not known from the Baltic amber, at any rate not from the Copenhagen collection, which at present totals about 200 specimens.

The Phoridae are one of the insect families which really need to be revised anew. Loew (1850) considered that he had at his disposal 11 species, which without further determination and description he referred in all cases to the genus *Phora*. Since then, Meunier has published a couple of small reports (1908c, 1909b), and in 1912 he gave a comprehensive treatment of the material he had access to. Finally, Brues (1923) published 2 species of *Dohrniphora*, and later (1939, p.411) a more comprehensive paper.

Platypezidae are a small holarctic family, in which only a single species is known from Baltic amber: *Oppenheimiella baltica* Meunier (1893, see also 1895). No material is found in Copenhagen. In general the biology of the family is only poorly known. Many recent species form dancing swarms, especially in shady and damp sites; they are also encountered in shrubberies and other low vegetation. Many of the now living species, at any rate, breed in the fruiting bodies of fungi.

Sciadoceridae are a small family which at the present day live in various regions in the southern hemisphere (Southern Argentine, Southern Australia, Tasmania, New Zealand). Only a single species is known from amber: *Archiphora robusta* (Meunier, 1905b, p.92; 1905b, p.92; Hennig, 1964).

Cyclorrhapha Acalyptratae. Before Hennig took up the topic in 1965, knowledge of these flies as amber fossils was indeed poor. In the subsequent years (1967b, 1969b and 1971b), Hennig not only revised the older literature, which was dominated in particular by Loew and Meunier, but also worked up what was available of new material in a long series of European and North American institutions. The total number of these flies is quite strikingly small compared for example with Empididae and Dolichopodidae, and a large number of recent families are still unknown from Baltic amber. Altogether, Hennig's studies included a total of 118 specimens, many of them from the Copenhagen collection.

By far the most common family is Heleomyzidae (19 specimens), the larvae of which are today saprophagous or are found in fungi. Of Sciomyzidae there are 14; the larvae of this family often live in water, and are reported to be malacophagous. Among the Diastatidae, larvae in wood mould, 12 specimens are known, but 10 of these were taken together in two pieces of amber. Likewise in the Camillidae (7), several specimens were taken in a single piece of amber. At the present day, the Camillidae are reported to breed in the nests of rodents, where the larvae live as saprophages; this can also have been the case in the Eocene. Of the other families, only 1-7 specimens are known: Cypselosomatidae, Calobatidae, Megamerinidae, Diopsidae, Psilidae, Dryomyzidae, Sepsidae, Lauxaniidae, Chamaemyiidae (Text-fig.26), Proneottiophilidae, Pallopteridae, Lonchaeidae, Odiniidae (Text-fig.27), Acartophthalmidae, Clusiidae, Chyromyidae, Aulacigastridae, Anthomyzidae, Asteiidae, Cryptochetidae, Carnidae, Milichiidae, Drosophilidae (Text-fig.28) and Chloropidae (Hennig 1969b), and Pseudopomyzidae (1971b).

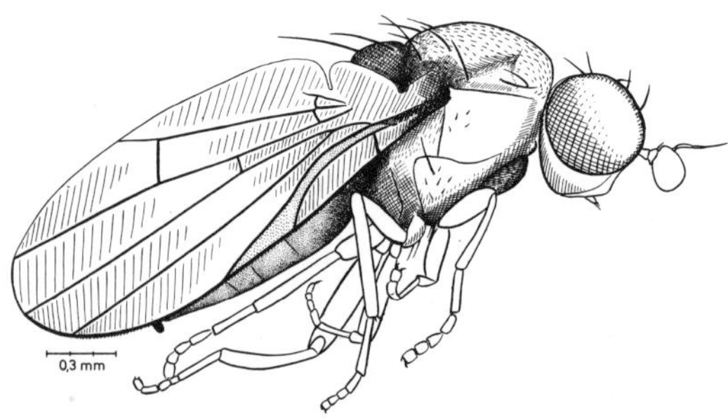

Text-fig.26. *Procremifania electrica* Hennig (Chamaemyiidae). (From Hennig, 1965, fig.148).

At the larval stage, the majority of Acalyptratae have been saprophages or associated with fungi, and especially the Lauxaniidae have any likelihood of being typical for the *Sciara* biotope. Many families have species whose larvae must be sought under dead bark of trees or in rotting wood; this is the case for Calobatidae, Megamerinidae, Lauxaniidae, Pallopteridae, Lonchaeidae, Odiniidae, Acartophthalmidae, Clusiidae and Diastatidae, but this need of course not always have been so for the amber species. A number of these larvae are predators; they are insects with biological contact with the later mentioned fauna from dead tree-trunks. Larvae of Drosophilidae and Psilidae live in the sap secreted by the trees, some Psilidae in the living roots and underground plant stems, Lonchaeidae, Anthomyzidae and Milichiidae also in living plant tissues, as well as Chloropidae, mainly breeding in leaf sheaths and blades in monocotyledonous plants.

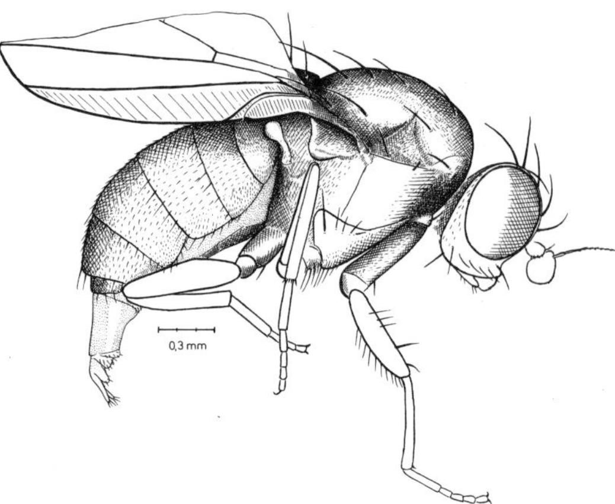

Text-fig.27. *Protodinia electrica* Hennig (Odiniidae). (From Hennig, 1965, fig.156).

A few families occur as parasites on Aphidoidea, Aleuroidea or Coccoidea. This is the case with Chamaemyiidae, which at the present day has species bound to specific representatives within the three host groups, and is also the case with Cryptochetidae, parasites in *Monophlebia* (Coccoidea). Both these families have had the possibility of breeding on the trunk of *Pinites*.

In the case of the imagines, they have in general been rather tied to the locality of their growth, on their own accord they have not flown much or far, but to a pronounced degree have stayed on leaves and shoots of vegetation within a narrow region. Not many of them have actually belonged to the amber forest, but far rather to the thickets and continuous groves surrounding this, in fact a number of them belong preferably to the meadow-like biotopes in the surrounding territory.

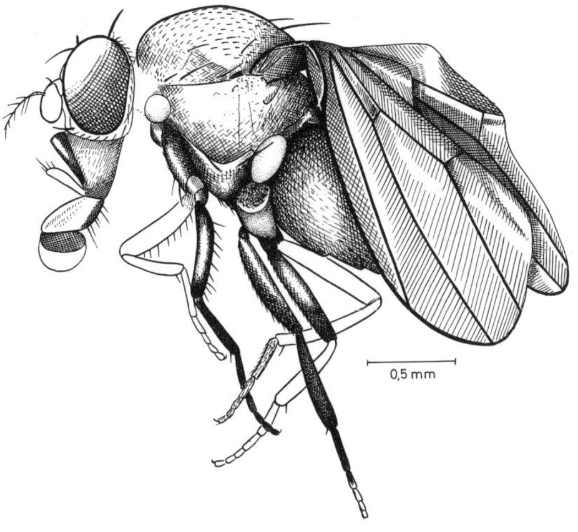

Text-fig.28. *Electrophortica succini* Hennig (Drosophilidae). (From Hennig, 1965, fig. 307).

Cyclorrhapha Calyptratae. Hennig has also worked on the Calyptratae (1966a), but here the material is even more scanty than in the case of the Acalyptratae. Hennig has found a single specimen in Loew's material which belongs to this nowadays very large group, whereas all the other material which could be identified among that previously published belonged to other families of flies. This one specimen belongs to Fabricius' old and well-known species *Fannia scalaris;* Hennig was at any rate unable to tell any difference which could justify establishing a new species for the fossil specimen.

Chapter F
The Fauna of Moss and Bark

Moss and liverwort have played a dominating role among the epiphytes of the trees. Lichen has also been present, but to judge from the content of plant remains in the amber, this has been of less significance for the fauna. There must also have been unicellular green algae and other microscopic plants in large numbers, as well as Infusoria, Tardigrada and other animal organisms of the same order of magnitude. As in the ocean, such organisms constitute the necessary basal stage in the food chains of the local fauna.

The fauna which has lived here has typically no dependence on any definite tree type, but is directly or indirectly associated with the epiphytes, which provide the fauna with shelter and protection, and which constitute a considerable proportion of its food.

Where the forest has constituted an uninterrupted whole, far from the brow of the woods and the open spaces, a vertical layering can be observed in it; conditions must have varied considerably from the base of the trees to the tree-tops, for example with respect to humidity and light intensity, and this must have been responsible for a considerable variation in the composition of the fauna. As a whole the forest has been a mixed forest, with a crown consisting of trees relatively open to the light, such as oak, pine and sequoia. The undergrowth has been rich in species of luxuriant, most often sub-tropical plants, which have been dense especially at the base, so that it has almost completely excluded growths of a herbaceous character. It is in this lowest layer that rotting leaves and twigs have collected so thickly that the Sciaridae, for example, have dominated here. Here in particular the liverworts have grown, to some extent also up the trunks. The mosses have also undoubtedly growth conditions here, but they have nevertheless been able to flourish higher up the tree-trunks. The moss fauna has very likely been most prolific in the *Sciara* region, and this is where most of the photophobic arthropods have been captured by the resin. The fauna has really been a forest floor type, but to varying degrees it has followed the moss upwards; the average size of the animals has gradually decreased, and the composition of the population has gradually changed. The growth of the moss upwards has decreased both in amount and in luxuriance, and has been more or less replaced by lichens; Psocoptera at this point becomes a dominant element of the fauna. The lowest placed colonies of plant lice have also probably been here, and most of the cicadas and spiders have no doubt lived in the crowns of the undergrowth.

Spring-tails (**Collembola**). This is one of the groups which has been worked on best in the available and relatively younger literature (Handschin, 1926). Handschin has had the opportunity of studying a material of 354 specimens, the greater part of them originating from Klebs' collection of Samland amber from the old Königsberg museum. He recognized a total of 8 genera in this material, all found at the present day, and he groups it into 12 species (none of them recent), although he points out that several of them (especially *Entomobrya* and *Orchesella)* must be regarded as collective species; the species-distinguishing characters, as employed in modern Collembola systematics, cannot be applied to the amber material.

The 12 species are found in the following numbers: 1 *Hypogastrura protoviatica* Handschin, 1 *Hypogastrura intermedia* Handschin, 3 *Isotoma protocinerea* Handschin, 4 *Isotoma crassicornis* Handschin, 32 *Tomocerus taeniatus* Koch & Berendt (1854), 228 *Entomobrya pilosa* Koch & Berendt, 8 *Lepidocyrtus ambricus* Handschin, 18 *Orchesella eocaena* Handschin, 16 *Sminthurus succineus* Stach (1922), 21 *Alacma plumosetosa* Handschin, 11 *Alacma setosa* Handschin and 11 *Alacma plumosa* Handschin. As can be seen, species with a well developed springing organ dominate very strongly, leading Handschin to the conclusion (p.336) that it is a case of a moss fauna which has lived in the forest floor and on the bark of trees, but not under dead bark. To this must be added that

also *Hypogastrura* (the genus with the weakest developed springing organ) is most often found in moss today, and is at any rate not typically subcortical.

At the present day, the 8 genera mentioned are individually distributed over very large land areas, especially within Holarctis, but also throughout other parts of the world. This tells us very little as to the temperature preferences of the Collembola of amber. However, Handschin concludes (p.339-340) that the amber forests have lain within a region in which all genera mentioned have been able to coexist. This has only been possible in warm temperate southern parts of the Palaearctic and Nearctic region and in the adjacent parts of the northerly subtropical belt. These are conditions of existence which coincide with what we have learnt are the requirements of the amber forest. The fact that the percentage for Symphypleona *(Sminthurus* and *Alacma)* is relatively great (almost 17%) suggests in Handschin's opinion that the fauna has belonged to the northerly subtropical region without acquiring typical tropical elements itself. That the majority of the Collembola found in amber have also existed far up into the temperate parts of Holarctis cannot be doubted, likewise that they have become mixed with more boreal genera and species, by which they have been replaced in the course of the Neogenic period; but tangible evidence for this is not likely to be forthcoming. The humidity requirements of the Collembola fauna have been considerable, and it has probably been found in particular near the root-neck of trees, not very high above the base of the *Sciara* zone.

The spring-tails in the Copenhagen collection have not been subjected to any particular study. There are altogether 356 specimens, distributed as follows: 13 Poduromorpha *(Hypogastrura?)*, 241 Entomobryomorpha (of these about 10 *Isotoma)*, 97 Sminthuridae and 5 indetermined specimens. Thus also here it is a case of forms which in by far the majority of cases have a well-developed springing organ, in other words a typical moss fauna. The very considerable percentage representation of Symphypleona, almost 30%, is remarkable. It may possibly be interpreted as expressing a climatic difference between South Scandinavia, where the Danish material is considered to have mainly originated, and West and Central Russia, where the Samland material must originate. What is more likely, however, is that it merely emphasizes the unreliability of percentage calculations and the uncontrollable fluctuations within the collected fossil material.

Mites **(Acarina)**, like spring-tails, are represented in the moss fauna. Of this heterogeneous group of animals, Koch & Berendt (1854, p.103) described 16 species, which they referred to 10 genera. Menge increased this number by just a few species (1856, and also with editorial notes in Koch & Berendt, 1854). These authors referred this classic material to the families Trombidiidae (here also including Tetranychidae), Bdellidae, Oribatidae, Sarcoptidae and Gamasidae. A great many features have been altered in this taxonomy since then. Sellnick (1931, p.148) has presented the most recent list of mites in amber, although he does not pay much attention to other groups than Oribatidae, and it is these others in particular which are in great need of revision. Following Sellnick, the number of known species of mites is increased to 90, but there is no doubt that even among the Oribatidae, a revision of the 400 specimens which are found today in the Zoological Museum, Copenhagen, will reveal much that is new.

According to Sellnick, the species hitherto known all belong to a forest fauna, a forest floor fauna, whose species have also been more or less common in those carpets of moss which have grown on the tree-trunks. A significant proportion of these mites are phytophagous, while others are parasites or predators.

Oribatidae are common in amber, the Copenhagen collection including 108 specimens, and 72 species are included in Sellnick's revision. They are typical inhabitants of moss carpets, and absolute vegetarians. It has been observed how some recent species collect fungal spores together with their palpi and consume them (Willmann, 1931, p.81). Their diet, however, consists mainly of green algae and other unicellular organisms, together with small fragments which they cut from the moss leaves by means of their chelicers, their functional "mandibles". Oribatidae are photophobic, and there is no doubt that only quite extraordinary conditions have driven them out of their hiding-place into the liquid resin.

Koch & Berendt established two species of Oribatidae, and both have since been rediscovered by Sellnick (1919). Of one of these species *(Oribates politus)*, Sellnick writes that he is unable to recognize any difference between this and the now-living *Ceratoppia bipilis* Hermann; he therefore establishes it as *Ceratoppia bipilis* forma *fossilis*. Similarly we have *Eremaeus oblongus* forma *fossilis* and *Nothrus horridus* forma *fossilis*. In his most recent publication on the mites in amber (1931), Sellnick mentions a further number of species as forma *fossilis*. Among the Oribatidae, therefore, there are a number of species in which the difference between fossil and recent individuals is so slight that it cannot be documented; in these cases, therefore, practically no evolution has taken place in the course of the last almost 50 million years.

Among the plant-sucking Tetranychidae there are a few species which have been described. These animals have always been found singly or at most with a couple of specimens in the same lump of amber, and normally they are hardly likely to have lived on the tree-trunks, at any rate not on the trunks of the thick-barked amber trees. They can have been chance guests, brought with the wind from the leaves of other trees or perhaps from the lianas and epiphytes of higher phylogenetic rank than mosses.

A quite small species, described by Koch & Berendt under the name *Acarus rhombeus* (1854, p.110), and by these authors referred to the family Sarcoptidae, but which rightly should be referred to Trombidioidea (Erythraeidae?), is exceedingly common in Danish amber (96 pieces of amber, often with several specimens). At times, half-a-score or more of specimens can be found in a small piece of amber; Bachofen-Echt (1949, p.63) even mentions one piece with more than 30 specimens. Nevertheless, in spite of the number it hardly belongs on the trunk of the amber tree, at any rate not direct, as assumed by Bachofen-Echt. A characteristic feature is that it is commonly found together with considerable amounts of detritus originating from insect larval burrows. There is much to suggest that it has been exceedingly common in the larval burrows, both in the bark of the trees and in the wood, and probably also in the fruiting bodies of the Polyporaceae, where its most important source of nourishment appears to have been the broken-down larval excrement, or more exactly the microscopic fungi growing on this. The resin which has been flowing out has occasionally carried along some of the contents of the larval burrows on to the surface of the trunk, where it now produces a cloud of detritus in the amber, containing gnawed chips, mites, and other possible items (Schubert, 1939, p.38). This mite has thus lived (at any rate also) higher up on the trunk than is typical for the other mite fauna. The enclosed nature of the biotope has been an adequate guarantee against the animals drying out.

The other mites found in amber are more or less certainly predators, and in most cases they have been insect parasites at one larval stage. They thus represent a higher link in the food chain than the Oribatidae, in addition a very complicated link, since many of them, at various early development stages, have depended on animal forms for food or transport, being themselves as adult mites a source of food for these animal forms. The free predator stages live mainly on Collembola, Oribatidae and the small worms mentioned below. As predators, however, they are not very typical representatives of the fauna of the tree-trunks, but like the corresponding faunal element in the present, must be assumed to have been far more common in the forest floor.

Trombidiidae are common; the Danish amber contains 138 specimens, and Bdellidae are likewise common (28 specimens in Copenhagen), while Gamasidae (10 specimens) are less common. Uropeltidae are not known from amber. Wheeler (1915, p.121) describes and illustrates a mite which is attached firmly to the leg of an ant *(Lasius schiefferdeckeri* Mayr) (Text-fig.43). In all probability this is a case of a parasitic form and not of any Uropeltidae. Hydrachnidae: although it is of course impossible to obtain any real impression of the frequency of this family in the amber territory, it is nevertheless a fact that it has existed, as larvae have been found parasitic on a few specimens of caddis flies from amber (Bachofen-Echt, 1949, p.64). Wood ticks (Ixodidae) from amber have been described by Weidner (1964, p.143).

Moss scorpions **(Pseudoscorpionida)** (Plate 6,D) are among those predators which lived to the greatest extent on the biotope's own microfauna: small worms, spring-tails and mites. Today they live in obscurity, many of them in the upper layer of the forest floor between rotting leaves and moss cushions, or they exist under the loose bark of dying and dead trees and in dry rot; by means of the last-mentioned biotopes a number of them have established a more or less permanent relationship to bird's nests, animal lairs and the nests of bees or ants. They have undoubtedly led a similar existence in the amber forest. Those genera specially associated with the moss vegetation of the bark appear to have been relatively common, the natural explanation for this no doubt being that they came into contact with the amber tree and its flowing resin more easily than the more exclusive forest floor forms or those forms living a subcortical existence. However, none of the pseudoscorpions have had a special association with the amber tree. A number of species have no doubt been more common in the moss forming carpets around the foot of the tree, and have met their fate in the form of flowing resin, pressed out of the moss in the same way as has probably been the case with so many of their prey. They have undoubtedly been restricted completely to the *Sciara* zone.

The number of species of Pseudoscorpionida, and undoubtedly also the number of individuals, has been far greater in the amber forests than in the present temperate forest at the same latitudes, but the centre of gravity of their distribution also at the present lies within the tropics and subtropics. All now-living families are represented in amber, and this is likewise the case with a number of the recent genera. A few genera appear to have been declining since the Palaeogenic; for example relatively many specimens of the genus *Pseudogarypus* have been found in Baltic amber, while at the present day it is only known by two rare species in the southern parts of North America. According to Max Beier (1947), 21 species of moss scorpions have been found so far; 4 species were described already by Koch & Berendt (1854, p.94), and several were described at the same time by Menge (1855). Menge's material has since been checked by Beier (1937). The Copenhagen collection contains 15 specimens, 8 of them having been referred by Max Beier to 8 different genera and species, all of them known from the publications mentioned. Altogether, these data suggest a richly varied fauna of Eocene Pseudoscorpionida.

Beetles **(Coleoptera)** likewise include many genera which belong to the moss fauna of the tree trunks, no doubt mainly as predators. Like wood-lice and millipedes, most of these beetles are found mainly in the forest floor and in the moss surrounding the root-neck of the tree-trunk, the biotope of the *Sciara* community.

Carabidae. Numerous genera of this family have been named in the classic literature, but some must be regarded with distrust, as they do not normally belong to more or less closed forest biotopes (e.g. *Clivina*). The great majority of these beetles belong to the forest floor fauna which is often encountered in the moss around the tree root-neck, and only a few of them belong to the typical tree fauna, which obtains its food between the leaves of the epiphytes, under the scales of the tree bark and in its fissures, in short in the zone of the Psocoptera. These species are climbers, and are unlikely to have been found in particular on the amber tree, but probably to an equal extent on the numerous surrounding oak trees.

In this special fauna of tree-trunk climbers, the genus *Dromius* has been dominant. It is by far the most common in amber, and this is in full agreement with its mode of life at the present. Menge (1856, p.23) recognized 2 common species in his material, and compared these with the recent *Dromius agilis* and *Dromius quadrimaculatus*, although not identifying them with these. Giebel (1856, p.70) gave a thorough redescription of *Dromius resinatus* (Germar), based on the type to Germar's *Lebina resinata* (1813), and there is much to suggest that this species is indeed the dominant one in the material of amber fossil carabids. Germar compared the species with *Dromius quadrimaculatus*, but among other characteristics Giebel was unable to recognise the cross marking which Germar considered could be seen. He found the upper surface apparently shining brassy-yellow, most probably due to the fine layer of air which the microscopic pubescence had retained during the embedding in the resin. A comparison with the Copenhagen material suggests that this is the case. The great relative incidence of the genus (or the genus-group) in amber appears already from the work of Menge, in which 17 out of 26 Carabidae belong to the genus. However, the possibility cannot be overlooked that Menge may have included other genera, for example *Metabletus* and *Lebia.* The corresponding figures in Klebs (1910) are 20 *Dromius* out of 73, and in the Copenhagen collection 3 out of 6, one of which is a *Dromius* larva, which morphologically deviates only slightly from the recent forms. *Dromius* has presumably gone through its entire life cycle on tree-trunks and branches higher up than the *Sciara* zone, in contrast to the majority, at any rate, of the genera mentioned below.

In the list of Carabidae from Baltic amber which was published by Klebs (1910) and worked on by Edm. Reitter, in addition to the 20 *Dromius* there are mentioned 1 *Acupalpus,* 1 *Agonum,* 1 *Amara,* 1 *Apristus,* 1 *Balius,* 1 *Bradycellus,* 3 *Calathus,* 2 *Dyschirius,* 1 *Lagarus,* 1 *Lebia,* 1 near *Lebia,* 7 *Metabletus,* 3 *Nebria,* 1 *Ophonus,* 4 *Platynus,* 3 near *Platynus,* 1 near *Polysticus,* 5 *Pterostichus,* 1 near *Pterostichus,* 8 *Trechus* and 6 undetermined. They are mainly typical forest floor forms, which today have a very wide distribution with preponderance within Holarctis. Many are also found in decaying wood and under dead loose bark, and a few (e.g. *Lebia*) perhaps also higher up on the living tree trunk. A number of these genera are also mentioned in the older literature, which also mentions *Carabus* (Gravenhorst, 1835, p.92), *Cymindis* (Burmeister, 1831, p.1100), *Clivina* (Berendt, 1845, p.56), *Chlaenius* (Berendt, 1845, p.56, and Helm, 1896c, p.224), *Harpalus* (Berendt, 1845, p.56) and *Bembidion* (Helm, 1896c, p.224), and in the Copenhagen collection there is also a *Tachys* s.l., described by Erwin (1971) as *Tarsitachys bilobus.* Only a few species have been described: *Cymindoides sculptipennis* Motschulsky (1856, p.25), *Agatoides carinulatus* Motschulsky (1856, p.26), *Protoscalidion rugiae* Schaufuss (1888, p.266), *Trechoides fasciatus* Motschulsky (1856, p.26) and *Dyschiriomimus stackelbergi* Yablokov-Khnzorian (1960, p.94) (Text-fig.29), in addition to

the previously mentioned *Dromius resinatus* (Germar). Further, Giebel has described *Bembidion succini* (1856, p.64).

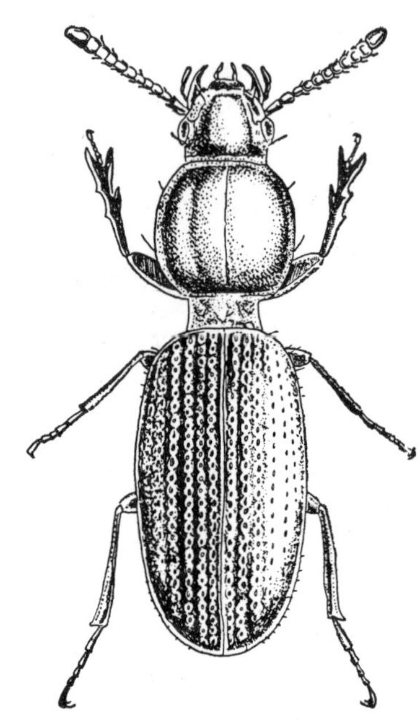

Text-fig.29. *Dyschiriomimus stackelbergi* Yabl. (Carabidae). (From Yablokov-Khnzorian, 1960, fig.2).

Bachofen-Echt considers it remarkable that *Calosoma* has not been found in amber (1949, p.105). If it is appreciated, however, that the amber tree usually has had its base in the humid darkness of the *Sciara* community, where the typical prey of the *Calosoma* species, large Lepidoptera larvae, has been absent or only very scantily represented, the above absence must be regarded as quite natural. The forest forms of *Calosoma* are found in clearings and the edges of the forest with thickets of hazel, birch and willow, for example, and with at any rate some herbaceous vegetation. The recent species in Holarctis are in addition generally more boreal in their characteristics than the amber tree at its northern boundary.

In connection with the Carabidae, two findings must be mentioned which are very remarkable. Both are tiger beetles

(Cicindelidae), and both are imagines. One is *Tetracha*, according to Walter Horn (1906, p.329) identical with the recent *Tetracha carolina* (L.). *Tetracha* is usually regarded as a sub-genus of *Megacephala*, which at the present day has a very wide distribution in the tropics and subtropics in all parts of the world on both sides of the Atlantic and Pacific, while the species *carolina* is limited to America, primarily the southern U.S.A., the West Indies and Central America. Thus, in the Palaeogenic part of the Tertiary, the genus *Megacephala* must have lived in the Baltic region as a link in a world-wide distribution. The possibility cannot be rejected that *Tetracha carolina* has existed as a species and has been widespread on both sides of the Atlantic; numerous amphi-atlantic beetles are found today, the recent distribution of which must be based on a spreading far back in the Tertiary era or actually even earlier, and there is nothing to hinder *T. carolina* having perished on one side of the ocean, while continuing to exist on the other side. Even if the amber specimen, in spite of Walter Horn's expertise, should turn out to be taxonomically deviant from the recent form, the relationship between them is nevertheless so great that the variation cannot be ascribed any fundamental significance.

It is surprising, however, that this large and powerful insect, with its home in open territory, should have been captured in the resin. Further, the genus flies very little, but seeks its prey (various insects) on foot and at night (Houlbert, 1912, p.105). The specimen captured in the resin must have lived in the extreme edge of the amber forest and by chance while hunting have passed the foot of an amber tree producing resin, just like many of the forest floor's own hunters. The larvae live in earth burrows in the same way as the larvae of most other Cicindelidae.

The other remarkable find of Cicindelidae is an imago of *Collyris* or a form very closely related to this. The specimen, which was found in the ground in Salling, North Jutland, belonged to the Copenhagen collection, but is most regrettably missing. A study of the piece was rendered somewhat difficult by extensive opacities, but the general body build of the animal, together with the completely clear regions around the head and prothorax as well as front legs, mouth parts and antennae, established the determination mentioned as certain.

Collyris lives today in India and Further India, together with the adjacent regions of Indonesia; the closely related *Tricondyla* has a further spread throughout Indonesia, while *Ctenostoma* has its home in tropical America. The two last-mentioned genera differ from *Collyris*, among other things by the imagines being unable to fly, but otherwise they are morphologically and biologically close. *Collyris* is the most primitive of the genera. Its home is in localities with open forest and bush vegetation, where as a sunshine-seeking insect it hunts its food in particular among the flower-seeking insects. The larva live in burrows just like other Cicindelidae larvae, but not in the ground. They are found in the pith of finger-thick branches, through an opening in which they catch passing insects, not least ants. By this mode of existence, recent species cause considerable harm to coffee trees (species of *Coffea*, Rubiaceae).

Staphylinidae have lived then, just as they do now, more or less constantly in the moss cushions, and have thus to a certain degree become included in the fauna of the tree-trunks, although they have had their existence particularly in the *Sciara* zone. Many of these species have lived on the local small animals, but just as in the present, a not insignificant proportion have probably lived mainly on fungi or detritus. There are large number of genera, but the most common are *Tachyporus*, *Anthobium*, *Lathrobium*, *Atheta*, *Bryocharis*, *Medon*, *Philonthus*, and *Scopaeus* (Klebs, 1910, Edm. Reitter det.). In the Danish collection there is an unexamined material of 30 specimens, including 4 larvae

(Lathrobium, Anthobium (?), *Tachyporus* and *Micropeplus).* A revision of the generic determinations, both of the existing classic material from the Baltic and of the Danish collection, has been carried out by Dr. V. Puthz, Germany. The list of the Copenhagen material reviewed (received 1974) comprises the genera *Tachyporus, Eusphalerum, Rugilus, Oligota, Carpelimus, Sepedophilus,* prope *Lathrobium,* Proteininae g.sp., in addition to some pieces which Dr. Puthz did not consider could be identified at present.

The families **Scydmaenidae and Pselaphidae** have a mode of existence which much resembles that of many staphylinids. They are typical inhabitants of the *Sciara* zone. Most of them are essentially forest floor animals, and are therefore often found in the moss around the base of the tree trunks; in addition imagines swarm in the dusk and also in this way may have been exposed to the risk of being captured by the flowing resin. The genera *Euconnus* and *Stenichnus* (Scydmaenidae) together with *Euplectus* and *Bythinus* (Pselaphidae) appear to have been particularly common (Klebs, 1910, Edm.Reitter det.). L.W. Schaufuss (1890a, p.564) has described a number of Scydmaenidae from Helm's collection: *Cryptodiodon corticaroides, Cyrtocydmus laticlavus, C. carinulatus, C. capucinus, C. titubans, Semnodiocerus halticaeforme, Palaeomastigus helmi, Hetereuthia elegans, Palaeothia tenuitarsis, Heuretus coriaceus* and *Electroscydmaenus pterostichoides.* Also described are *Clinidius balticus* C. Schaufuss (1896, p.51) and *Scydmaenoides nigrescens* Motschulsky (1856, p.27).

Among the Pselaphidae, numerous species have been described by L.W. Schaufuss (1890b, p.113) belonging to the genera: *Greys, Tychus, Bryaxis, Parabryaxis, Bythinus, Monyx, Deuterotyrus, Hagnometopias, Batrisus, Cymbalizon, Tyrus, Ctenistodes, Dantiscanus, Pammiges, Pantobatrisus, Nugaculus, Nugator, Euplectus, Hetereuplectus* and *Faronus.* Further, C. Schaufuss (1891,

p.53) described *Bryaxis patris* and Motschulsky (1856, p.26) *Euspinuides glabrellus* and *Tmesiphoroides cariniger.* As can be seen, L.W. Schaufuss regards the great majority of species as belonging to extinct genera.

The Copenhagen collection contains 66 Scydmaenidae (2 of them larvae); they are being studied by Professor W. Suter, U.S.A. A material of 16 Pselaphidae has not been studied so far.

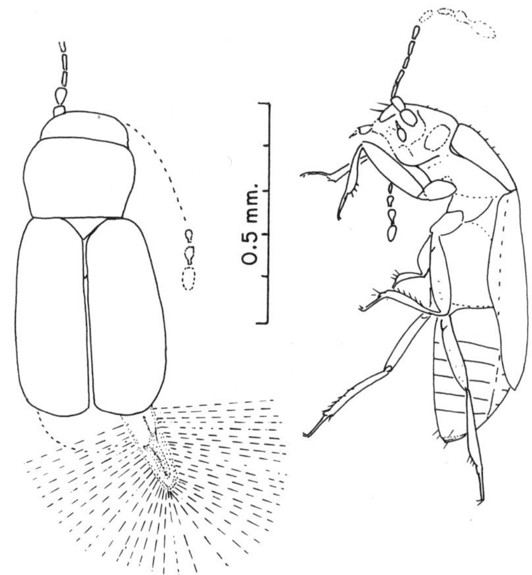

Text-fig.30. *Microptilium geistautsi* Dybas (Ptiliidae), dorsal and lateral views. (From Dybas, 1961, fig.4).

The quite small **Ptiliidae** are represented in the Copenhagen collection by 5 pieces of amber, but one of them contains 3 specimens. These are beetles which are extremely common in the forest floor at the present day, especially near tree roots and windfalls, where the moisture is best retained. There can hardly be any doubt that this family has been far more common in the amber forest than the few finds would indicate. Dybas (1961, p.1) has described the species *Microptilium geistautsi* (Text-fig. 30) and Parsons (1939, p.62) the species *Ptinella oligocaenica.*

Silphidae. Originating from the same biotopes as the beetle families mentioned above are a number of small Silphidae,

apparently all belonging to a single species, *Ptomaphagus germari* Schlechtendahl
(1888, p.483), referred by Reitter (in
Klebs, 1910) to *Nemadus*. It has presumably been breeding in the nests of
mammals and birds and has no doubt
itself often been present on carrion. It is
to be expected that on the whole its biology has been related to that of the
recent species. The Klebs collection has
8 specimens, and the Copenhagen
collection 15.

Lagriidae is known from only a few
specimens, either related to *Statira*
(Smith, 1868, p.184; Klebs, 1910) or to
Lagria (Klebs, 1910). The Copenhagen
collection contains a single larva, a rather
large species, undoubtedly belonging to
the Statirinae. The larva has presumably
belonged to the forest floor fauna.

Fleas (Siphonaptera) are found in the
Baltic amber to the extent of two specimens, described as *Palaeopsylla klebsiana*
Dampf (1911, p.238) and *Palaeopsylla
dissimilis* Peus (1968, p.62) (Text-fig.31).
No other Tertiary fossil fleas are known.
Their phylogeny is discussed by Hennig
(1969c). It is not surprising that both
species, which differ morphologically
from each other to quite some extent,
should be referred to *Palaeopsylla*, as in
the present this is actually one of the
genera most exclusively associated with
small insectivores, which were the most
dominant group of burrowing mammals
in Europe in the Eocene.

The larvae of fleas feed on detritus, and
have most probably had their biological
origins in the *Sciara* zone of a tropical or
subtropical forest, where their mode of
life has presumably developed more or
less like that of the *Sciara* larvae. The
transition of the imaginal fleas to parasi-
tism on mammals, however, must in all
probability (just like the amber forest)
be much older than the Eocene. Marsupials, bats and insectivores have today
their exclusive groups of fleas, respectively Macropsyllidae, Ischnopsyllidae
and Hystrichopsyllidae (to which group
Palaeopsylla belongs). The separation

between them is older than the estab-
lishment of *Palaeopsylla*, and the mammalian groups mentioned have not
exchanged species since. The biological
and morphological separation between
these 3 families of Siphonaptera must lie
far back in the Mesozoic. As many small
rodents evolved later, and often overtook burrows from insectivores, they
also overtook parasites of the Hystrichopsyllidae family. From this point they
have then developed significant elements
of their own flea fauna, and this has
hardly been much older than the Tertiary.

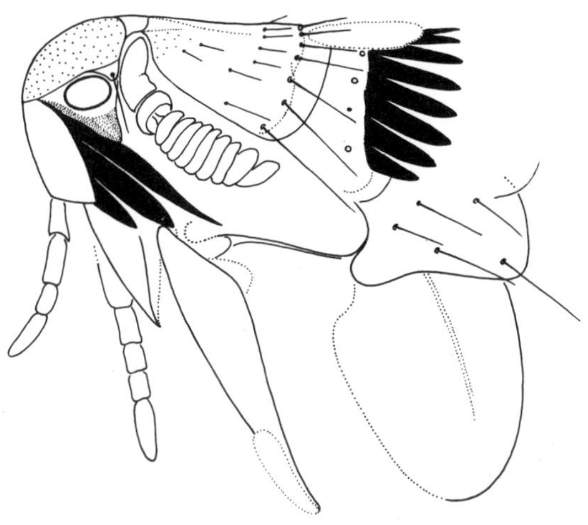

Text-fig.31. *Palaeopsylla dissimilis* Peus
(Siphonaptera), head and anterior part of
thorax. (From Peus, 1968, fig.1).

It must thus be assumed that Macropsyllidae have still been living in the
amber forest as late as the Eocene, in
spite of the increasing rarity of the marsupials, and that Ischnopsyllidae have
been very common. However, fleas in
amber will always be a great rarity.

Strepsiptera. Menge in 1866 published
a description of *Triaena tertiaria* (Text-
fig.32), a specimen of hitherto unknown
family of the Strepsiptera. As the name
Triaena was preoccupied, it was altered
in 1886 by Grote in a short note (Canad.
Entom. 18, p.100) to *Mengea* after the
finder, so that the species is now called
Mengea tertiaria (Menge). Later, several

specimens were found (a total of 6). Keilbach (1939) has given an overall account of all this material. In addition, recent representatives of the family have now been found in the Western Mediterranea.

Mengeidae differ on significant points from the other Strepsiptera, and present in particular a series of primitive characteristics. Particularly striking is the observation that the female (at any rate in the recent species) is not endoparasitic but a wingless larviform creature living in the soil. The larva is known as a parasite in Lepismatidae, but whether other insects which are free-living in the soil are included in the choice of host animals, or have been included in earlier times, is not known.

It is likely that the host for *Mengea tertiaria* has been an animal living in the soil, and among the host possibilities present in the forest floor of the Eocene might be mentioned Thysanura (Machilidae rather than Lepismatidae), cockroaches, earwigs and possibly crickets; in addition there has been a throng of ants and many termites. Cicadas have not been likely host animals, as in the main they must have lived in the partly rather high-placed crowns of the undergrowth, as well as having been relatively lightly represented in the amber forest. Other Aculeata than ants were rare in the amber forest, which as a whole has been poor in sunshine and flowers. At first sight, one might be inclined to consider that the Strepsiptera in the Eocene were at a phylogenetically young stage of development. But this view must be rejected, in this case as in the case of many other insect groups. Considered as a biotope the amber forest was exceedingly unfriendly to most of the other Strepsiptera. Solitary bees were almost completely lacking, and social, higher bees dominated; the Hemiptera were relatively few. We know from the Danish Mo-clay, however (Larsson, 1975) that within the same regions there have been landscapes open to the sunshine and with a very rich fauna, including Pentatomidae and cicadas. Alone for technical

reasons, however, the Strepsiptera have had practically no chance of appearing among these fossils.

The finds of imagines made hitherto all originate from the East Baltic amber.

Text-fig.32. *Mengea tertiaria* (Menge) (Strepsiptera), male in dorsal view. (From Ulrich, 1927, fig.17, redrawn).

Rhipiphoridae (Plate 11,A). Menge (1856, p.23) mentions from his collection of amber fossils "ein stück mit 7 rötlich gelben larven, die den auf blumen lebenden und sich an bienen anhängenden meloe larven ähnlich sind; aber am ende der zweigliedrigen tarsen befinden sich nur zwei klauen. Der leib ist spindelformig, mit kopf zugespitzt, taster viergliedrig; hinterleib aufwärts gebogen, spitz zulaufend, ringsum abstehend behaart." Apart from the observation of a doubly segmented tarsus, which must be based on an error of observation, there can be no doubt, following this description, that this has been a case of a newly hatched brood of one or other of the Rhipiphoridae.

The literature mentions *Rhipiphorus* (Berendt, 1845; Menge, 1856), but there has only been a slight chance that this genus was present in the amber forest, as it lives on Aculeata. Reitter has determined two specimens in the Klebs collection' (1910) as belonging to *Pelecotoma*, the larvae of which are said to live as predators on other beetle larvae, which they seek out in rotted wood. This is the most primitive life form in the family, and the genus has quite definitely had the possibilities of existence in the amber forest. Finally, Stein (1877, p.29) has described *Rhipidius primordinalis*. Also this genus, the larva of which is endoparasitic in cockroaches, has had rich opportunities for existence in the locality. Menge's brood of larvae could thus belong to one of the two last-mentioned genera, both of which could have existed in the forest floor.

The Copenhagen collection contains a single Undetermined imaginal specimen of Rhipiphoridae (Plate 11,A).

Meloidae. A find of *Lytta* is mentioned by Burmeister (1832, p.635) and Berendt (1845, p.56, as *Cantharis*), but must be regarded as uncertain. A first stage larva of a Meloidae is present in the Copenhagen Collection (Text-fig.33).

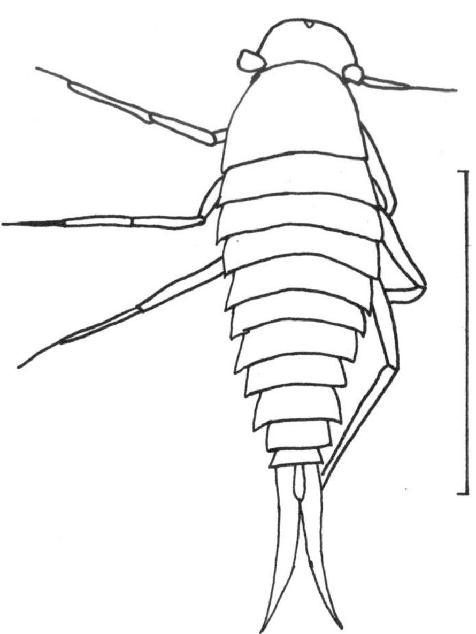

Text-fig.33. First stage larva, probably of a species of Meloidae, dorsal view. Scale: 0.3 mm. (Original drawing from specimen in Copenhagen collection).

Embioptera. This remarkable order of insects with archaic features is known from a few isolated specimens. Pictet (1854) has described the apterous male of *Embia antiqua* (Text-fig.34), the later *Oligotoma antiqua* (Pictet) Hagen (1856), and *Haploembia antiqua* (Pictet) Davis (1939b, p.562), all based on the same specimen. Recently Ross (1956, p.76) redescribed Pictet's type, combined with a supplementary apterous male and a late instar juvenile both belonging to the same original material. Ross altered the generic name to *Electroembia*. An additional specimen is known from Bachofen-Echt (1949, p.77), however, not described, only pictured (fig.68), and also this specimen seems to be a wingless male. The order is not represented in the Copenhagen collection.

Ross comments on the apterism in the male Embioptera as follows: "Apterism in male Embioptera is characteristic of modern species which live in ecologically marginal life zones of the world. The apterous condition is almost universal in regions experiencing prolonged dry seasons. Thus male apterism in *Electroembia* strongly suggests that the early Tertiary environment of the species suffered a long dry season in the summer month (the season of sexual maturity in modern North-temperate species)." However, this seems hardly to have been the reason for apterism in the amber forest.

Earwigs (**Dermaptera**). This order consists mainly of nocturnal insects, creeping about in their search for food, which consists both of dead organic matter and soft plant parts, as well as small animals of various kinds, e.g. plant-lice. There is no doubt that at night time a number of the species have crawled upwards in the vegetation, also on the irregular trunk of the amber tree, which has both provided food and possible shelter for the coming day. Judging from the rarity of the group in amber, it must be presumed that they most commonly have been present under the loose bark on fallen, rotting trunks and in the forest floor itself, as is also the case today.

The earwigs from the East Baltic amber, mainly originating from the Klebs collection, have been described by Malcolm Burr (1911, p.145), and nothing has been published on them since. Burr's material comprised a total of 13 specimens: 8 males, 2 females and 3 larvae. Like so much else, the relatively many larvae suggest that the production of resin by the amber tree has been most active at the start of the annual growth season. The excess of males must be interpreted as a relatively greater activity in males than in females.

In this material Burr describes 4 species and refers them all to *Forficula* (Forficulidae), although only one species *(Forficula praecursor)* is typical of the genus. Two of the larvae deviate strongly from Forficulidae, for one thing because of the construction of the feet, but because of their young age they cannot on the other hand be referred with certainty to any other genus. One is possibly a *Labidura*, the other a *Pydidicrana*.

The Copenhagen collection contains two pieces of Danish amber with earwigs, both of them *Forficula* in Burr's comprehensive sense; one is probably a male of *Forficula klebsi* Burr. At any rate one of the specimens occurring in the Mo-clay also appears to be *Forficula klebsi*, or rather a closely related species. The males are in considerable preponderance also in the Mo-clay.

Cockroaches **(Blattoidea)** (Plate 6,A) from the Baltic amber have been treated by R. Shelford in two studies. In the first (1910, p.336) he worked up the material from the Klebs collection, in other words exclusively East Baltic amber. In the second (1911, p.59) the material belongs to the British Museum, and originates partly from Samland, partly from Miocene deposits at Stettin. Shelford was unaware that this latter amber material was on a secondary deposit, and regarded the fossil insects it contained as having lived in the actual Miocene. For this reason he gave one of the species the unfortunate designation *Ceratinoptera miocenica* (1911, p.64).

The Stettin amber has undoubtedly been brought to this site by water courses, but it need not be East Baltic in origin; possibly it is more likely to have come from the Sudeter region, where Palaeogenic amber forest may also have grown (Oder-Neisse drainage area).

Shelford has recognized a total of 29 species, all of which he referred to now-living genera. A single recent species, *Eutyrrapha pacificia* (Coquebert), is included. This lives today in South America, Africa, Madagascar and Polynesia. Also the extinct species belong to genera which at present are found first and foremost in the tropics and sub-tropics, more rarely in warm temperate regions. By far the most common in amber are Phyllodromiidae, which today have genera and species in all tropical and sub-tropical regions, in Australia and far south in South America. The genus *Ectobius* is found widespread in Africa and Europe. It is remarkable that the more common of the two species in amber, *Ectobius balticus* (Germar & Berendt) appears to be more closely related to the recent *Ectobius lapponicus* (L.) than any other species.

Cockroaches live more or less on the same materials as earwigs, but are much more rapid in their movements. This is no doubt one of the reasons why they are also much more common in amber. Also among cockroaches, the males are much more common in amber than the females (91 ♂♂ against 31 ♀♀), most likely due to their greater activity. Shelford's material includes about 75 specimens of larvae. Here, too, this must be interpreted as a result of a period with strong production of resin coinciding with the most important reproductive period in the cockroaches, i.e. a coincidence of seasonal phenomena.

The Danish amber contains 31 specimens of cockroach, a single specimen having been determined by K. Princis as *Ectobius balticus*. A further 3 imagines probably belong to this species. 22 of the insects are lavae of various ages, a number are quite young, two of them newly hatched in the same amber piece.

Thysanura. These primitive wingless insects are not particularly rare in amber. Up to about the turn of the century, more than 30 species had been described. However, Silvestri (1913) revised this entire material of far more than 100 specimens, and reduced the number of species to 10, as he based his work exclusively on sexually mature animals of such a quality that important determinative characteristics were present. According to Silvestri, the total material contained two relatives of our silver-fish (Lepismatidae) (Plate 7,A), both belonging to now extinct genera. One of the species, however, *Lampropholis dubia* (Koch & Berendt) Silvestri, has nevertheless a close relationship with the now-living *Isolepisma*. On the other hand, *Lepidothrix pilifera* Menge is systematically isolated within the Lepismatidae, and according to Silvestri is more primitive than any now-living species. The 8 klipspringers (Plate 7,B) (Machilidae: 7 *Machilis* and 1 *Praemachilis*) are systematically very close to present species. In fact, Silvestri was unable to specify any real differences between the fossil *Machilis diastatica* (Olfers) Silvestri and the recent *Machilis polypoda* Linnaniemi.

At the present day, Thysanura can be found throughout the tropics and temperate zones, and on a very great number of different biotopes. The recent forms, therefore, do not give much guidance in evaluating the biology of the fossil forms found in amber. A great number of living species, however, are associated with a moist forest floor and moss, and it must be assumed that those animals trapped in the resin of the amber tree have lived as plant or detritus eaters on this biotope around the base of the tree-trunk. This *Sciara* zone in the amber forest has apparently been more favourable for Machilidae than for the more hidden Lepismatidae.

The Copenhagen amber material includes 31 specimens, only a very few of which are Lepismatidae.

Silvestri's Thysanura material also included a single **Diplura** namely *Campodea darwinii* Silvestri (Campodeidae), which resembles completely the now-living *Campodea staphylinus*. A further single specimen is found in the Gyllenhal collection in Uppsala, Sweden (Selberg, 1963). This small, pale and extremely light-shy animal must have been captured under conditions similar to those for the Thysanura and Symphyla.

Symphyla is a small order of arthropods whose members are widespread, and a few species are also found today in the fauna of the Baltic. It is very light-shy and needs a very humid atmosphere. It therefore lives well hidden in the forest floor and under cushions of moss. However, it may also be found under the moss a little higher up on the root neck of the tree-trunk, provided the air humidity is high enough. Of this animal group, only two fossil specimens are known, both in Bachofen-Echt's collection of amber (Bachofen-Echt, 1942, p.397).

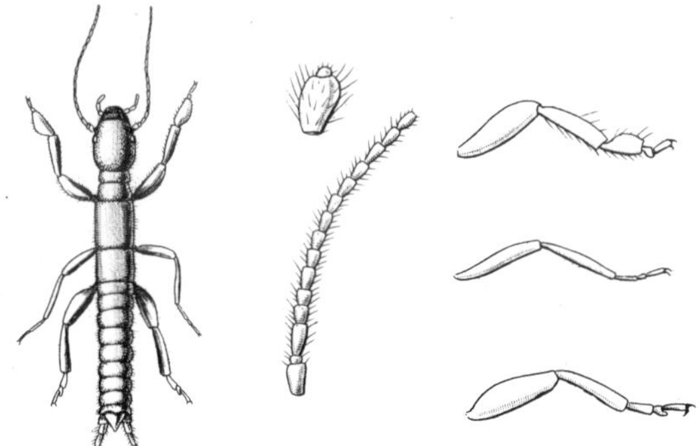

Text-fig.34. *Electroembia antiqua* (Pictet) (Embioptera), apterous male. (From Berendt, 1856, pl.V, fig.7).

Millipedes **(Diplopoda)** are found to only a slight degree on the surface of the bark and in the moss cushions, and in fact surprisingly few specimens are known. A few Juliformia and Nematophora are known, and a few more Polydesmoidea (Plate 7,C) but not many more than were already known from the publication by Koch & Berendt (1854).

From the same publication we know *Glomeris denticulata* Menge (p.12) (in 1856, p.6, mentioned under Isopoda), and Bachofen-Echt mentions a single find of *Polyzonium* (1942, p.399), which is so close to the now-living European species that he was unable to tell the difference. In general, here as everywhere else, the development which has taken place later among arthropods is surprisingly slight.

The only Diplopoda which are anything like plentifully represented are the small brush-clad Polyxenidae (Plate 6,C), which are typical bark forms. Some few species are already known from Koch & Berendt and from Menge, one of them having a very great resemblance to the now-living European *Polyxenus lagurus* L., probably being identical. According to Attems, another species is identical with the present-day South African *Schindalmonotus hystrix* Attems (Bachofen-Echt, 1942, p.397).

Bachofen-Echt was in close personal touch with the specialist K. Attems during the preparation of the above-mentioned publication on millipedes in amber. He points out that present-day relatives of the Early Tertiary Baltic millipedes are scattered over very large regions of the globe, and that some groups occur today in isolated populations with very great distances between them. This is the case, for example, with the order Colobognatha, to which *Polyzonium* belongs. He regards these populations as the remains of ancient stocks with very extensive distribution. These ideas are in good agreement with the impression obtained from the vegetable kingdom, where a rich development occurred in the Cretaceous period, and at the same time a spread over enormous land areas.

Bachofen-Echt also draws attention to the fact that by far the greater proportion of the millipedes found in amber, and this holds for the Chilopoda too, are immature animals, often only newly hatched. This agrees with the findings for bugs, cicadas, cockroaches and many other animals, and appears to confirm that a pronounced seasonal change has existed in the amber forest.

The Copenhagen collection of amber fossils contains 6 Polyxenidae and 7 other Diplopoda.

Also centipedes (**Chilopoda,** pars) are rare in Baltic amber. Lithobiidae (Plate 8,B) are the most numerous; a number of geophilids are known and only few scolopendrids. The Copenhagen collection contains a single scolopendrid, 9 lithobiids and 3 geophilids. The largest specimen mentioned in the literature is a scolopendrid, but this is not complete, only about 50 mm of the animal being preserved. Bachofen-Echt (1942, p.400) estimates from the size of the remains that the full length has been about 70 mm. Such large species are not found north of the Mediterranean region at the present day. These families of chilopods are above all forest floor animals, larger specimens do not move around very much higher up on the tree-trunk. In view of their special biology, Scutigeridae will be discussed later.

Scorpions (**Scorpionida**). Menge's collection included 2 specimens, one of which he described (1869) under the name *Tityus eogenus*. The other specimen was too poor to be of scientific use. No scorpions have been mentioned from amber since then, and the Copenhagen collection has thus no specimens. Morphologically, the species found is of quite a modern type. It is unlikely that scorpions, as warmth-loving animals, should have been rare in the amber forest or in the surrounding territory, at any rate in the more hilly regions, where they have had so many hiding places available in the day-time. Food has been plentiful in the humus layer of the forest floor, in the moss cushions and in the many fallen and more or less decayed tree-trunks: beetles, millipedes, woodlice. The fact that two specimens have been preserved in amber may be the result of sheer chance; following a nocturnal search for food they may have sought shelter in a deep moss-covered crack in the bark at the foot of an amber tree, and there have been caught by the flow of resin next day.

Wood-lice (**Oniscoidea**): two species are described, *Oniscus convexus* and *Porcellio notatus*, both Koch & Berendt (1854, p.9 and 10). In addition the editor of this publication, A. Menge, in an accompanying note, names two, possibly three species of *Porcellio* (*Porcellio granulatus* Menge and *Porcellio cyclocephalus* Menge), as well as a trichoniscid, not determinable, from his own collection. All these species should probably be referred to quite different genera. Menge supplements the list with the genus *Ligia* (1856, p.6). Following the extremely brief description, it is not possible to check this determination, which cannot be correct. It is possible, however, that the fossil belongs to the family Ligidiidae, which at present day also has representatives in the moist forest floor. There is no more recent study of amber Oniscoidea. Bachofen-Echt appears to depend exclusively on this material, although his figure (1949, p.41) probably originates from a specimen in his own collection. The Copenhagen collection contains a further unexamined material of 4 specimens, which according to N. Thydsen Meinertz (personal communication) all belong to the genera *Trichoniscus* or *Trichoniscoides* (Plate 8,C), or are at any rate very close to them.

All oniscoids are free-lying in the amber, which is without notable content of earth and plant particles. On the other hand, they are all clouded on one side by "milky amber" or "mit Schimmel bedeckt", as Koch & Berendt express it. Animals such as these can quite easily creep up on the lower parts of moss-covered and fissured trunks of amber trees when the weather is damp. If they are disturbed, or if they encounter unsurmountable hindrances, they simply let themselves fall, and this is presumably how they have got into the amber.

Included in the fauna whose home has been in the moss on the trunk of the amber tree among other places, are a number of oligochaetes and nematodes.

The known **Oligochaeta** probably all belong to the Enchytraeidae. In spite of being exceedingly common today in a damp forest floor, just as they must have been at the time of the amber forest, only a very modest number of specimens have been found in amber. Menge (1866, p.8) described the species *Enchytraeus sepultus*, which he found in 3 specimens, Bachofen-Echt (1949, p.41) mentions and illustrates a single, and the Copenhagen collection contains a further single specimen. This sparsity is only natural, however, for the terrestrial species of the group are light-shy and very sensitive to drying-out; they emerge only occasionally above the ground surface. However, when the humidity is high enough, they can be found a little way up the tree-trunks near the base of the moss carpets, but they definitely only belong to the *Sciara* zone. Just like other residents of the biotope (e.g. springtails and mites) they seem to have a tendency to work their way out of the moss when it is under pressure, and it is probably the gliding mass of resin which, on its way down, has pressed the small animals out of their hiding places and then captured some of them. In fact, they lie freely in the amber and not together with branches of moss or particles from the forest floor. The Enchytraeidae from amber are quite small animals.

Of **Nematoda,** there are both free-living Anguillulidae and insect-parasitic mermitids. Anguillulidae from amber are mentioned for the first time by von Duisburg (1863, p.31). In 1866, Menge (p.7) described the species *Anguillula pristina* and *Anguillula capillacea*, each from its piece of amber, but both of them occurring in very large numbers (respectively about 30 and about 50 specimens). The Copenhagen collection contains a single piece of amber in which there are a couple of quite small, partly spirally twisted specimens, apparently also belonging to this group, some genera of which live mainly in moss cushions. Also the animals mentioned by Menge have probably lived in moss cushions on the trunks of the amber trees, and have bored out of these as a result of the pressure from the downward flowing resin. To judge from flow lines in the amber, pictured in Menge's paper, the nematodes have moved for some time after being captured, and in both cases their movements appear to have had a common main direction. Menge provides no information as to the direction of this movement in relation to the flow of resin, and the incidents cannot be reconstructed today. The direction, however, suggests that the animals have reacted to a common stimulus, and one is tempted to assume that the light-shy animals have moved in towards the dark trunk and thus almost at right-angles to the direction of the flow. Future investigators of similar cases should take this into account.

Mermitidae are parasites in their 2nd larval stage, by far the majority being insect parasites. Menge (1866, p.5) has described the species *Mermis matutina* as an example of this family, the description being based on 3 specimens, all of which had just worked their way out of the abdomen of a chironomid. The Copenhagen material consists of 2 pieces of amber, and in both cases the host here is also a chironomid. In the one piece the parasitic worm still has its posterior end enclosed in the abdomen of the host, in the other piece the two animals are completely separate, but here too there cannot be any doubt as to their association. There are indications that all five parasites belong to one and the same species. Also at the present day, Chironomidae are often attacked by mermitids. Infection takes place at the larval stage of the host.

Snails (**Gastropoda**). Among the phytophagous amber fossils are land snails, but they are rare; years elapse between finds, and there are still none in the Copenhagen collection. With one exception the specimens are all lung snails (Pulmonata). Most of the species have been described by Klebs (1886, p.366-394). Important comments on this study and on a contemporary publication by Helm (1884a, p.125) are given by Sandberger (1886, p.137). Finally, Bachofen-Echt (1949, p.180) mentions a *Helix* which is determined by Dr. Adensamer as most closely related to a now-living African species, but without mentioning which. This *Helix* has apparently not been described, but according to Bachofen-Echt is to be found in Berlin. Naturwiss. Museum. Among the amber fossils found is also *Electrea kowalewskii* Klebs, belonging to the terrestrial family Cyclostomatidae (Prosobranchia).

The amber fossil snails are small, only a few mm in diameter and height. Most common are Zonitidae (*Hyalina (Conulus) alveolus* Sandberger, *Hyalina gedanensis* Klebs and *Microcystis kaliellaeformis* Klebs) and Pupidae (*Vertigo hauchecornei* Klebs and *Vertigo künowii* Klebs). In addition there is *Parmacella succini* Klebs, which is referred with some doubt to Helicidae, and *Balea antiqua* Klebs (Clausiliidae). However, Sandberger very strongly doubts the last generic determination, which is based on a not fully developed specimen. These fossils are snails which are typically found in the forest floor among rotting plant remains and in moss cushions, but do not creep up the tree-trunks, and it is likely that they are captured from the moss at the foot of the amber tree. A characteristic feature of bark-creeping snails is that they prefer trees with a smooth, naked trunk, e.g. beech and ash, where the population of algae they rasp off is rich and easily accessible. There are grounds for believing that trees of this ecological group have been quite common. The amber tree, on the other hand, with the rough bark indicated by the morphology of some aphids, e.g. *Germaraphis*, has not been attractive, and it is presumably for this reason that only forest floor forms have been captured, and no bark-living forms.

The snails found belong to genera which have a very wide distribution, and the great majority of the genera also occur today in the Baltic area. If, however, the nearest relatives of the fossil species are sought among now-living species, they are not found in the European fauna. Some are found in North America, especially in the eastern U.S.A., others in East and South East Asia. Thus, geographically they appear as a parallel to magnolias and other plants. If *Balea antiqua* is correctly determined generically it belongs to an ancient genus which today has one species in Europe and a few on New Zealand and on the distant island Tristan d'Acunha.

Water-bugs (**Heteroptera**, pars). A few amber fossils, most probably captured by the resin on the moss and livermoss which have formed carpets at the foot of the amber tree, are of quite chance character. Very little is known as to actual Hydrocorisae in amber, and this is naturally explained by the mode of life of these animals, for they only rarely leave their element, and their entire life history takes place in water. Berendt (1830) mentions a water scorpion *(Nepa)*, which however may have been captured on the river bank. Far more incomprehensible is the occurrence of three wingless *Corixa* larvae (Corixidae) in a single piece of amber (Bachofen-Echt, 1949, p.168). This may signify a sequelae to a brief violent rise in water level, perhaps a spring phenomenon. A single specimen of Belostomidae has been found in Mo-clay, one of the oldest known finds of this family.

There is rather more known about the pond-skaters (Plate 8,A) of the amber territory (Gerroidea), which are not actual water animals, but as it were live exclusively on the water surface, and (for example while overwintering) in the very nearest terrestrial surroundings. These finds are however modest. Germar & Berendt mention a larva of *Halobates* (1856, p.19, fig.8), and Menge's collection includes the foot of a *Hydrometra* (1856, p.20). According to Bachofen-Echt (1949, p.172) a total of 3 species are known from Baltic amber, belonging

to the genera *Gerris* and *Metrobates.* These generic determinations must be accepted with great reservation. The Danish amber collection includes 2 larvae, found in one and the same piece. In the Mo-clay about 16 adult specimens have been found so far, including several belonging to a very large species.

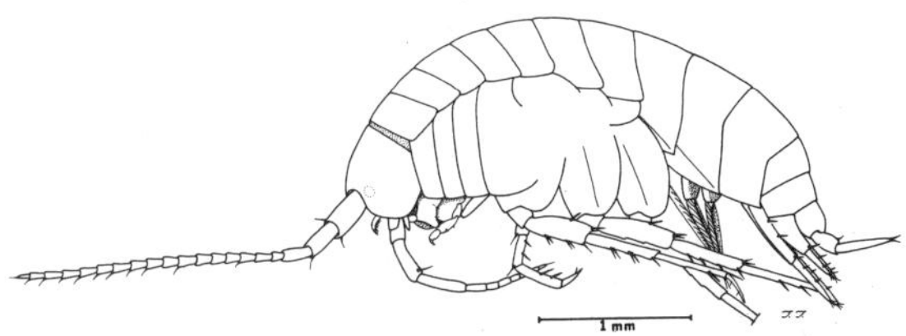

Text-fig.35. *Palaeogammarus danicus* Just (Amphipoda). (From Just, 1974, fig.1).

Finds of beach-fleas (**Amphipoda**) should be just as incomprehensible as that of the *Corixa* larvae mentioned. The family appears to be represented by only a single genus, of which 4 specimens in all are known to date. The first description was by Zaddach (1864, p.1), who named the animal *Palaeogammarus sambiensis.* This piece of amber, however, seems to have long since disappeared, at any rate Lucks did not succeed in finding it for comparison with his later find; but Zaddach's study is very carefully done and well illustrated. The next specimen was described under the name *Palaeogammarus balticus* (Lucks, 1927, p.1). Of this material there were originally two specimens in the same piece of amber, but for the sake of the description it was necessary to grind away the poorer of them. The subsequent fate of this piece, which was privately owned in the town of Zappot (Bachofen-Echt, 1949, p.43), is unknown. The Copenhagen amber collection now has a fourth specimen, *Palaeogammarus danicus.* Just, 1974 (Text-fig.35). This author refers the genus to Crangonycidae (Bousfield, 1973). While the first pieces were Prussian amber, this last piece is of Jutlandic origin.

Palaeogammarus has apparently been a lively jumping animal, with the chance, when particularly unlucky, of finishing up in the resin at the root of trees which have been near the shore of the fresh water where the animal has lived. But here too a temporary rise in the water level may have played a part. Its great rarity in amber seems to depend on the considerable distance which has existed between the normal biotopes for amphipods and the amber tree.

Also the few finds of water beetles (**Dytiscidae**) from amber depend to a high degree on chance, and several of the determinations are uncertain and inadequate. Imagines are mentioned only by Menge (1856, p.23: *Agabus*?) and Helm (1896c, p.224: Dytiscidae). There is greater interest in Weidner's description of larvae (1958), and the determination of these appears to be certain, even though the author notes his own reservation in the one case: "*Rhantus* (?) spec." (p.53). The other species was first mentioned as a thysanour under the name *Glesseria* (Berendt, 1845, p.56), and since described as *Glesseria rostrata* (Koch & Berendt, 1854, p.872) (Text-fig.36). Handlirsch (1908, p.717), however, was already aware that it was not a case of a thysanour, but of a larva of a dytiscid related to the recent *Hyphydrus*, at any rate belonging to Hydroporinae. Weidner (p.58) has given a redescription of this originally misunderstood animal, which is now found in the amber collection in Geolog. Staatsinstitut, Hamburg. These larvae have probably been captured by the resin on their way from the forest pool where they have developed, during their search for a suitable site for pupation.

The Danish amber collection does not include any Dytiscidae, but there are specimens from the Mo-clay, although they are not common. Among these will be mentioned only one particularly large specimen (the size of a now-living *Hydaticus*), the upper surface of which is almost complete.

Whirligig beetles (**Gyrinidae**) are not known from amber or from Mo-clay in the Danish collections. The genus *Gyrinus* is mentioned both by Berendt (1845), Menge (1856) and Helm (1896c), but only Motschulsky describes a species: *Gyrinoides limbatus* (1856, p.26). Bachofen-Echt (1949, p.106) mentions in addition *Orechtochilus*, which lives in running water. See also Hatch (1927).

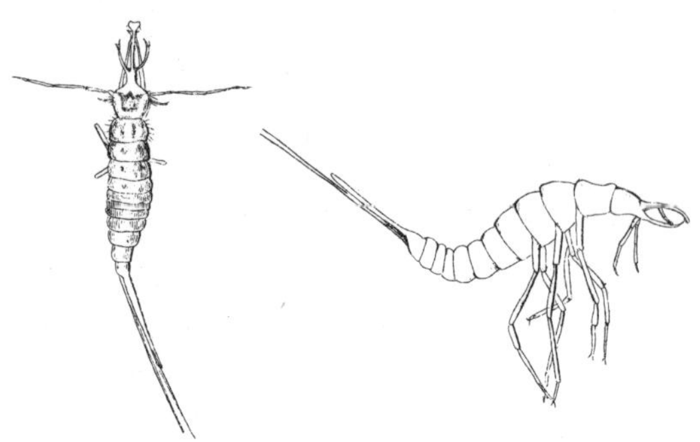

Text-fig.36. *Glesseria rostrata* Koch & Berendt (Dytiscidae, Hydroporinae), reconstruction, originally described as a thysanuran. (From Berendt, 1854, pl.XVII, fig.154).

Lacewings (**Neuroptera**, pars) (Plate 9,A). In addition to the already mentioned Osmylidae and Sisyridae, amber has been found to contain representatives of Berothidae, Coniopterygidae, Hemerobiidae, Neurorthidae, Psychopsidae, Ascalaphidae and Nymphidae, the last three of which have been discussed by MacLeod (1970). The earlier descriptions originate from Hagen, 1856, Enderlein, 1910, and Krüger, 1923. *Chrysopa* has been mentioned on several occasions in the earliest literature on amber fossils, but according to MacLeod (1970, p.148) no definitely certain chrysopid has as yet been described from Baltic amber. However, Schlüter (1975, p.155, fig.2a and b) mentions that at least one specimen of Chrysopidae has been found in Cenoman amber from north-western France.

It is striking that present-day Neurorthidae, today a family with a very limited number of species, is the one most richly represented in amber, with 24 out of the 57 specimens which have been available to MacLeod. This material comprises only imagines, no larvae. Very little is known of the group's recent biology, but it may be taken as certain that at least one species has been common in the amber forest. On the other hand it is unlikely that the larva has had any special relationship to the amber tree, as in that case, just because of the incidence, it should be found among the amber fossils. This is the case, for example, in *Propsychopsis* (Psychopsidae), relatively common in amber; both larvae and imagines are found, and in approximately equal numbers. Psychopsidae live as larvae under loose bark, bark fragments and the like, from which they surprise their booty of small insects. This family today is primarily indigenous in Australia, only a few species being found in Africa, India and China. In Australia they are found particularly on *Eucalyptus* with coarse bark (Tillyard, 1926, p.315). Also the larva of *Pronymphes mengeanus* is known (Hagen, 1856; Nymphidae). Nymphidae larvae

are slow animals, which lurk after booty under loose bark or dead leaves. The larvae of *Neadelphus protae* MacLeod (1970; Ascalaphidae) (Text-fig.37) has most probably lived under similar conditions.

Berothidae, Hemerobiidae and Coniopterygidae live both as larvae and imagines on plant lice and scale insects in particular. Among amber fossil species might be mentioned the hemerobiid *Prophlebonema resinata* Krüger (1923), *Prolachlanius resinatus* (Hagen, 1856) and *Prospadobius moestus* (Hagen, 1856) together with the coniopterygids *Coniopteryx timidus* Hagen (1856) and *Archioconiocompsa prisca* Enderlein (1910). The taxonomy of this last-mentioned family is discussed by Enderlein, 1930. Many species of these lacewing families live today on the tree-trunks of various forest trees, and it is surprising that the known amber material does not include larvae identified with certainty. *Germaraphis* must have been one of their favourite forms of food.

There is a Copenhagen materal of 11 imagines, now being studied by Dr. Ellis G. MacLeod, U.S.A. One piece of amber includes 3 small larvae which appear to be newly hatched specimens of a small neuropterous insect, possibly a psychopsid (Text-fig.38).

It might be of interest to mention the considerably larger material of lacewings found in the Early Eocene Danish Moclay (Larsson, 1975). Here there are impressions of wings, often particularly well preserved. Even though this material has not undergone any finer sorting, it may be said that both psychopsids and hemerobiids are commonly represented. Any possible occurrence of chrysopids is unknown.

Lady-birds (Coccinellidae). Although these are aphid and scale insect eaters, only a few lady-birds are known from amber. *Coccinella* is mentioned by Berendt (1845, p.56), *Scymnus* by Helm (1896c), and Menge (1856, p.23) mentions a larva without other determination than the collective *Coccinella*.

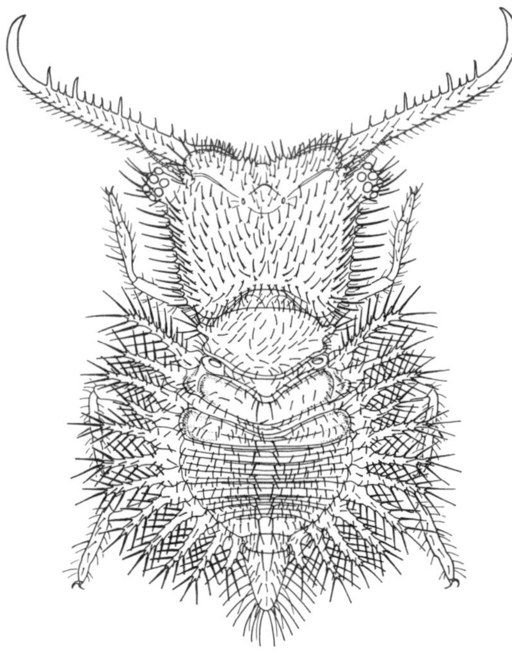

Text-fig.37. *Neadelphus protae* MacLeod (Neuroptera), reconstruction of larva in dorsal view. (From MacLeod, 1970, fig.2).

Klebs (1910) has 12 specimens in his collection: 2 *Coelopterus*, 2 near *Pharus*, 2 near *Platynaspis*, 4 *Scymnus* and 2 non-determined. The Copenhagen collection contains 4 non-determined specimens. Reitter's determination of Klebs' material appears probable; it is natural to find in amber those small species which are in fact mainly encountered on the tree-trunks. They have probably fed on scale insects more than on aphids.

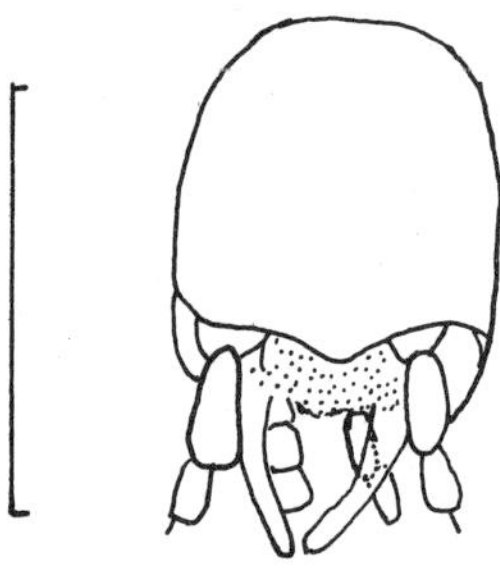

Text-fig.38. First stage larva of ?Neuropteron, head in dorsal view. Scale: 0.25 mm. (Original drawing from specimen in Copenhagen collection).

Lepidoptera. The following is mainly based on Rebel's study of amber moths of East Prussian origin (1934, 1935, 1936 and 1937), together with Margaret R. MacKay's examination of some microlepidopterous larvae from Danish amber (1969). In addition, Menge (1856) has some remarks on pupae and larvae and describes the cases of several Psychidae. Kuznetsov (1941) describes a number of species from eastern amber: *Electrocrania immensipalpa* (p.19), *Tillyardinea eocaenica* (p.23), *Martynea rebeli* (p.27), *Dysmasiites carpenteri* (p.29), *Scardiites meyricki* (p.32), *Proscardiites martynovi* (p.34), *Palaeoscardiites mordvilkoi* (p.37), *Glesseumeyrickia gerasimovi* (p.43), *Prolyonetia zeckerelli* (p.45), *Paraborkhausenites innominatus* (p.50), *Oegoconiites borisjaki* (p.53), *Symmocites rohdendorfi* p.56) and *Electresia zalesskii* (p.63). Most recently, species from eastern amber have been described by Skalski (1973a and b, 1974). For further information, referen-

ces should be made to coming studies by Dr. Skalski, Poland, including an investigation of the Copenhagen, mainly West Baltic material (51 imagines and 8 larvae), but also including a revision of older literature, which appears to include many taxonomic errors. Such, one Skalski paper being of great importance to taxonomists, has already been published (1976); however too late to be discussed in the present book.

Of the many Lepidoptera found, all belonging to small species, many have lived free as larvae, although most often in self-spun cases. To a major extent their habitat has been on freely exposed parts of tree-trunks, and they have probably in most cases lived on lichens. This is the same biotope in which most of the Psocoptera are found. Other Lepidoptera have been associated with the lower, moss-grown regions of the tree-trunks in the *Sciara* zone. A number have been leaf-miners in leaves or needles, most often as quite young larvae, following which they have passed the rest of their life in the forest floor between leaf fragments they have spun together. Finally, a number have lived under loose bark and in rotted wood. In most cases a probable domicilary relationship to the amber forest can be demonstrated. However, a far greater variety of forms has undoubtedly existed.

In the Psychidae, numerous types of larval cases are known, and Menge (1856) in particular has laid great emphasis on their description. In all he has been familiar with 13 specimens, which he refers to presumably 7 species. Rebel's material (1934) includes a single larval case. No imagines of this family are known. Quite small fragments of bark and wood are common, often even dominating among the foreign bodies which are spun into the larval cases, and in some specimens fragments of leaves are found from deciduous trees or conifers. It is probable that they have lived on shoots and branches and have been spun fast to parts of branches and trunks during resting periods and during pupation. In a couple of specimens the stellate hairs of oak are a very important constituent of the material of the larval case; this suggests that these species, at any rate, have terminated their larval period in the course of spring. They have probably been lichen feeders, some on oak and other deciduous trees, others on conifers, probably including *Pinites*.

Rebel mentions 2 species of Tortricidae, which he refers to the genus *Prophalonia* (1935), but which according to Skalski (1973a, p.342) actually belong to Tineoidea. In 1934 Rebel described and illustrated a tortricid larva, which Kuznetsov (1941) refers to Oecophoridae. Menge has had 15 imagines, 4 pupae and 7 larvae for study, but gives no taxonomic analysis, only reporting that in his opinion the pupae belong to 2 species, the larvae to 4. With regard to the one pupa he states that it lies in the amber with "zerkauten oder als koht ausgeworfenen holztheilen" (1856, p.29). It is probable that most of the Tortricidae have lived on various deciduous trees in the forest, not least on oak, and their larvae may have been captured by the resin while lowering themselves down to the forest floor. Some species, however, have no doubt been breeding on the amber pine. Menge's pupa mentioned above would suggest this; it must have been pressed out of its pupal cave by outflowing resin as was often the case with the mite *"Acarus" rhombeus*. Just like the now-living *Laspeyresia coniferana* (Ratzeburg), it may be assumed to have lived by mining in the bark of *Pinites* and perhaps other conifers. The recent species breeds "apparently in particular in decaying trees, it reveals its presence by resin flow and accumulation of excrement on the trunk. Pupation where it feeds" (Larsen, 1917, p.130, translated from Danish). The tortricid described by Kuznetsov, *Electresia zalesskii*, is otherwise regarded as actually being related to *Laspeyresia*.

Rebel (1934, 1935) describes a total of 11 species of Oecophoridae, and refers 2 of them to *Depressarites* and 9 to *Borkhausenites*. MacKay (1969) describes a larva which in her opinion

most probably belongs to the same family, Skalski (1973, p.154) imago to *Epiborkhausenites obscurotrimaculatus*, and Kuznetsov (1941, p.50) *Paraborkhausenites innominatus*. The recent Oecophoridae do not fly far and in fact keep to the lowest part of the tree-trunks. Their larvae live mainly in rotten wood or under dead bark, *Tubuliferola flavifrontella* Hb. (and possibly others) in a case made from lichen or leaf fragments, on tree-trunks or on the forest floor. That so relatively many Oecophoridae have been found in amber is therefore quite natural, since they have had excellent conditions for breeding in the *Sciara* zone of the forest, with all the fallen and rotting trunks of many different trees, as one may assume has undoubtedly been the case. However, it is hardly likely that any species has been breeding on the living amber tree.

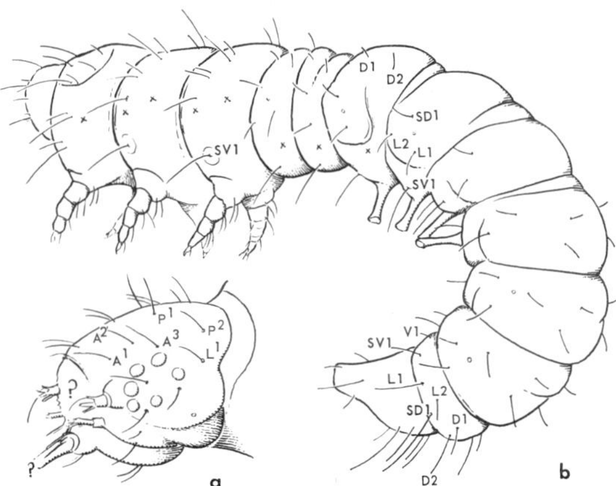

Text-fig.39. ?Plutellidae, first instar larva, a: head, b: thorax and abdomen, both in lateral view. (From MacKey, 1969, fig.4).

Rebel (1934, p.16) mentions quite briefly a larval case of a microlepidopterous insect, which may have belonged to the Coleophoridae. The determination is doubtful. Rebel (1934) classifies an imago in Elachistidae, referring with reserve to *Anybia*.

Yponomeutidae. Rebel refers 7 species of his complete material (1934, 1935) to this family: 1 *Scythropites*, 4 *Epino-*

meuta and 2 *Argyresthites.* No larvae are mentioned. Most of the yponomeutids found may be imagines of species which have been breeding in the numerous bushes and small trees of the forest undergrowth. Some recent genera, however, e.g. *Cedestis, Dyscedestis* and *Ocnerostoma,* breed on pine shoots, where the larvae live in webs between the young pine needles. These genera belong to Argyresthiinae, and *Argyresthites* from amber has possibly had a similar life pattern.

MacKay (1969) doubtfully refers 2 larvae from Danish amber to Plutellidae (Text-fig.39). As a number of plutellids (e.g. *Cerostoma)* live at the present day in leaf rolls on different deciduous trees, especially oak, their presence in the amber forest is highly probable.

Rebel's 2 *Architinea* species and 1 *Tineolamina* (1934) must presumably be referred to Tineidae. Skalski (1974, p.97) describes *Tineosemopsis decurtatus* (Text-fig.40). Menge (1856) claims to have 66 imagines, but as he neither mentions oecophorids nor yponomeutids, most of this material has probably belonged to other families than in fact Tineidae. Menge mentions a larval case which from his description appears to have a great resemblance to the case of the recent *Diplodoma* (Lyposinae), the larvae of which live in a triangular case covered with sand grains or the like, on the moss at the base of tree-trunks. Many recent tineids live as larvae in the fruit bodies of *Polyporus* species and in rotten wood (containing mycelium?), while one species, *Infurcitinea argentimaculella* Stt., lives in tubular galleries on lichens on walls and stones. It is therefore likely that the weak-flying tineids have been more common than the scanty material would indicate.

Presumably belonging to Incurvariidae is Rebel's *Incurvarites* (1 species) and to Adeliidae *Adelites* (1 species). As completely young larvae, Incurvariidae live by mining in leaves, while Adeliidae bore in seeds or flowers. In both families, the older larvae live in flat cases on the ground. The probability of their occurrence in the amber forest is high.

Micropterygidae. Tillyard (1926, p.410) mentions a micropterygid, not further described, from Baltic amber, and Cockerell (1919, p.193) describes *Micropteryx parvetus* from Burmese amber. Otherwise, only a single specimen is known from Rebel's material, which he names *Micropteryx proavitella* (1935, p.185). This weakly flying family may be assumed to have had particular favourable breeding conditions in the rich moss and liver-moss growths of the *Sciara* zone, and it may be anticipated that several specimens will be found.

Text-fig.40. *Tineosemopsis decurtatus* Skalski (Tineidae). (From Skalski, 1974, fig.3).

In Middle Cretaceous amber from north-western France (Cenoman, about 100 million years old), Schlüter (1975, p.157, fig.5) found about 150 insect scales, very similar to scales of recent micropterygids. Although the identification is not absolutely certain, they belong in Schlüter's opinion to a member of the suborder Zeugloptera.

Psocids (**Psocoptera**) (Plate 9,B,C) are found everywhere wherever trees grow, creeping or running around on the bark of the tree-trunk, where they feed upon microflora, moss, lichens and fungi; in contrast to collemboles and mites, they live in fact on the lichen-covered areas of the trunk, rather than in the obscure *Sciara* zone. From Enderlein's thorough monograph (1911, p.279) we know 29 species from the Baltic amber, distributed into 6 families. This study is based on 400 pieces of amber. None of the species are recent, but the great majority belong to now-living genera, *Psocus* and *Caecilius* dominating among them just as in our day. Only one of the amber forest families is unknown in the present, and has probably become extinct, but there is no certainty about this, as tropical forests continue to provide faunistic surprises. An example of this is the genus *Archipsocus*, which was originally described from a species from amber (Hagen, 1882, p.222), but since then one recent species has been found in Further India, one species in Brazil and Paraguay and one species in East Africa.

A number of psocids can be found at the present day on trees of greatly differing systematic position, while others to a greater or lesser degree prefer and breed on a definite species of tree. It is natural that such a characteristic should leave a trace on the frequency with which the individual species of psocids are captured by the resin, and it may be assumed that species which have had a special preference for the amber tree, have the chance of being particularly numerous in amber. In addition, many species seek together in colonies in the same way as, for example, the recent *Psocus longicornis* Fabricius. Species whose larval stages are found in amber are among the most common. As an example, Enderlein (1911, p.296) mentions *Copostigma affinis* (Pictet), the most common species in his material (65 specimens). As another interesting example he mentions *Archipsocus puber* Hagen. In addition to normal winged individuals, the recent species of this genus have a number in which the wings are reduced, in some being merely scale-shaped vestiges. These species construct

extended webs, often several square metres in area, under which the many members of the colony find protection. In Enderlein's material of *Archipsocus puber*, 1 out of 7 specimens was typically brachypterous. Among the 6 Copenhagen pieces there are 3 macropterous, 1 brachypterous, 1 as good as apterous and 1 larva in its final stage.

Among other species which have lived in great numbers on the amber tree, and perhaps have been indigenous to it, might be mentioned *Psocus picteti* Enderlein, *Epipsocus ciliatus* (Pictet & Hagen), *Caecilius debilis* (Pictet & Hagen), *Caecilius proavus* (Hagen), *Caecilius prometheus* Enderlein and *Amphientomum paradoxum* Pictet.

Another remarkable observation, made by Enderlein, was that the phylogenetically older groups were only represented by strikingly few specimens by comparison with the young groups. For example, of the phylogenetically young suborder Isotectomera (families Thyrsophoridae, Psocidae and Caeciliidae), there were found 16 species with a total of 323 specimens, while of the older Heterotectomera (the rest of the families) there were 12 species and in all only 77 specimens. If these numbers are compared with the numerical relations among the present day tropical forms, approximately the same result is obtained. This appears to show that the psocopterous fauna of the tropical belt (and perhaps its entire insect fauna) may have been subjected to only insignificant changes since the end of the Mesozoic. This also shows that at that period the Psocoptera had long since reached their present development.

An evaluation of the relationship of the individual fossil forms with those of the present-day gives Enderlein (1911) an occasion for the following zoogeographical considerations. Many of the genera occurring in amber have today a widespread distribution in Holarctis and are represented here by many species. A further evaluation, however, shows that these species do not have particularly close relations among the recent species in Western Eurasia, whereas a very large number have their closest relationship with species and genera in the present-day tropics and subtropics, most in South and East Asia, others in equatorial parts of America and Africa.

The Copenhagen material contains a non-described material of 87 pieces, and here too Psocidae and Caeciliidae are among the most common families. Some of the species seem to belong to other genera than those known to Enderlein. There are a number of Lipo-scelidae (in addition to *Sphaeropsocus*), and it is remarkable that some (apparently in particular one wingless species) are usually found in amber together with the mite mentioned above (*"Acarus" rhombeus* Koch & Berendt) and plentiful detritus from the tunnels of insect larvae.

Grasshoppers (**Saltatoria**) are not common in amber, and only the long-horned grasshoppers (Ensifera) are represented. The primitive family Gryllacridae is represented by 3 species which have been described by Chopard (1936). These belong to the extinct genus *Prorhaphidiphora*, all based on only a single individual. They have neither stridulating organ nor auditory organ, and are usually predators, living in trees and similar vegetation. Zeuner (1936) has described 2 species of Tettigoniidae: *Eomortoniellus handlirschi* and *Lipotactes martynovi*, the nearest recent relatives of which live on the Sunday Islands; both these species are represented by several individuals, a number of them larvae. A considerable weakness of the amber material of grasshoppers is that a large part of it consists of non-describable larvae; thus, the subfamily Phaneropterinae is only known by one single quite young larva (Bachofen-Echt, 1949, fig.73, p.83-84). This suggests, however, that the larval development has been early during the season. Also the tettigoniids mentioned are tree-climbers.

Among crickets (Gryllidae), mainly detritus eaters, 5 genera have been described, each with one species (Germar

& Berendt, 1856; Chopard, 1936), and each belonging to its own sub-family (Text-fig.41). Of these *Stenogryllodes* is extinct. The others have all their nearest recent relatives in the tropics and subtropics of the old world.

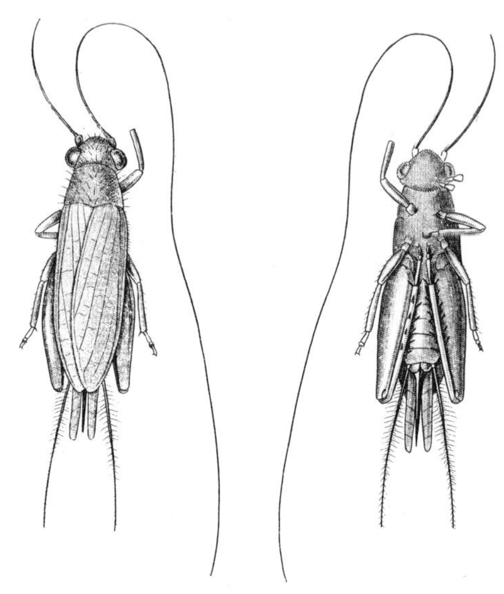

Text-fig.41. *Gryllus macrocerus* Pictet & Hagen (Ensifera), female in dorsal and ventral views. (From Berendt, 1856, pl.IV, fig.8).

The known grasshoppers in amber are all rather small, and as mentioned there are many larvae among them. The largest imagines which Bachofen-Echt has seen reached a length of 30 mm (1949, p.83). Larger species have lived, however, and Bachofen-Echt has found grasshopper legs in amber which would suggest that the original bearer has been at least 50 mm long. The ability to autotomy of a limb is definitely an important reason why these animals are rare in amber, since such limbs are not particularly unusual, but this is hardly the most important reason. Grasshopper species which climb trees are also common to-day, especially in hot climates; however, they almost always keep to deciduous trees, and the amber tree, at any rate, has been a conifer. It must therefore be considered probable that grasshoppers in particular have lived on oak and similar trees, and have only found themselves on amber trees by accident, for example when the wind has been blowing or when they have jumped the wrong way and once more have taken flight. This is probably a contributory cause of the very great preponderance of larvae.

The Copenhagen collection contains 11 non-described specimens.

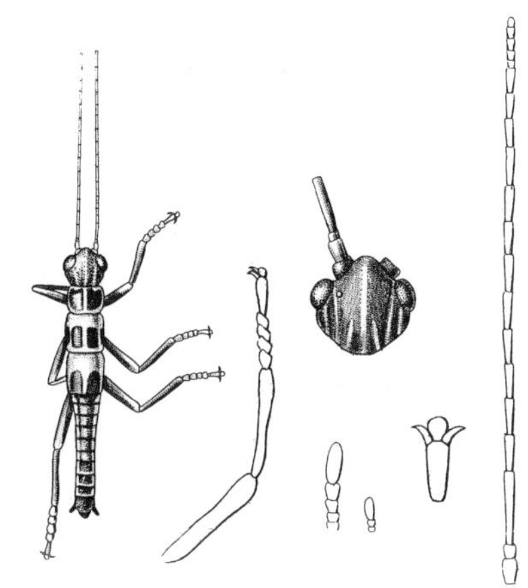

Text-fig.42. *Pseudoperla lineata* Pictet & Hagen (Phasmoidea), dorsal view, and some details. (From Berendt, 1856, pl.IV, fig.10).

Phasmids (**Phasmoidea**) are plant eaters, and in the amber territory have presumably lived in shrubbery and oak brush. They have only occasionally been carried on to the amber tree and thus on to the resin. A single species, *Pseudoperla lineata* Pictet & Hagen (Text-fig.42), is relatively common, and has been recorded by Bachofen-Echt as larvae of various sizes, right up to 30 mm (1949, p.85). It has probably lived in vegetation of somewhat greater height than that usually preferred by the members of the family, or it has been able to stand more shade. Its recent relatives live in the tropics of South America. One undetermined specimen is found in the Copenhagen collection.

Praying mantids (**Mantoidea**). These have lived on the same biotope as most of the phasmids, but have been predators. They are even more uncommon in amber than the groups mentioned above, and are only known as larvae. Bachofen-Echt (1949, p.92) considers that he can identify at least 2 species in his material of 3 specimens. The Copenhagen collection contains a single very young larva. The occurrence of these larvae in amber appears to be quite by chance.

The fauna of the tree-trunks also includes a fauna of free-living animals, often quick-moving, in other cases trap-constructing, which by day or night have hunted the permanent fauna or to an even higher degree all those Diptera and other flying insects which have merely sought rest and hiding during the passive part of their diurnal rhythm. This type of fauna includes the already-mentioned Coccinellidae, which must presumably have fed especially on quite small Homoptera, a number of grasshoppers, *Dromius*, and many of the Neuroptera found. During the day, there have otherwise been above all small birds and ants which hunted for food on the tree-trunks. Birds have left innumerable traces in the amber in the form of half-eaten insect remains. Among ants, the lively Formicinae and Dolichoderinae have been present in particular, with their established tracks to the aphid colonies, undoubtedly an important goal for their wanderings, while at the same time they have also carried insect booty home to their nests. Scorpion flies with their leg-traps appear to have less significance. There have been hunting spiders all day round, using many different hunting methods, active forms during the day, trapnet spinning particularly during the reduced light of morning and evening. The particularly long-legged forms, Opilionida and Scutigeridae have been proper nocturnal animals. Many hymenopterous insects have searched for food for their larvae, or they have occurred as parasites on the native fauna.

Ants (**Formicidae**) had already reached their complete specialization at the time of the Baltic amber, but nevertheless only quite few fossils are known which are older than the Eocene, and only from the Upper Cretaceous. Sharov (1957, p.943) described the wing of *Cretavus sibiricus*, which may possibly be referred to Formicidae. Wilson, Carpenter & Brown (1967, p.1) described *Sphecomyrma freyi* from New Jersey amber, on the basis of 2 worker ants in the same piece of amber, belonging to an extinct and particularly primitive sub-family (Sphecomyrminae).

There have been many species of ants among the insects which lived on the tree-trunks in the amber forest, both on the amber tree as well as on oak and other species. Basic to our knowledge of this fauna is the publication by G.L. Mayr (1868) and the later study by W.M. Wheeler (1915) of about 10,000 individuals, the great majority originating from the Königsberg Geological Institute and from Professor Richard Klebs' private collection. While this material consisted almost exclusively of East Baltic amber, the Zoological Museum collection in Copenhagen today contains by comparison a still undescribed material of about 425 individuals, by far the great majority originating from South-West Scandinavia.

It is characteristic of the East Baltic material that a few individual species occur in extraordinarily large numbers of individuals in relation to the total, and there can be no doubt that these species have played a considerable role in the life of the amber forest. The same is the case for the western material, but there seems to be some difference in the relative incidence of the dominant forms between the two localities of origin.

Formicinae occur in Wheeler's material in 32 species distributed over 13 genera, and he and earlier authors (Mayr, 1868; André, 1895) have together examined 3827 individuals within this sub-family. Interest is aroused in particular by *Formica flori* Mayr (1310 specimens) and *Lasius schiefferdeckeri* Mayr (1172 specimens) (Text-fig.43; Plate 10,C). In both cases, the recent fauna contain species which are so close to those mentioned that it depends in fact on the per-

sonal opinion of the morphologist whether he wishes to consider them as one and the same species or different. Wheeler's opinion (p.18) is that while the recent form is no doubt different from the fossil one, it must nevertheless be regarded as a direct descendent.

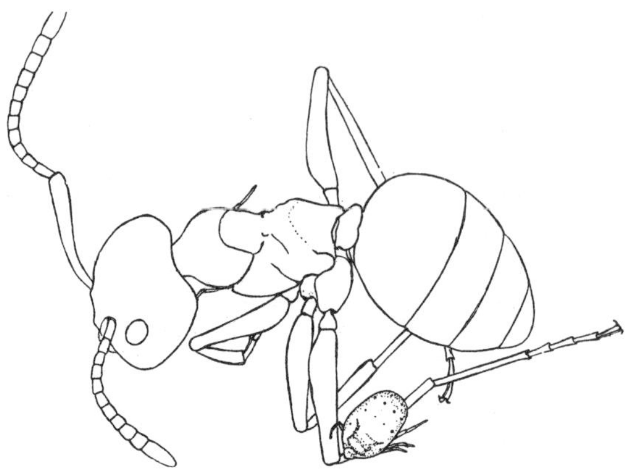

Text-fig.43. *Lasius schiefferdeckeri* Mayr (Formicidae), worker. A parasitic(?) mite is attached to the left hind tibia. (From Wheeler, 1915, fig.58).

The "double" which *Formica flori* has in the recent fauna is *Formica fusca* L. (the black slave ant). *Formica fusca* is a pronounced boreal species with holarctic distribution. In Southern Europe it occurs only in mountainous territory; in North America its southern boundary lies at South Carolina. If *Formica flori* has also corresponded biologically to the modern species, which is likely, it must be presumed that its greatest distribution has been north of the amber territory, and here could only have been found in the vicinity of the boundaries of the amber production, both vertically and horizontally. In all probability, therefore, it has only had a possibility of common distribution with the amber tree in the warm temperate zones of that period, corresponding in fact to the most northerly distribution of the amber tree. Its distribution in common with typical pines, on the other hand, must have exceeded by far the northern boundary of the amber-resin producing forest.

Formica species have always had their nests in the ground or in tree stumps and old wind-falls, and *Formica fusca* is one of the species which make the most modest demands on the nature of the environment; they spend a very considerable part of their time in the vegetation, however, also high up in the crowns of the trees. *Formica flori* has hardly been particularly associated with the amber tree, but here as elsewhere it has had its permanent routes up to the top in its search for food: insects, aphid honey and perhaps also other plant juices.

Lasius schiefferdeckeri Mayr (Text-fig. 43; Plate 10,C) corresponds to the recent *Lasius niger* (L.) to such a high degree that the ant-specialist Stitz (1939, p.275) considers that they are identical. The distribution of *Lasius niger* is on the whole coincident with that of *Formica fusca*, it stretches not quite so far north and somewhat further south. This too is very modest in its ecological requirements. Its food consists mainly of aphid honey and other sugar-containing homopterous excrement, which it find up in the vegetation or in special "aphid stalls", which it creates around aphid colonies near the ground, often even root-sucking aphids. It is also to some degree carnivorous. Its nests are most often complicated underground residences, but it can also set up a home in tree-stumps and dead branches. From our knowledge of the biology of recent forms, it may be assumed that a relatively greater percentage of *Formica flori* existing in the Eocene have been captured in the amber resin than of *Lasius schiefferdeckeri*, particularly because of their greater hunting territory.

In both the Eocene species, a number of winged individuals are known of both sexes, although to a very great extent males. In addition, a few de-alate *Formica* females are known. Also this suggests that they have often had their nests near the amber trees. The recent species swarm in mid-summer, although *Lasius niger* is often earlier on the wing than the *Formica* species. Should this also have been so for the fossil forms, it must be assumed that in *Formica flori* the swarming has just touched the period in which the amber trees have had their optimal resin secretion, while the very great number of *Lasius* males in amber suggests that this form, also during the Eocene, started swarming somewhat earlier than *Formica*.

Both *Formica fusca* and *Lasius niger* set up new colonies independently, without forcing other ant societies to help, and corresponding conditions may be assumed to have existed in the species found in amber. In both, however, already established nests are utilized as a basis for other ant species which are themselves unable to start new colonies unless on the basis of the work of other species. *Formica fusca* is used for example by *Formica truncorum* Fabricius, the corresponding form in amber being *Formica phraëthusa* Wheeler, two workers of which are known. *Lasius niger* is used among other forms by *Lasius umbratus* Nylander, whose partner in amber is *Lasius nemorivagus* Wheeler. *Lasius umbratus* leads an almost entirely underground life, and has therefore very little opportunity of being captured in resin. After the nuptial swarming, the de-alate fertilized females attempt to penetrate the *Lasius niger* colonies, where their adoption has the result that the queen of the host colony becomes eliminated. *Lasius nemorivagus* is only known by one specimen, a de-alate, probably fertilized female, so this too apparently leads a hidden existence. *Lasius umbratus* is known as cultivating underground aphids, as is also the case to a lesser extent with *Lasius niger*. *Lasius nemorivagus* (probably also other Eocene *Lasius* species) may be assumed to have had similar underground aphid cultures. There is thus a reasonable probability that plant roots have harboured a living aphid fauna, but of this we can hardly have any direct evidence.

Both *Lasius* and *Formica* are common in Danish amber, but in contrast to what was the case in East Baltic amber, *Lasius* is here somewhat more common than *Formica* (78 *Lasius* and 43 *Formica*).

Prenolepis henschei Mayr occurs relatively frequently among the other Formicinae in Wheeler's East Baltic amber material (611 specimens). The corresponding recent species is *Prenolepis nitens* Mayr, which lives in forests and needs much more warmth than the recent Formicinae mentioned above. As a whole, the genus is found in warm climates. Numerous males and a number of fertilized females are known. In Danish amber, the genus *Prenolepis* is relatively less common than in East Baltic amber (11 specimens). The same is the case with *Camponotus*, which is widely distributed today, but which is almost entirely thermophilous (3 Danish specimens compared to 129 East Baltic specimens), and the Indo-Australian *Geomyrmex* (3 Danish specimens and 161 East Baltic specimens). On the other hand, the genus *Plagiolepis*, which today lives in the warm regions of the Old World, appears to be relatively common in Danish amber (19 specimens compared to 112 East Baltic specimens). This may be due to territorial differences in the amber forest of west and east. The recent *Plagiolepis pygmaea* Latreille preferentially has its nests under stones and in cracks on the hillsides, and avoids continuous forest and luxuriant vegetation. Its social organization most often contains rather many individuals, and may include many root aphids. However, care must be taken not to attribute too much significance to these figures, as they are not only small, but may depend on the varying attitude of the collectors to the material. Thus, in the East Baltic material, it is the large species which dominate percentwise, in contrast to the Danish material, where the small species are relatively more common.

Within the Dolichoderinae, the genus *Iridomyrmex* dominates strongly in Wheeler's material *(Iridomyrmex goepperti* (Mayr) with 5428 specimens, *Iridomyrmex geinitzi* (Mayr) with 1289 specimens and the genus as a whole with a total of 6870 specimens). Today, the genus is very common in the tropics, both in the Old World and the New World, and it has apparently found an exceedingly satisfactory environment in the amber territory, the subtropical region of which has been within its optimum region of distribution. These ants are particularly active, not only making nests in the ground, but often in hollows high up in the trees, both in dead and in living plant parts. For example, some always make their nests in natural cavities characteristic of certain living plants (e.g. *Iridomyrmex myrmecodiae* Emery in *Myrmecodia* and *Hydnophytum*). Many *Iridomyrmex* societies have a plastic mutual balance, as they form suburban colonies which to some degree maintain contact mutually and with the mother colony, and can unite once more when circumstances demand this. The two species mentioned, with their strong representation in amber, must be assumed to have been practically omnivorous. Aphid excrement has had very great significance, as can be seen not least from a piece in the Danish collection, which contains a confused mixture of *Iridomyrmex* workers (about 10) and at least as many aphids of the genus *Germaraphis*, living on the bark of the amber tree, and including at least one wingless parthenogenetic imago (or last-stage larva?) and larvae of varying sizes. Just as at the present day, some *Iridomyrmex* species have presumably cultivated root aphids. The number of known winged specimens from amber is relatively small (16 ♂♂ and 8 ♀♀).

The Danish amber includes 59 specimens which must be referred to *Iridomyrmex*, i.e. about the same number as *Lasius* (78 specimens), and only slightly more than of *Formica* (43 specimens). Even though the figures are small, they may be considered as striking, as the East Baltic material contains more than five times as many *Iridomyrmex* as *Lasius* (6870 specimens against 1257 specimens). In this case, the size of the animals and their general appearance cannot have played any role for the human selection. This might indicate that the site of origin of the Danish amber, probably Southern Scandinavia, has been closer to the northern boundary of the thermophilous genus *Iridomyrmex* than the average site of origin of the Samland amber, presumably central regions of Russia. This may indicate that *Iridomyrmex* is somewhat more thermophilous than the amber tree. It may however also be due to landscape differences between East and West. A species determination of the Danish material might be of great interest.

Also common in the East Baltic amber is *Dolichoderus tertiarius* Mayr (494 specimens); this species corresponds to the recent *Dolichoderus quadripunctatus* (L.). *Dolichoderus* is as a whole a thermophilous genus, but the strongly dominating *Dolichoderus quadripunctatus* in particular is among those species which penetrate furthest north in the warm temperate regions. Nor does the relative frequency of the genus decrease anything like as strongly from the East Baltic amber to Danish amber (36 specimens), as was the case with the more tropical *Iridomyrmex*. It does not build earth nests, but breeds in hollow spaces in dried branches and tree-trunks, especially of deciduous trees. It is therefore likely that *Dolichoderus tertiarius* has had its nest in oak or other common deciduous trees in the amber forest, and only had food-seeking tracks on the bark of the amber tree. In addition to flower nectar, *Dolichoderus* lives on the excrement of Homoptera, which it licks from the leaves, but it does not seek direct contact with aphids.

The reason the other species of Formicinae and Dolichoderinae are far less common than the few mentioned (some of them represented by only a single or a few specimens) is presumably that they have been correspondingly far less common on the trunk of the amber tree. On the other hand this does not necessarily indicate that they have been less common in the amber forest. The selection the amber fauna represents would undoubtedly have been quite different in composition than that now known, if it had been oak, for example, and not a conifer, which produced the amber.

Myrmicinae are represented by 30 species in amber, i.e., about as many as those of Formicinae, but in the East Baltic amber there are 16½ times as many specimens of the last-mentioned subfamily as of the first-mentioned. The difference in the Danish amber is much less pronounced. Of the total Danish material, Myrmicinae constitute about 14%, but in the East Baltic material only about 2%. The probable reason for the relative rarity of Myrmicinae lies in the presence of special features in the mode of existence of these insects, and not in the actual incidence. They often prefer more dry biotopes and are usually not such pronounced forest animals. The difference between the western and eastern populations may be due to factors of climate and especially terrain.

The societies of the Myrmicinae are most often far less rich in individuals than those of the Formicinae or Dolichoderinae. They are in general light-shy individuals, living close to the ground and with very little tendency to climb the vegetation, while most Formicinae and Dolichoderinae are lively animals with a large radius of action. Two species of *Monomorium* and a single *Leptothorax* are relatively frequent in Wheeler's material (26, 28 and 34 specimens), and the same is the case for a single species of *Sima*. *Monomorium* is also common in Danish amber (13 specimens), and strangely enough this seems also to be true for the obscure-living, xerothermic genus *Aphaenogaster* (11 specimens).

Wheeler (1915, p.8) finds it strange that the large thermophilous genus *Crematogaster*, which in fact is mainly to be found in trees and bushes, where it also has its nests, is not found at all in amber. In his opinion this is because during its spread this form, probably like so many other forms, never reached the amber forest territory. Here, however, it must be considered that the number of species known today from amber is exceedingly small in relation to the fauna which, from our knowledge of the recent subtropical forest, must have existed in the amber territory. Many, also quite common species, must necessarily remain unknown, but may also turn up in the course of further research. Most Myrmicinae are insectivores, but a number of them have become specialized as granivores.

The Myrmicinae known from amber belong almost without exception to thermophilous genera, most of them represented today by ordinary species in the Mediterranean regions, but they have their main distribution in the subtropics and tropics.

Ponerinae, regarded as the most primitive of all known ants, are represented by even fewer finds than Myrmicinae, with a total of 111 in Wheeler and only 5 in the Copenhagen collection. *Ponera atavia* Mayr (Text-fig.44) is by far the most common species. It is closely related to *Ponera coarctata* Latreille, the main distribution of which today is in the Mediterranean region, but which is also one of those species in the sub-family which have penetrated furthest north in the warm temperate region. Ponerinae are insectivores, forming small sub-terranean nests, where the worker caste plays a far minor role than is otherwise the case among the ants. It is also characteristic that Wheeler's 42 *Ponera atavia* include only one worker but 9 females and 32 males (although the few Danish specimens nevertheless include 2 workers), and the species *Euponera succinea* (Mayr) is known by 24 females and neither males nor workers. The situation is different in the case of *Bradoponera meieri* Mayr, known by 17 workers and a single ergatoid female, *Prionomyrmex longiceps* Mayr (1 ♂ and 9 ♀♀) and *Procerapachys annosus* Wheeler (2 ♂♂ and 6 ♀♀). These figures could suggest that the last-mentioned species moved more freely in the vegetation, and that there may have been closer relationship between the amber tree and these three species in particular, while especially the first mentioned species have been captured by the amber resin during their mating flight. The variation in the figures may also be due to the genera

making different demands on degree of humidity and temperature, so that some of them have only been breeding outside the actual forest regions and have come into contact with the resin only during their brief swarming.

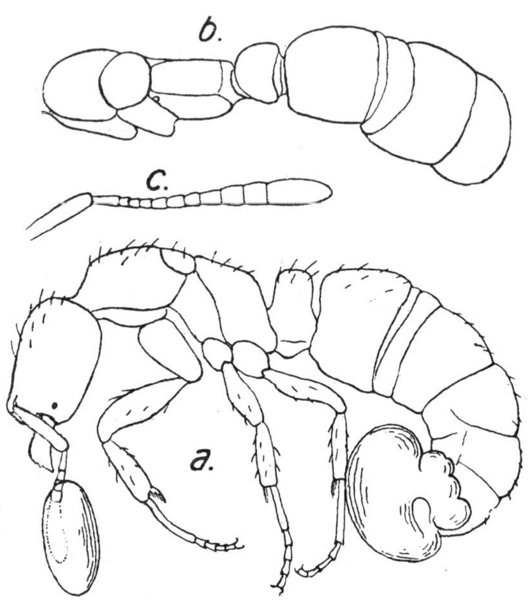

Text-fig.44. *Ponera atavia* Mayr (Formicoidea), a: worker in profile, b: same in dorsal view, c: antenna. (From Wheeler, 1915, fig.9).

According to Wheeler (1915) Dorylinae, the wandering ants of the tropics, are unknown in amber, and he makes no comment on the species mentioned by Handlirsch (1908, p.881), *Anomma (? rubella* Sav.) Smith (1868). There is in fact probably no reason for continuing to quote this old report. At any rate, it must be admitted that Dorylinae, if they have lived here, do not appear to have had any contact with the amber resin. The wanderings of Dorylinae have been only periodical events. They are determined to a high degree by their food requirements and are thus most common when there are many larvae in the insect society. In addition they are to a high degree dependent on external conditions. They never appear when the sun shines hot and the resin is flowing most strongly, but when the evening coolness appears or dense cloud layers reduce the power of the sun. This subfamily appears to be more tropically inclined in its climatic requirements than the amber territory has been able to satisfy, at any rate in the European region. Its absence must be regarded as a confirmation that the climate of the amber territory has at most only touched that of the tropics.

The ants in amber can be divided into two main groups, a thermophilous (e.g. represented by *Iridomyrmex*) and a boreal (among others represented by *Formica flori* and *Lasius schiefferdeckeri*), but otherwise with all possible transitions. Wheeler (1915, p.15) is of the opinion that these two extreme types mark definite periods in the course of the amber forest's existence, at first a warmer one with dominance for thermophilous species, which during the cooling in the Oligocene were ousted by the steadily more boreal fauna. He sees an important argument for this point of view in the exceedingly great numbers in which several of the typical boreal species appear. He supposes the amber to have been transported from the site of formation by streams, and the site of deposit is for him exclusively the then subtropical Samland. If amber with boreal and subtropical species appeared simultaneously, then according to his mode of thought this would imply that the sites of formation were geographically separate, and that the boreally formed amber containing boreal insect species must have undergone a far longer journey. This presupposes a certain degree of plasticity in the ecology of the amber tree. However, this had to involve a correspondingly greater loss for the boreal amber than for the subtropical amber. He does not find it likely that under such conditions the boreal species could occur in such relatively large individual numbers alone on the basis of vertical distribution. On the other hand, if the amber collections had originated from the same restricted geographic region during the course of a longer period with decreasing temperature and

changing fauna, a rich representation of both thermophilous and boreal species would seem to him natural. He does not discuss the possibility that ecological plasticity of the species might play a role.

In principle, there can be no objection to Wheeler's argumentation, but nevertheless it cannot be correct. The excretion of amber resin in the forest (and thus, as a result, the presence of amber fossils), has been bound to a limited climatic region, as earlier pointed out. In the north, this has extended a little into the warm temperate regions of that time, more precisely to the most northerly line for the possible existence of the amber tree. From this northern boundary the region of the amber forest comprised the subtropical and probably also the tropical belt, in so far as forest could grow there with regard to other climatic factors. The amber forest of North and Central Europe has undoubtedly been under development long before the Eocene (section 3), and was fully established at the transition to the Tertiary. It is probable that the moderate fall in temperature during the Palaeocene left a trace on the fauna, and it is possible that during this a number of boreal species have reached the amber forest. In addition, there may have been a pause in the production of resin in the most northerly regions during this period. It is important that the boreal ants of the amber have at all times had their southern boundary south of the northern boundary of the production region of the amber resin, and that in addition most of them have formed exceedingly densely populated societies. This has made it possible for them to become ordinary amber fossils without regard to the fact that their main region of distribution must have been outside the amber region. It is also remarkable that *Formica rufa* L. (or a substitute for this) is not known from amber, in spite of its very great colonies. It has a southern boundary which is only slightly but sufficiently further north than that of *Formica fusca*, so that its entire region of dispersal has apparently been outside the amber region. In addition, it is a pronounced inhabitant of the spruce forest, and even though spruce trees were present in the amber territory, they have not constituted any coherent biotope as we know it from the boreal regions.

Insect societies have been a common event in the amber forest, and older fossil finds show that at any rate termites and ants have had social forms of existence back in the Mesozoic, on lines which have been related to those of the present day. Also some of the bee societies we have learned a little about from amber, seem to be so developed and rich in individuals that they can hardly be very young. Much also suggests that in the Eocene amber forest, these insect states have already had a fully established cooperation with myrmecophilous and termitophilous insects. Our knowledge of this, however, is necessarily limited to those forms which at any rate occasionally moved around outside the boundaries of their hosts' nests. We have no precise information in the case of many of these small insects. They belong to genera which from our knowledge of recent forms can equally well be encountered on biotopes outside the insect states, and which for that matter are independent in their mode of existence. The interplay between *Germaraphis* and *Iridomyrmex*, however, is a fact, which shows that aphid honey was already involved as a permanent item in the nutrition of at any rate some ants.

Bugs (**Heteroptera**, pars). The fact that ants in the Eocene, exactly as today, like drug addicts oblivious to all else, could become addicted to the gland secretions from certain myrmecophile bodies, is illustrated in an example provided by Wasmann (1932, p.1). There is on Java today a bug, *Ptilocerus ochraceus* Montad, with this seductive characteristic. It lives in the bushes, where it is sought after by the ant *Dolichoderus bituberculatus* Mayr, which avidly licks the secretion up, while all the time the bug impales one ant after another and sucks them dry. The desiccated bodies of the ants can be found on the ground beneath the bug's resting place. In a piece of amber from East Prussia Wasmann found a bug which was not identical with *Ptilocerus* but which was quite the same type and with corresponding glandular areas. In the same piece of amber there lay "unterhalb der Wanze in einer Entfernung von 5-7 mm auch zwei Leichenhaufen von Ameisen, an denen die Köpfe mit den Fühlern meist unversehrt sind, während der ausgesogene Rumpf zusammengeschrumpft ist." Wasmann named the bug *Proptilocerus dolosus;* the ant was *Dolichoderus tertiarius* Mayr.

Paussidae are actually carabids which are strongly modified and specialized myrmecophiles, both at the larval stage and as imagines. At the present day they are found only in tropical and subtropical regions, and may therefore be regarded as important evidence for the existence of a hot climate in the amber forest. However, the family is far from common in amber, and is not found in the Danish material. One significant reason for its rarity is probably that these beetles do not usually move outside the ants' nests, and thus have not often had the occasion to come into contact with the resin. It must therefore be reckoned that the family has been far more common than the finds might suggest. Paussids and ants have been captured together in a single piece of amber, but this is no absolute proof of their coexistence. The specimens are *Eopausus balticus* Wasmann and *Formica flori* Mayr, probably a tropical and a boreal species.

Wasmann (1929) gives a detailed account of the known amber fauna of paussids, as well as of the older literature: Menge (1856), Motschulsky (1856), Stein (1877), C. Schaufuss (1896), Klebs (1910, a total of 4 specimens) and Wasmann himself (1926a, 1926b, 1927, 1928). A total of 7 genera are known: 3 species of *Arthropterillus*, 12 *Arthropterus*, 1 *Cerapterites*, 2 *Protoceraptus*, 1

Arthropterites, 1 *Paussoides* and 1 *Eopaussus*. All are extinct with the exception of *Arthropterus*, whose recent species are all found in Australia. Wasmann argues strongly for the idea that the Paussidae consist of 4 branches, each with its own origin among the Carabidae. This is vigorously opposed by Kolbe (1925), who is convinced of their monophyletic origin.

Digger-wasps (**Sphecoidea**) are, next after ants, that group of aculeate wasps which are most common in amber. However, relatively few specimens are known, and only few have been described. Brischke's list of genera, found in collections belonging to Menge and Helm, is the most recent and also the most important literature available, but it is hardly completely reliable. Brischke's list (1886, p.278) comprises a total of 40 specimens which practically all can have found the food for their larvae on the plants of the amber forest, including the amber tree itself. 25 of the digger-wasps found in amber (24 *Crossocerus* and 1 *Mellinus*) at the present time prey on flies, and in the early Tertiary forest it has been most likely Dolichopodidae and Empididae which occurred in enormous numbers. From what we know of the recent life-habits of the genera, a number of the digger-wasps must be assumed to have preyed on aphids, some of which they could find on the bark of the amber tree, but no doubt also on many other plants. This is the case with the genus *Passaloecus*, of which Brischke has found 8 specimens in the material mentioned. This may also be the case with 3 specimens which Brischke has referred to *Psen* or possibly *Mimesa* (which, however, preys on cicada larvae). Thus also in this way, *Germaraphis* probably has played a part in the animal communities of that time. Brischke refers the last two specimens of the collection, although with some doubt, to *Gorytes*, which preys on the larvae of cercopids, and *Cerceris*, which preys mainly on weevils. The occurrence of these two wasps is apparently more of a chance nature. In addition, Cockerell (1909a) has described 2 *Crabro* species (flies as larval food) and a *Pison* species.

In Copenhagen there is an unanalyzed material of 3 Sphecidae, but in addition to this 10 undetermined Aculeata, some of which at any rate belong to Sphecidae. The Copenhagen collection also contains a species belonging to Ampulicinae (O. Lomholdt det.), a thermophilic subfamily, probably using cockroaches as larval food. They do not form nests, but use any available cracks and holes.

The historical material of spider wasps (**Pompiloidea**) is very uncertainly determined, Burmeister (1832, p.636) mentions the genus *Pepsis*, Brischke (1886, p.278) *Pompilus*. The Danish collection contains a single pompilid. The group prey almost exclusively on spiders for larval food. Their nests are constructed typically of clay, and must be exposed to the sun or at least freely exposed if they are to dry and harden quickly enough. The actual amber forest has therefore hardly provided the group with great possibilities for satisfactory existence. From the few findings it is possible to recognise its existence and its complete specialization, and it has probably been common on open and more arid regions outside the continuous forest.

Also hornets (**Vespoidea**) are free-living predators, collecting food for the nest, and they too are very weakly represented in amber. Bachofen-Echt (1949, p.133) considers that the total material comprises about 6 exclusively social species; here he must have included unpublished pieces from his own collection. Notes by Gravenhorst (1835, p.92) and Brischke (1886, p.78) are highly uncertain. Menge (1856, p.26) describes *Vespa dasypodia*, a form with reddish-yellow bristles on the underside of the front tarsus "ähnlich wie bei einigen lebenden Andreniden." Cockerell (1909a) has described a species of *Palaeovespa*. The group is not represented in the Copenhagen collection.

Scolioidea are aculeate wasps which do not collect food for their larvae, but occur as ectoparasites on host animals in their natural environment, in all cases earth-dwelling larvae. Bachofen-Echt (1949, p.132) mentions a *Scolia* species, parasitic on chafer larvae, and Brischke (1886) a *Tiphia* (doubtful). Mutillidae, which are parasitic on other aculeate wasps, is known by a few specimens; Menge (1856) mentions 6 specimens, Brischke 3 *Mutilla.* Three specimens of this family are found in the Copenhagen collection. Myrmosidae are found in 7 species of *Protomutilla*, described by Bischoff (1915) and one (Text-fig.45) by Brues (1932). They lay eggs in nests of other aculeates. Among Sapygidae, Brischke mentions a single *Sapyga.* Scolioidea are thus weakly represented in the Baltic amber fauna.

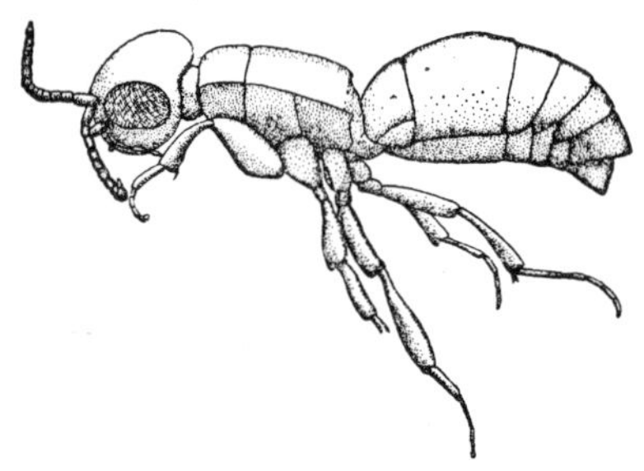

Text-fig.45. *Protomutilla megalophthalma* Bischoff (Hym., Myrmosidae). (From Brues, 1932, fig.88).

The cuckoo wasps **(Chrysidoidea)** are mentioned by Menge (1856), and Brischke (1886) mentions the genera *Chrysis* (aculeate parasite) and *Cleptes* (parasite in Tenthredinidae). Brues (1932, p.153) discusses in detail the exceedingly strange *Protochrysis succinalis* Bischoff (1915) and describes *Omalus primordinalis.* There are 2 specimens in Copenhagen of this very scantily represented group.

Bethyloidea. Dryinidae are a very remarkable family not very rich in species. The members are parasitic in various cicadas, among others Fulgoridae and Jassidae, which are so relatively well represented in the amber forest. They are endoparasitic in the cicada larvae, in their abdomen, but finally appear as an outer cyst by a moult. Pupation is in the ground or vegetation. As a result of its mode of existence, the larvae (sometimes there are several) causes the host to lose its power of reproduction. Brues (1932, p.146) describes 3 *Taumatodryinus,* 2 *Lestodryinus* and 1 *Neodryinus.* The Copenhagen collection contains 2 specimens.

Bethylidae is a relatively large family, also well represented in amber. Brues describes (1932, p.116, but partly already in 1923), a total of 27 species distributed over 17 genera, one of which, however, he feels very uncertain about *(Epyris).* Of the other 16 genera, 7 are extinct *(Uromesitius, Palaeobethylus, Palaeobethyloides, Bethylopteron, Parapristocera, Protopristocera* and *Pristapenesia),* almost all primitive types, of which *Uromesitius* and *Bethylopteron* show that great similarity with the Chrysidoidea, that the posterior abdominal segments form a tube. Of the now-living genera *(Eupsenella, Prosierola, Perisierola, Artiepyris, Misepyris, Holoepyris, Isobrachium, Rhabdepyris* and *Laelius*, i.e. apart from *Epyris),* 2 are today cosmopolitan, while 7 are more limited in their occurrence; 6 are neotropic, 5 are palaearctic, 1 African, 1 Indomalayan and 1 Australian. It is actually the Australian genus which is the dominating one in the Baltic amber. There is an unusually large representation from South America and Australia. The Copenhagen material contains 24 specimens. The family is also known from Burmese amber, 4 species, each with its own genus, having been described by Cockerell (1917a, 1917b, 1920). Bethylidae are above all parasites in coleopterous and lepidopterous larvae.

Brues (1932, p.113) describes 2 species of a quite small family Embolemidae belonging to the recent genera *Embolemus* and *Ampulicimorpha.* Their biology is unknown, although they have been observed together with ants in the recent fauna.

Proctotrupoidea are at the present day one of the very large groups of parasitic wasps, as they have also been in the amber forest. The Danish collection of 348 specimens is now being studied by Dr. Lubomir Masner, who is also studying a considerably larger material of East Prussian origin. A discussion of this group prior to the publication of these studies would seem impracticable. The older literature (Brues, 1923) mentions numerous species of Scelionidae which are parasites in insect eggs, 6 species of Ceraphronidae, apparently above all parasites in braconids and chalcidids, themselves primary parasites in aphids and scale insects, 1 Diapridae *(Paramesius)*, 1 Belytidae *(Belyta)*, probably a parasite in flies, and 6 Proctotrupidae, the recent species of which are mainly parasites in beetles.

Masner (1969a, p.105) has published some considerations regarding the subfamily Ambositrinae (Diapridae), at the present day widespread throughout Africa, Madagascar, Australia and America, and which is also found in 2 or 3 species in Baltic amber. One of these amber fossil species is very common and also very closely related to the African *Ambositra famosa* Masner. Masner "considers *Ambositra* to be a genus of southern origin which spread from the hypothetic Gondwana across the African continent northwards to Europe." It is however more likely that also this little group has been subject to the same biogeographic laws as flora and fauna in general, and that it has had an exceedingly wide distribution at the transition between Mesozoic and Caenozoic. Where and when it actually originated is hardly possible to determine.

Chalcidoidea. In the case of this very large family group of parasitic wasps, the taxonomic treatment of the amber material has been completely neglected, only a single species having been described. Determinations by Menge (1856, Pteromalinae), Helm (1896c, *Pteromalus)* and Brischke (1886, *Perilampus)* can hardly be taken quite literally. Duisburg (1868) illustrates a *"Proctotrupes"* which according to Bakkendorf (1948, p.213) "shows a mymarid, which seems to be identical with *Petiolaria anomala* Blood & Kryger." The recent type specimen was caught in an old pine-wood.

The Copenhagen collection contains 137 Chalcidoidea (O. Bakkendorf det. to families). Of these, Torimidae are represented by 2 specimens. Most Torimidae are parasitic in gall-forming Diptera and Hymenoptera, a possible biotope in the amber forest, but other recent species live in seed, among others of conifers. There are 5 Eurytomidae, whose present mode of existence varies considerably, many as larvae being phytophagic, others parasitic; 4 Pteromalidae: parasites or hyperparasites in many different insects; 29 Encyrtidae: parasites in many different insects, but especially Homoptera and Lepidoptera; 4 Eulophidae: very wide-spread parasites; 11 Aphelidae: parasites in aphids and scale insects. This material has been sent to Prof. M.J.P. Mackauer, Canada, for study; 11 Trichogrammatidae: egg parasites in many insects; 22 Mymaridae: exclusively egg parasites; 10 Mymarommidae: exclusively egg parasites.

Evanioidea. Of this small, very characteristic group, Brues (1923, 1932) has described a total of 7 species, 4 Aulacidae and 3 Evaniidae. The Aulacidae include 2 *Pristaulacus*, a genus, with a very wide distribution in the present, being absent only in the Ethiopian region. The other two species belong to *Micraulacinus*, also recent, but today occurring exclusively in Australia. The recent aulacids are known as parasites in various wood-boring larvae (ciricids, buprestids and cerambycids).

The 3 evaniids were all referred by Bruce to the subgenus *Parevania*. While *Evania* used in the widest sense of the term has a world-wide distribution, *Parevania* in the present is limited to South Africa and Eastern India. In the older literature *Evania* is mentioned by Burmeister (1831, p.1100) and *Brachygaster* by Brischke (1886, p.278).

Evaniidae are parasites in the egg capsules of cockroaches.

Thus, in the Eocene, both these families have not only had favourable breeding conditions in the amber forest, but probably over almost the entire globe. Geographically, the group clearly marks the seeming randomness with which the fauna has in general survived the cold wave of the late Tertiary and the Quaternary. *Evania* has had a very great ability to survive, and subsequent ability to spread. The same holds for *Pristaulacus*, which for incalculable reasons has been excluded from Ethiopia, which is no common phenomenon. *Parevania* has been able to survive in South Africa as well as in Eastern India, while *Micraulacinus* has disappeared everywhere except in Australia.

The Copenhagen collection contains 3 Evanioidea.

Ichneumonoidea. The largest of these families, Ichneumonidae, belongs to those insect groups which from a taxonomic point of view are those most poorly studied in the Baltic amber. Brischke (1886, p.278) mentions a series of genera (*Porizon*, near *Mesochorus*, *Tryphon*, near *Mesoleptus*, *Cryptus*, *Pezomachus*, *Hemiteles* and *Phygdeuon*), and Giebel (1856, p.155) describes *Pimpla succini*. In addition, *Bassus* is mentioned in Burmeister (1832, p.626) and in Keferstein (1834, p.332). However, these generic determinations must all be accepted with great reservation. The Copenhagen collection contains 132 Ichneumonoidea, of which at least 27 are Ichneumonidae, a relatively small number (Plate I).

The position is quite different in the case of the other very large family, Braconidae. Here Brues (1932) has had a large material from Königsberg for study, representing numerous subfamilies with a total of 127 species. Spathiinae: 1 *Cantharoctonus* (today South America). Hormiinae: 1 *Hormiellus* (today Formosa), 1 *Prochremylus* (Text-fig.46) (extinct). Doryctinae (today mainly Holarctis): 5 *Doryctes*, 1 *Dory-*

ctomorpha. Hecabolinae (only Holarctis): 1 *Polystenus*, 1 *Promonolexis* (extinct). Rhogadinae (widespread): 2 *Rhyssalus*, 1 *Palaeorhyssalus* (extinct), 1 *Digastrotheca* (extinct), 2 *Coeloreuteus* (Africa), 6 *Clinocentrus*, 1 *Semirhytus*, 1 *Rhogas*. Cheloninae (especially Holarctis): 2 *Phanerotoma*, 8 *Ascogaster*, 1 *Diodontogaster* (extinct). Helconinae (widespread): 1 *Chelonohelcon* (extinct), 3 *Electrohelcon* (extinct), 1 *Austrohelcon*, 1 *Eumacrocentrus*, 1 *Gymnoscelis*, 5 *Aspicolpus*. Diospilinae (particularly Palaearctis): 9 *Microtypus* (Text-fig.47), 1 *Diospilites* (extinct), 1 *Taphaeus*. Blacinae (mainly Holarctis): 5 *Blacus*, 1

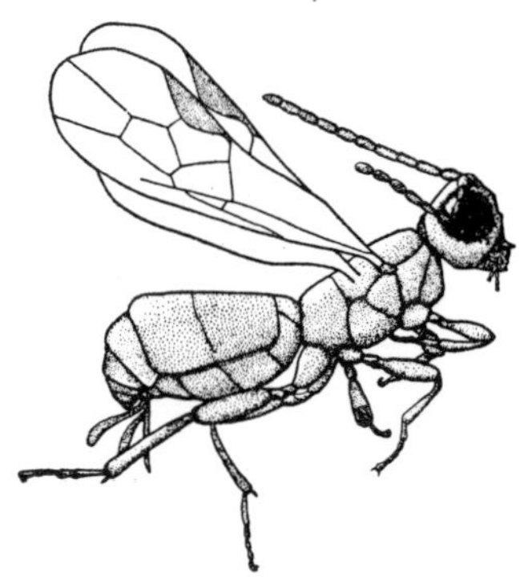

Text-fig.46. *Prochremylus brevicornis* Brues (Hym., Braconidae). (From Brues, 1932, fig.12).

Electroblacus (extinct), 2 *Neoblacus*, 1 *Pygostolus*. Ichneutinae (Holarctis): 2 *Ichneutes*. Leiophroninae: 1 *Parasyrrhizus* (extinct). Cardiochilinae: 1 *Eocardiochiles* (extinct). Agathidinae: 1 *Microdus*, 1 *Snellenius* (Indomalaya). Neoneurinae (mainly Holarctis): 1 *Elasmosomites* (extinct). Microgastrinae (mainly Holarctis): 1 *Miracoides* (extinct). Meteorinae (widespread): 4 *Meteorus*. Euphorinae (widespread): 1 *Microctonus*, 1 *Onychoura* (extinct). Aphidiinae (mainly Holarctis): 1 *Ephedrus* (very common in the present amber collection), 2 *Aphidius*, 1 *Propraon* (extinct).

During the preparation of this material, Brues has expressed surprise over various aspects. He has felt the lack of now

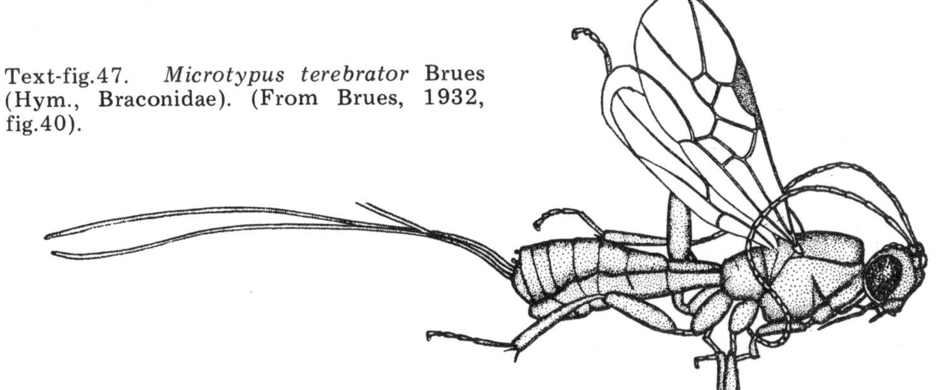

Text-fig.47. *Microtypus terebrator* Brues (Hym., Braconidae). (From Brues, 1932, fig.40).

common subfamilies, for example he has not seen a single specimen of the subfamily Braconinae, not even among the many pieces which were too badly preserved to form the basis of a description. Likewise, he was surprised not to take a single find of the related family Alysiidae.

His table of the amber material (p.110) shows that of the 28 now-living genera there are 18 in Palaearctis, of these 15 also in Nearctis, where a total of 17 live, but this tells us nothing as to the distribution of these genera in the North-South direction, which would have been desirable. There are 12 neotropic genera, not less than 10 of which are common with Nearctis. Of the 9 African genera, only *Coeloreuteus* is endemic. Of 11 Indomalayan genera only *Hormiellus* and *Snellenius*, and of 11 Australian genera only *Austrohelcon* and *Neoblacus* are recorded as endemic, and of these, according to the text, *Hormiellus* is actually indigenous to Formosa (Brues, p.26).

Braconidae are parasitic on insects of practically every systematic grouping, but lepidopterous larvae seem to be the most sought host. Aphidiinae are specialists, breeding in aphids. The Danish collection contains about 100 Braconidae, 19 of which are Cheloninae and 5 Aphidiinae (the latter sent to Prof. Mackauer, Canada, for study).

Apart from the above, Brues (1932) has dealt with the small families Stephanidae (3 species of the extinct *Electrostephanus*), Megalyridae (3 species of the extinct genus *Prodinapsis*, the nearest now-living relatives of which must be sought in South Africa), and the completely extinct family Pelecinopteridae (1 *Pelecinopteron*).

Leather-winged beetles (**Cantharoidea**). These beetles are mentioned in the classic collections (Berendt, 1845; Menge, 1856; Helm, 1896c), as well as in Klebs (1910).

Cantharidae are all found as imagines on open inflorescences such as Umbelliferae, for example. Here they have fed on small insects and no doubt also sucked nectar, and have themselves been booty for the tiger beetle *Collyris*, among others. The genera *Cantharis* and *Rhagonycha* have been dominant, and although they have mainly lived in open spaces and on the forest edges, and not in continuous forest, they are in no way completely rare. Nor have the larvae of these genera belonged in the forest; they are all soil dwellers, living on both vegetable and animal detritus. They are not known from Baltic amber.

The picture is somewhat different with the larvae of *Malthodes* and *Malthinus*, genera which at the imaginal stage are both quite common in amber. In the present these larvae are found commonly under bark scales, in cracks and in the moss on tree-trunks, where at any rate *Malthinus* larvae are carnivorous, and even though hardly any species has belonged exclusively on the amber tree, it must be considered strange that so far there is no report of their being found in amber at this stage of development.

Camillo Schaufuss (1891, p.58) has described the 6 mm long *Cacomorphocerus cerambyx* from Coll. Helm, Yablokov-Khnzorian the 7 mm long *Malchinus kryshtofovichi* (1961b, p.95). Klebs' collection contained 2 *Absidia*, 12 *Cantharis*, 3 near *Cantharis*, 4 *Malthinus*, 5 *Malthodes*, 15 *Rhagonycha*, 5 near *Rhagonycha* and 2 *Silis*. Only few cantharids are known from the Danish collection.

Lampyridae: Coll. Klebs had 3 *Luciola* (fireflies), which most probably have been very numerous on the favourable localities. In addition, Berendt mentions *Lampyris?* (but this has probably also been a *Luciola)*. In Klebs, Lycidae are represented by 3 *Dictyoptera*, 1 near *Dictyoptera* and 1 near *Lygistopterus*. Lycids have mainly lived in dead and rotting tree-trunks, and have probably been far more common than the few finds suggest.

A single larva belonging to one of these last-mentioned families has been seen in a private collection. It was, however, unobtainable, and may today possibly exist as a piece of jewellery.

Scorpion flies (**Mecoptera**) are rare in Baltic amber. According to Carpenter's latest publication (1954, p.31), 2 species have been described in the genus *Panorpodes* (Text-fig.48), which occurs today only in East Asia (Japan, Korea), and 2 species of *Panorpa*. In the same study Carpenter describes 3 species of *Bittacus*, but already 1854 (p.379) Pictet has described *Bittacus antiquus*. According to Carpenter (1954), the determination of *Bittacus validus* Hagen (1854, p.92) is uncertain. Recently, Carpenter has studied the type specimen and states in a note (Psyche 82, 1976, p.303) that *B. validus* Hagen is not a mecopterian, but belongs to the Trichoptera. *Bittacus antiquus* Hagen (1856) is a different species from *Bittacus antiquus* Pictet,

but synonymous with *B. fossilis* Carpenter (1954, p.36). The Copenhagen amber contains a single undetermined scorpion fly. Altogether we know about a score of specimens. In addition a few specimens are known from the Danish Eocene Mo-clay.

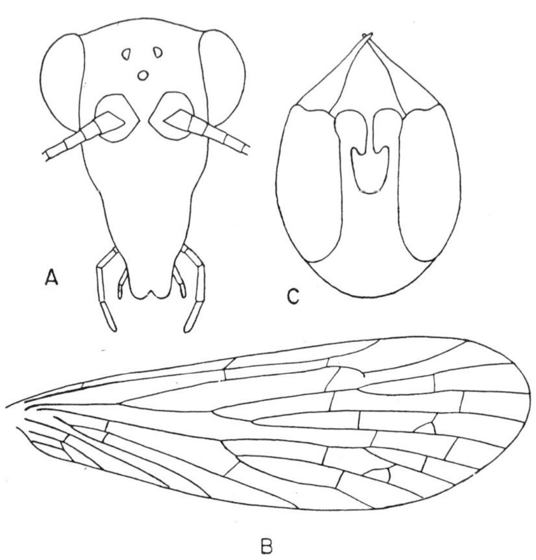

Text-fig.48. *Panorpodes brevicauda* (Hagen) (Mecoptera), a: head in frontal view, b: forewing, c: genital bulb of male. (From Carpenter, 1954, fig.1).

Scorpion flies live both as imagines and larvae on animal matter, mostly small animals, living or dead. The imagines are found in low vegetation, herbage, bushes and small trees, and often use their slender and prehensile tarsi in capturing their living prey. They are poor fliers, only changing their situation unwillingly, but prefer to hunt for a longer interval from one chosen site, most often bushy. They have only rarely landed on the trunk of an amber tree. The larvae are slow animals, superficially resembling lepidopterous larvae; they are found in moss and plant detritus and do not occur among the amber fossils; they have inhabited the *Sciara* zone.

Spiders (**Araneida**). The classic literature here, as almost universally, is Koch & Berendt (1854), with important comments by Menge, the publisher of the study. Menge's own studies are also important (1856, 1869). What we know of the occurrence of this order of arthropods in amber is due above all to Alexander Petrunkevitch, who in several major publications examined as much material as he at all could have access to. After having first reviewed what was available of Baltic amber in the museums of the U.S.A. (1942, 1946 and 1950), he made a collecting trip to a number of European museums and private collections in 1949, and the product of this trip, published in 1958, is today the basis for any serious discussion of Tertiary spiders.

The work of Petrunkevitch was of considerable significance for the organization of the material of Danish amber fossils, as his visit resulted in a coordination of existing public collections as well as in a greatly increased interest in an extension of this collection. As a result, we have today in the Zoological Museum, Copenhagen, a collection of about 500 specimens over and above the modest number Petrunkevitch has studied. It is much regretted that he did not live to utilize this result of his trip to Europe. Provided West Baltic (Danish-Scandinavian and possibly West German) amber and East Baltic (Samlandian) amber are not of quite the same origin, but have originated in somewhat different landscapes and climates within the region of the European amber forest, this special West Baltic collection will acquire increasing interest.

According to Petrunkevitch's last account (1958), the material of spiders in the Baltic amber includes representatives of 37 families, 6 of which are now extinct. The material includes 140 genera and 267 species; only few of the genera and none of the species can be recognized in the fauna of the present. The spider fauna of the amber has only very little in common with the present European fauna, indeed with the Holarctic fauna as a whole, and then only through genera which also have a great geographic distribution over several continents.

On the other hand, the great majority are closely related to forms which today are found in the tropics and subtropics. As examples, Petrunkevitch (1950, p.259) mentions South Africa, Malaya, Australia and South America, to which may be added all the sub-tropical and tropical region of America and of Asia, New Guinea, New Zealand, Azores, St. Helena, Kerguelen Islands, Madagascar and Seychelles.

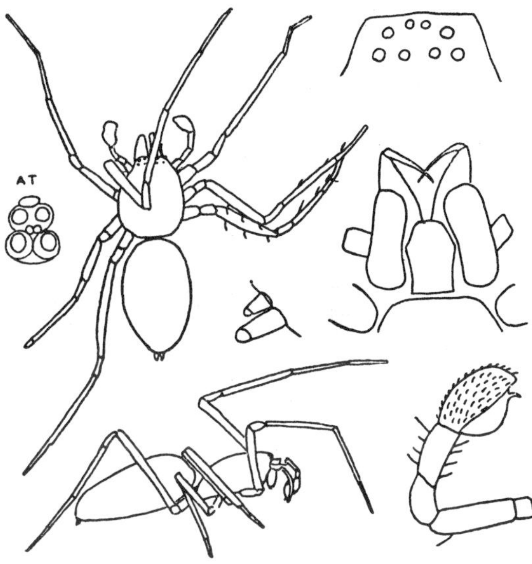

Text-fig.49. *Myro extinctus* Petr. (Agelenidae), male, dorsal and lateral view, and various details. (From Petrunkevitch, 1958, figs. 42-47).

Among those families especially common in amber might be mentioned Agelenidae (Text-fig.49) (e.g. species related to the recent *Tegenaria*, the house spider), Araneidae (orbitelous spiders, e.g. *Araneus*, the garden spider), Clubionidae, Eusparrasidae, Gnaphosidae (Drassidae), Linyphiidae, Salticidae (jumping spiders), Segestriidae and Theridiidae (Text-figs.50, 51). These are all families within which, at the present day, at any rate a proportion of the representatives are associated with vegetation, and which are often found creeping on branches and trunks, and whose retreat is hidden under loose bark and in fissures. A number of them, for example the salticids, do not spin traps, others must have had their webs in bushes and small trees around the trunks of the larger trees, between lianas, in the crooks of the branches or as fine carpets over the moss. The spiders thus seem to have been widespread from the forest floor to high up in the treetops. Most of them have had their period of activity during the hours of darkness, when in particular they have constructed their very different types of net. Others, above all the Salticidae, have hunted freely on trunks and branches in full sunshine, and have been passive at night. The booty has probably consisted by far of gnats and other flying insects, caught during their seeking after resting places. Fine spinning threads are found, not rarely almost undamaged in the amber, sometimes together with entangled insect remains, which have quite clearly been the work

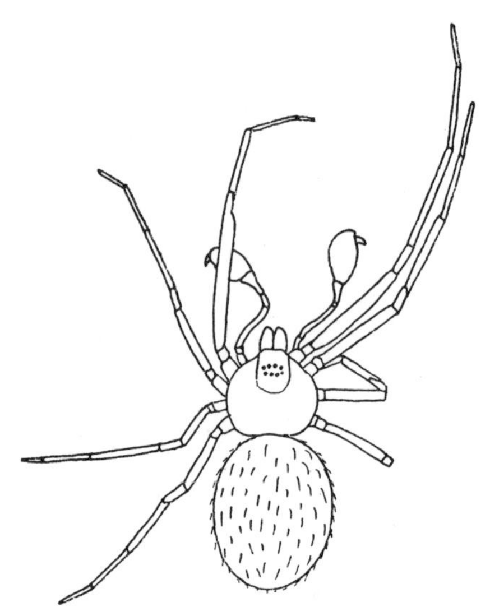

Text-fig.51. *Eosmysmena moritura* (Petr.) (Theridiidae), male. (From Petrunkevitch, 1958, fig.108).

of spiders. These often well-preserved fragments of webs are evidence of the thin-flowing resin, and thus of the considerable increase in temperature which has at any rate at intervals characterized the amber forest. Menge (1856, p.9) gives a quite detailed account of the spider's webs he found in his collection.

It is remarkable that groups like Theraphosinae (bird spiders) and Lycosidae (hunting spiders), which are restricted to the ground or to heaps of stone or fissures in cliffs, are very limited in numbers, also those forms which as newly hatched have a spreading stage as gossamer. The *Sciara* zone has obviously been unsuited to the earth-dwelling and generally rather xerophilic forms. Many of them have also no doubt been too strong to remain captured by the resin.

Autotomy is not unusual among the spiders captured by the resin, and loose legs in amber are quite common. For example, Bachofen-Echt (1949, fig.50, p.54) illustrates a leg of one of the large hairy bird spiders (Mygalidae), a family which is otherwise almost unknown in Baltic amber.

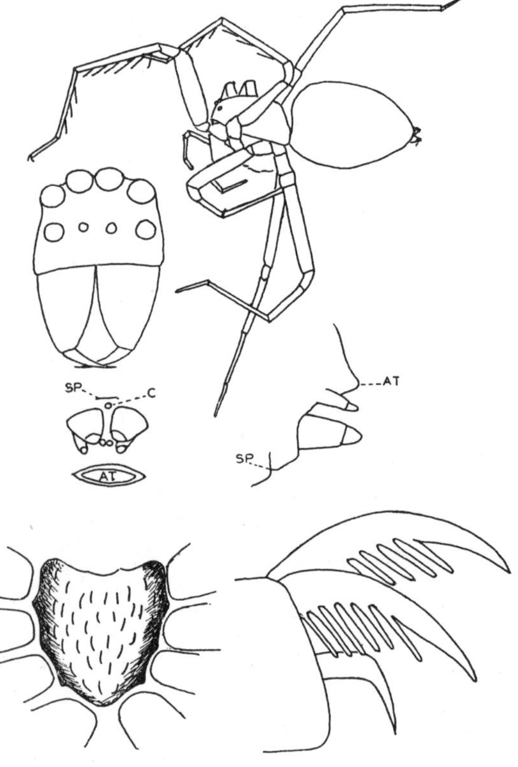

Text-fig.50. *Esuritor aculeatus* Petr. (Pisauridae), female and details of same. (From Petrunkevitch, 1958, figs.52-57).

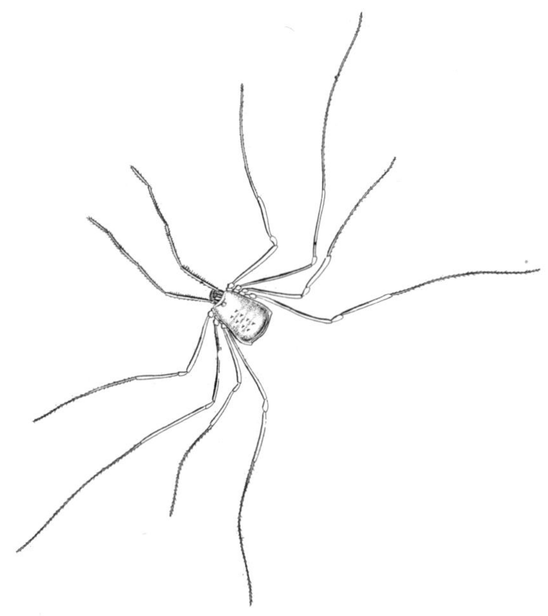

Text-fig.52. *Nemastoma tuberculatum* Koch & Berendt (Opilionida), reconstruction. (From Berendt, 1845, pl.XI, fig.97).

Harvestmen **(Opilionida)** (Plate 10,A). Koch & Berendt in 1854 (p.97) describes 7 species of harvestmen (3 of them belonging to the genus *Nemastoma)*, and Menge (1856) in addition described 4 species (including 1 *Nemastoma)*. Since then, no studies have been made of the Opilionida occurring in amber, in spite of their considerable incidence; for example there are 27 undetermined specimens in the Copenhagen collection, mainly Phalangidae *(Megabunus, Oligolophus* and *Licinius* or related forms) as well as a single *Nemastoma* (N.T. Meinertz det.). Part of the reason for the poor utilization is presumably in the nature of the material, which often presents quite considerable technical difficulties, for example one side of the animal is almost always "milky" or "mouldy", in spite of otherwise clear amber, and the long legs are almost always more or less defective.

The *Nemastoma* species (Text-fig. 52) appear to be those which provide the best amber specimens for scientific studies. They are relatively short-legged forest floor forms, apparently captured during their nightly wandering over the exposed part of the root-neck of the trees, or when they have been sitting concealed during the day in crevices at the same place. In the present, this genus has no fixed breeding period, at any rate not under boreal conditions. There are adult individuals all the year round, and this has probably also been the case in the amber forests, at any rate there are relatively many adult animals in the amber material.

Most of the other harvesters belong to genera which at any rate at the present day, especially at night-time, tend during their hunting to creep high up in bushes and trees. These species must have been very exposed to capture in the resin, and we know that this has in fact taken place. However, they also have the characteristic that their legs break off very easily from the body at the coxal joint by autotomy, when anything holds them firmly. For this reason, therefore, the amber is found to contain many loose legs which have belonged to harvesters, and which cannot be determined. The animals have let themselves drop to the ground as soon as they felt they were trapped, and only the leg has remained. Many of the individuals actually trapped in fact lack one or more legs, which have been lost on previous occasions. Furthermore, there is the disadvantage in this group of species that it is mainly young animals which are found in amber. At the present day, these species are tied in their breeding to a rather constant annual rhythm, as adult animals only appear rather late in the summer. If something similar has been the case in the amber forest, and this is most likely, the adult animals have not been generally common until the time when the growth period of the amber tree has almost terminated, and when it must therefore be assumed that the flow of resin has been weak or has perhaps completely terminated.

Scutigera is the only genus of centipedes **(Chilopoda,** pars), which really are indigenous on the tree-trunks of the amber forest. Koch & Berendt (1854, p.14) already describe 2 species (under the name *Cermatia* Leach), and in his comments, Menge adds that he too possesses some pieces. Three very defective specimens are found in the Copenhagen collection.

Scutigera is very long-legged, and from our knowledge of existing species we must assume that just like harvesters it has roamed around as a nocturnal predator hunting insects and small spiders. During this activity it has had excellent opportunities to come into contact with the resin, and then has undoubtedly also often been captured. When it is nevertheless so rare in amber, it is due above all to the fact that just like the harvesters it has released itself by autotomy from those limbs which stuck fast. The few specimens known have been so particularly unfortunate that in spite of their long legs they have been caught in the resin by the body itself. None of them have a full complement of legs, and some have in their haste cast off all their legs. As a further deterioration of the material is the fact that in some cases the birds have plucked large pieces out of the bodies, before they were covered by the next flow of resin.

Chapter G

The hidden Fauna of Tree-Trunks

In each natural forest there is hard competition between the individual plants; this starts already at the time the seed plants begin to germinate and lasts all their lives. Poor placing, not least in relation to older and stronger members of their own species, too poor supplies of nourishment, exposure to wind and unsatisfactory ability to bind the water in the soil, all reduce the possibility for the plants, and many must necessarily go under in the course of time. In addition, even the strongest and most favourably placed trees, in their high age, enter into a period of senility, where advancing weakness increases their susceptibility to external attacks in always increasing degree, and which with certainty leads to their death and destruction. Apart from the slowly acting natural competition, there may also be natural catastrophes such as forest fires, hurricanes and storm floods, exerting their influence on the survival of the forest vegetation, and finally, parasitic plants and plant-sucking insects such as aphids and scale insects have a strong influence on the vitality of the trees.

However, even those trees which have been mistreated in competition or by the hand of nature, have the greatest significance for the remaining life of the forest, for that weakness to which they are subjected, and which ends in death, plays throughout its whole course a decisive role for their suitability as a nutrient for a series of parasitic plants, especially fungi and bacteria, and for a very rich insect world. These insects can develop a primary attack on the weakened trees (for example many bark beetles), or they can appear following a primary attack by fungi, above all by species of *Polyporus* (for example the death-watch beetles). In many cases, however, it is not possible to decide the route along which an insect group has conquered the biotope in the first place. Only few of those insects that live in the bark and wood of the tree-trunks attack completely healthy trees.

Tree fungi (Polyporaceae) constitute a very important factor in the life of the forests. The significance of their life functions can be followed in the forests of today, and there is important evidence that they have played a similar role in the amber forests. The species of *Polyporus* which are known from amber have very close relations in the forests of today, and in fact Conwentz has even established them as forma *succinea* to recent species *(Trametes pini, Polyporus mollis* and *Polyporus vaporarius)*. However, these determinations can hardly be regarded as absolutely certain.

The influence of the Polyporaceae on the condition of the trees is violent, but insidious. The commencement of the attack by a spore germinating, takes place typically from a bark which is already weakened, e.g. after violent sucking by aphids, or injured by broken branches, but it is also possible for apparently completely healthy trees to be attacked. From the surface of the tree, the fungal hyphae penetrate into the heart wood, which is their most important region for growth. The fungus destroys the wood by its life functions in such a way that the wood gradually changes its nature and becomes red rot or white rot, determined by the nature of the tree and the fungus; it absorbs moisture and develops a more or less spongy consistency, and the resistance of the trunk to storm is considerably reduced. The outer appearance of the bark and the sapwood, apparently healthy, can for several years hide the true state of health of the tree, and for that length of time maintain the transport of sap in an almost normal manner. When the fungus has achieved suitable maturity, it sends its hyphae to the surface, where the characteristic spore-filled fruit bodies grow out, often on apparently unharmed trunks, and only now can the attack be obviously recognized.

The destruction produced by the tree fungus is furthermore supplemented by the gnawing of innumerable insect larvae, for which the fungal-attacked rotted wood is the ideal substrate for nourishment. Many of these larvae are unable to break down cellulose of the wood without this pretreatment, which is a result of the life processes of the fungus. The fact that many insects (to some extent the same as those in the rotting wood) attack the fruit bodies of the fungi, naturally signifies nothing for the state of health of the trees already sentenced to death, but is part of the picture of the fauna on this biotype, which has been quite common in the amber forest, and to which also the amber tree apparently has had to contribute.

Many of these insects, whose larvae live in more or less fungus-rotted wood, appear to originate from forms which already earlier in the history of the earth were adjusted to life in the living or rotting fungal tissue. This appears relatively clearly, particularly from details in a series of life pictures of clavicorne and heteromere beetles (Cucujoidea).

Ciidae. The genus *Cis* is mentioned for the first time by Berendt (1845, p.56), but apart from this, there has been only little mention of this family (Helm, 1896c, p.228). In Klebs, only 3 specimens are mentioned, and in the Danish collection there are 5 specimens, one of which at the moment is studied by Dr. John Lawrence, Australia. In addition, there are a few larvae in a single piece of Copenhagen amber.

It is quite obvious that the family is only very weakly represented in Baltic amber, which would seem surprising. Above all, its members are associated with the Polyporaceae, in whose fruiting bodies most of them breed; others are found in dead wood, which has been rotted and broken down by the life functions of the fungi. The family prefers to breed in deciduous trees, but there are nevertheless species which also breed in fungi on conifers, or which even prefer these. Their at any rate apparent rarity may be based on this food preference, but it may also be due to the fact that the family seeks out fungi on such greatly weakened trees that the ordinary flow of resin has stopped. It is also possible that geographic and climatic conditions play a role.

Mycetophagidae are a family which very characteristically belong to just this biological group, and our collections show that it has been common in the amber forest. Klebs mentions (as the first) 30 specimens, of which Edm. Reitter has determined 2 to *Litargus*, 2 to *Mycetophagus* and 17 to *Typhaea*. Out of Klebs' material Abdullah has later described the extinct genus *Crowsonia* with the species *C. succinium* (1964, p.334). In the Copenhagen material of 25 imagines, the following have been identified: 3 *Typhaea*, 3 *Pseudotriphyllis*,

2 *Mycetophagus*, 3 *Triphyllis* and 12 *Crowsonia* — or forms closely related to these. One is therefore inclined to assume that the majority of those specimens determined by Reitter as belonging to *Typhaea* have belonged to *Crowsonia*, which at that time was undescribed, but which appears to have been the dominating genus in the amber forest.

The Copenhagen collection also contains 9 larvae, which must be referred to one or several species of Mycetophagidae. This has its very great significance, because it documents that there actually have been Polyporaceae on the amber tree itself, and at such an early stage of the year that the tree has still retained its ability to produce a flow of resin. The mycetophagid larvae have a rather great mobility and occasionally appear on the under surface of the fruiting bodies, from which they can easily have fallen down on to the resin. It is quite unlikely that these many larvae should have found their way into the amber in any other way from the surroundings.

Circaeidae. Yablokov-Khnzorian (1961c, p.209) estalished this family for the new genus and species *Circaeus borisjaki* (Text-fig.53) in association with Mycetophagidae and Colydiidae. However, Crowson & Viedma (1964, p.99) doubt

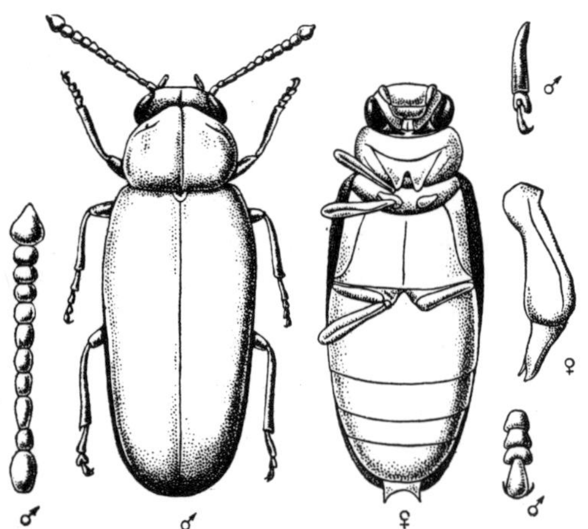

Text-fig.53. *Circaeus borisjaki* Yabl. (Circaeidae), dorsal and ventral view, and details. (From Yablokov-Khnzorian, 1961c, fig.1).

the taxonomic value of this family, making comparisons between *Circaeus* Yablok. and the heteromeran genus *Mycterus* Clairv. Its mode of life is unknown.

During the course of time, there has apparently been some difficulty in keeping the families **Aderidae, Anthicidae** and **Pedilidae** separate in the rather summaric descriptions of the beetles found in amber; the families are actually quite close to each other and have a great mutually similarity in their appearance. They are relatively common in Baltic amber, and all have recent species, the young stages of which more or less live to a pronounced degree in fungi. Berendt (1845, p.56) mentions only *Anthicus*, and Menge (1856, p.21) Anthicidae, whereas Helm (1896c, p.228) mentions both *Euglenes* sp. and *Notoxus*. A considerably larger material is found in Klebs as Hylophilidae: 20 *Euglenes*, 21 near *Euglenes*, 12 *Hylophilus*; as Anthicidae: 1 *Amblyderes*, 12 *Anthicus*, 3 near *Anthicus* and *Euglenes*, 9 *Macratria*, 3 *Ochthenomus*, 1 *Pedilus*, 3 *Steropus*, 2 *Tomoderus*, and 9 unknown (Edm. Reitter det.). These older determinations must be taken with reserve. Most recently, Abdullah has carried out a careful description of single species belonging to Pedilidae: *Protomacratria appendiculata* and *P. tripunctata* (1964, p.333) and *Macratria succinia* (1965, p.38).

The Copenhagen collection contains 19 Aderidae (= Hylophilidae = Euglenidae) as well as 3 Pedilidae and 17 Anthicidae. In addition to this imago material, there are 4 larvae (Text-fig.54B) belonging to the family complex.

The material does not tell us much on conditions in the amber forest, for one thing because of the uncertainty of the determinations, and for another because knowledge of the biology of the recent fauna is very slight. The larvae found, which constitute a by no means insignificant part of the Danish material, nevertheless show that at any rate some species have had the chance of existing on the surface of the living amber trees also at that stage, and it is most likely that they either have been living in the fruiting bodies of tree fungi on the tree-trunks in the same way as Mycetophagidae or as secondary fauna in attacks by bark beetles.

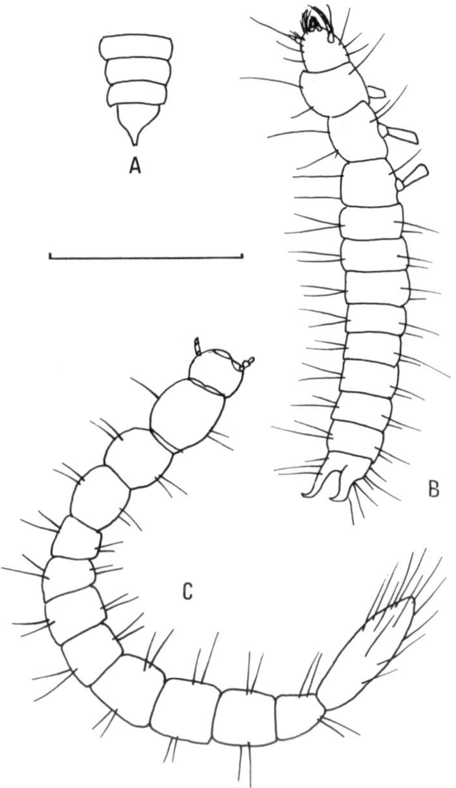

Text-fig.54. Various coleopterous larvae. A: abdominal end of *Osphya* (Serropalpidae); B: dorsal view of Aderidae; C. *Scraptia* (Scraptiidae). Scale: 0.5 mm. (Original drawings from specimens in Copenhagen collection).

Scraptiidae (Plate 10,B) are close to the above mentioned families both systematically and biologically. They are common in amber and appear to occur in several species. The family is mentioned for the first time by Helm (1896c, p.228), but apparently, several older mentions of *Anaspis* have referred to *Scraptia*. Klebs' collection contains 47 *Scraptia* and a specimen with close association to this. Recently, Ermisch (1941) has described *Scraptia pseudofuscula* (p.183) and *S. inclusa* (p.184),

and Abdullah (1964) has described *Palaeoscraptia elongata* (p.340), and *Archescraptia emarginata* (p.341). The undetermined Copenhagen material is relatively large: 32 imagines and a single specimen of the very peculiar larva (Text-fig.54C).

To judge from the biology of now-living species, the Scraptiidae of the amber forest have in part, but not exclusively, been associated with fungi on the trunks of the trees, not merely living trees, but also dead trees with loose bark. It is possible that they have mainly bred on deciduous trees, but the larva found could suggest that they have also been able to thrive in fungi on living amber trees.

Anaspidae, which are actually closer to Scraptiidae than Mordellidae, which they are most often ascribed to, have hardly had such direct association with the tree fungus as to fungal growth in general. In all probability, many older determinations are in fact erroneous. The first mention is by Berendt (1845, p.56). Klebs' collection contains 37 specimens, and the Copenhagen collection contains 16 imagines, but as expected no larvae.

Recently, Ermisch (1941, p.182) has described *Anaspis longispina* and *A. parallela*. According to Abdullah (1964, p.342), however, there are grounds for anticipating that a revised investigation of the type specimen of the latter species will show that it actually belongs to Scraptiidae. Abdullah has described *Anaspis (Silaria) parva* (1964, p.341).

Mordellidae. *Mordella inclusa* Germar (1813, p.14) is one of the amber insects which was earliest named. The genus and the family are also mentioned by Berendt (1845, p.56), Menge (1856, p.21) and Helm (1896c, p.228). Klebs' material includes 14 *Mordella* and 11 *Mordellistena*, and the Copenhagen material includes 6 *Mordella*, 9 *Mordellistena* (Plate 11,C) and 3 undetermined; this material has not been studied. Also Ermisch (1941) has described a mordellid species.

The larval biotopes of the recent Mordellidae are rather variable, but the most common and, in addition, most characteristic for the family, is heavily fungus-rotted wood from both deciduous trees and conifers; at times, the larvae are found direct in the fruiting bodies of the fungi. On the recent *Mordella maculosa* Näzen, Saalas (1923, p.275) writes that the imago is typically encountered on dry-rotted stubs of spruce, where the small fungus *Lenzitian saepiaria* grows. To begin with the larvae live in the fungus, and not until later penetrate in behind the fruiting bodies to the wood, which is invaded throughout by the fungal hyphae, and here pupation takes place later. Whether the example has general validity for the family cannot be said, but it suggests the possibility that the original breeding biotope of the family has been fruiting bodies of fungi in the same way as the families above, while the strongly transformed wood, which is invaded throughout by hyphae, has only a secondary significance, although a very important one.

In spite of their relatively great incidence, no amber fossil larvae of Mordellidae are known. The very slight mobility of the larvae is an adequate explanation for this; they do not voluntarily leave the medium in which they live. In addition, their attacks do not commence until the trees have been severely attacked by fungus, and are therefore hardly likely to have any longer a resin flow of significance.

Serropalpidae (Melandryidae). *Orchesia* is mentioned by Berendt (1845) and Helm (1896c) and *Hallomenus* by Berendt (1845). *Abderina helmi* has been described by Seidlitz in 1898 (p.577). In Klebs' list, a total of 44 Serropalpidae are mentioned: 5 *Abdera*, 1 near *Abdera*, 5 *Anisoxya*, 1 near *Carida*, 5 *Dircaea*, 3 *Eustrophus*, 1 near *Hallomenus*, 1 near *Hypulus*, 9 *Orchesia*, 3 near *Orchesia*, 1 *Phloeotrya*, 2 *Serropalpus* and 7 unknown. It is regrettable that this comprehensive and interesting material has not been described. The

imagines in the Copenhagen collection are as follows: 3 *Osphya*, 2 *Orchesia*, 3 *Abdera*, 1 *Melandrya* and 5 unknown, but in addition 6 larvae with the very characteristic structure for *Osphya* larvae (Text-fig.54A). This material has likewise not been studied. It is remarkable that Edm. Reitter does not mention *Osphya* from Klebs' material.

At the present day, the great majority of these genera breed in the fruiting bodies of the Polyporaceae, and to some extent in the strongly attacked wood behind these. Only in *Melandrya* are the larvae found by preference in the red-rotted heartwood produced by the Polyporaceae. However, biological details are missing for these insects, which are quite insignificant for the human economy. We therefore also lack a definite explanation why relatively many *Osphya* larvae are known in amber, while in the same material there are no larvae belonging to any of the other genera. It is conceivable that *Osphya* attacks the fungi at a time when the trees are relatively vital and producing a strong flow of resin, while the other genera do not breed until the trees are more weakened; the condition of the tree may possibly be explained by the character of the fungal bodies. The difference may also depend on a greater mobility in *Osphya* larvae than in the others. On the other hand, the number of *Osphya* larvae makes it rather unlikely that this should depend on chance alone.

Tenebrionidae. In addition to a few unlikely determinations (*Hopatrum*, Burmeister, 1832; Pimelidae and Opatridae, Helm, 1896c) there is an early mention of *Bolitophagus* (Berendt, 1845, p.56) as well as Diaperidae and Helopidae (Helm, 1896c, p.228). Klebs' list includes 12 specimens: 4 near *Helops*, 3 *Laena*, 1 *Lichenum*, 2 *Palorus*, 1 *Tribolium* and 1 *Uloma*, mainly animals which breed in tree fungi or in wood filled with mycelium. The Danish collection contains 3 specimens, which have not been further identified. Larvae are unknown. The material of Tenebrionidae is thus very modest. Biologically, they are related to Serropalpidae. When no larvae are known, the probable reason for this is the morphology of the larvae in addition to the relative rarity of the animals; contrary to the Mycetophagidae and for example *Osphya*, the very smooth click-beetle-like larvae with the weak legs do not appear on the lower surface of the fungi, where they would irretrievably fall down.

Klebs' collection of **Alleculidae** amounts to 17 specimens, which is a large number in relation to the modest mention which this group of beetles has previously received. Berendt mentions *Cistela* (1845, p.56), L.W. Schaufuss has described *Mycetocharoides baumeisteri* (1888, p.269) and Seidlitz *Isomira avula* (1896, p.102). Klebs' collection comprises 2 *Allecula*, 1 *Cteniopus*, 4 *Gonodera*, 1 near *Gonodera*, 1 near *Hymenalia*, 6 *Isomira*, 1 *Mycetochara* and 1 undetermined. The Danish collection contains only 2 pieces not further determined. The larva is unknown from amber. In contrast to the beetle families previously mentioned, the larvae of Alleculidae live practically exclusively in the wood which is rotted by fungi, and not in the fruiting bodies of the fungi. Contributing to the scanty number of the finds is that the family almost exclusively lives on deciduous trees. The find of *Cteniopus* is inexplicable, when the usual biology of this genus is taken into consideration.

Oedemeridae are mentioned by Berendt (1845) and Helm (1896c). In Klebs' collection, Edm. Reitter has found a single *Oedemera*, and the Copenhagen collection contains 2 specimens, all imagines. There is uncertainty with regard to the developmental biology of the genus, and possibly the finds are not placed at all in their correct biological relationship. It is known that certain recent species, including some of the most common boreal species, live as larvae in stalks of herbaceous plants, including *Typha*, *Circium* and *Eupator-*ium, which is an atypical although not unacceptable biotope in the amber forest. The majority of the genera in the family breed however in dead wood, almost like the Alleculidae mentioned above, and this life pattern is also followed for several of the *Oedemera* species. Also for the few known *Oedemera* in amber, the last-mentioned life pattern has undoubtedly been most probable.

Among the innumerable nematocerans (**Nematocera,** pars), which as imagines have been found in amber, numerous species have commonly bred in fungi, not least in Polyporaceae, and thereby probably as a result also on the amber tree. This is the case above all for Sciophilinae among the Mycetophilidae (see earlier for further details). A few mycetophilid larvae are found in amber. Normally, they pupate within the fungus, and advance by wriggling the abdomen halfways outside immediately before the metamorphosis. A few such exuvia are also found in the Danish amber; this emphasizes the probability that the amber trees attacked by fungus have been resin-producing, at any rate to begin with.

Among the relatively many small moths (**Lepidoptera,** pars) found in amber, some species have probably bred in tree fungi; this is particularly probable among Tineidae s.l.

Cleridae. The whole of the above-mentioned rich fauna in the fruiting bodies of the Polyporaceae has been hunted by Cleridae, Temnochilidae and Melyridae, three related families of beetles.

Among the Cleridae, Berendt mentions *Tillus* and *Opilo* (1845, p.56), but these generic determinations are not quite certain, however. In Klebs' collection there are mentioned 1 *Opilo*, 1 *Tarsostenus*, 1 near *Tarsostenus*, 1 *Tillus*, 1 *Trichodes*, 1 near *Trichodes* and 5 undetermined, a total of 11 specimens. The fact that half of the material has been unknown to Reitter, with his extensive knowledge of the Palaearctic beetle fauna, is interesting; Cleridae seems to have possessed more exotic features than the majority of other beetle families. The Copenhagen collection contains 5 imagines, which have not been identified further, but in addition there are 19 larvae, which possibly, with one exception, belong to one and the same species, at any rate they appear strikingly uniform. This common larva has been known to Bachofen-Echt (1949), cf. his fig.104, p.115: admittedly, he calls it a young larva of a leather-winged beetle (Cantharidae). These larvae belong to the sub-family Phyllobaeninae (Crowson det.); however, it has not been possible to associate them with any of the imagines which are known from amber. There is some possibility that it may have been a *Callimerus* or at any rate has belonged to this generic group, which at the present time is very common in South East Asia and Indonesia.

While the occurrence of imagines in amber appears to be by chance, as each species in Klebs is only represented by a single individual, the larvae must have belonged to a form which has been particularly associated with resin-producing trunks, at any rate with tree trunks. The typical working method of the clerid larvae is best known from the behaviour of *Opilo domesticus* in present-day roof constructions, where it seeks out wood-boring larvae in their galleries in the timber. When a region has been cleared of prey, the *Opilo* larva works out to the surface, after which it gnaws itself in at a new place, leaving wood dust externally on the timber construction in the form of a small "molehill." Thus it is relatively mobile. The larva of Phyllobaeninae has apparently not lived in dead or strongly affected trees, like the majority of other Cleridae, at any rate it has also been found on plants with relatively unweakened rise of sap. As it can hardly be conceived to have captured its booty from a fresh and untouched cambium, it must be presumed that it has had a corresponding mode of life in the *Polyporus* fungi, and in the wood damaged by them, or as accompanying fauna in attacks by bark beetles, where there has been rich booty in the form of beetle larvae, mycetophilid larvae and other insects.

Temnochilidae (Ostomidae) (Plate 11, B) are mentioned for the first time in Klebs (1910) with 4 specimens belonging to the genera *Calytis*, *Grynocharis*, *Lophocateres* and *Ostoma*. The Copenhagen collection contains 3 uniform but undetermined imagines. There are also 5 temnochilid larvae, which likewise have very uniform characteristics. Much suggests that one genus, possibly merely one species, has had a mode of existence corresponding to that of the above mentioned Phyllobaeninae among the Cleridae. The Temnochilidae are encountered at the present both as imagines and as larvae, mainly in wood-boring insect galleries or under loose bark, where they occur as predators.

Melyridae live when larvae as predators in the galleries of other insects under bark and in wood during decay, but many are also encountered under the bark scales in a similar manner to the larvae of *Malthodes*. The Danish material contains a similar relatively large number of larvae as in the Temnochilidae: 5 imagines and 6 larvae. It is therefore possible that also among these beetles there have been some which at any rate in part have sought their prey in the fruiting bodies of the Polyporaceae. In the East Prussian material of Klebs, there were the following representatives of Malachiinae: 2 *Apalochrus*, 1 *Attalus*, 1 *Calotes*, 4 *Ebaeus* and 3 *Malachius*, and of the Dasytinae: 1 *Cerallus*, 3 *Dasytes*, 1 near *Dasytes*, 1 *Dasytina*, 6 *Haplocnemus*, 2 *Microjulistus* and 1 near *Psilothrix*, and of the Melyrinae: 3 *Melyris* and 1 *Zygia*. As the Dasytinae are the group with the most forest-like features, it is quite reasonable that it is in considerable preponderance. Earlier, Berendt (1845) among others mentioned *Malachius*, *Ebaeus* and *Dasytes*.

While the above mentioned insects, mostly beetles and fungus gnats, have lived especially in and on the fruiting bodies of the fungi, the following group lives typically in the fungus-attacked wood of the tree trunks; only a few of them have in addition representatives which prefer fungal tissue alone.

Death-watch beetles (Anobiidae). This is one of the active beetle families on this special biotope. The larvae of many species live in wood of different conifers, other in deciduous trees, and here in particular in the red-rot of oak, and this material must have been very plentiful in the amber forest. For a number of species, the systematic position of the tree is of minor significance. A few species deviate from the typical mode of existence, as their larvae have to be found under dead bark, where they live almost in the same way as *Hylurgops* larvae; this is the case in certain *Ernobius*. Others again are found breeding in thin, dry branches and twigs, a few recent species even in the stalks of fresh cones of spruce. The sub-family Dorcatominae deviates as a rule from the typical Anobiidae in their mode of existence, as many of them breed in the fruiting bodies of tree fungi. There is much to suggest that this is the original biotope of the death-watch beetles. The adult beetles do not attract much attention during their flying time, which in most cases is found within the hours of reduced daylight, but for example the recent *Dryophilus* occurs commonly in full daylight on the flowers of conifers.

From these biological data, these insects should not be expected with any frequency in amber. Nonetheless, the death-watch beetles are among the most commonly occurring beetle families in amber. To a large extent this is probably because the death-watch beetles, just like so many other insects which are dependent on the Polyporaceae, have commenced their attack at a time when the outer layer of the trees was still fully healthy, so that the tension of the sap and as a result the production of resin dependent

on this, was still more or less normal. Their actual incidence may even conceivably have been still greater than the number of amber fossils suggest, corresponding to the very rich possibilities for breeding which must have been present on many different species of woody plants.

Amber fossil Anobiidae are often mentioned in the older literature, but the more precise determination is uncertain, and no species have been described. Klebs' collection contains 236 specimens (Klebs, 1910) and the Copenhagen collection contains 36 specimens. In Klebs, the Anobiidae constitute 3.7% of all beetles, while in the Copenhagen collection they constitute 5.6%. The Anobiinae in Klebs' collection are represented as follows: 89 *Anobium*, 11 near *Anobium*, 11 *Dryophilus*, 1 near *Dryophilus*, 13 *Ernobius*, 11 *Gastrallus*, 1 near *Hedobia*, 16 *Lasioderma*, 1 near *Nicobium*, 10 *Xyletinus* and 1 near *Xyletinus*, and of the Dorcatominae: 1 *Coenocara*, 1 near *Dorcatoma*, 14 *Mesocoelopus*, 2 *Mesothes*, 12 *Rhadine* and 12 *Theca*; in addition there are 10 undetermined. The Copenhagen collection contains 4 *Anobium*, 1 *Dryophilus*, 8 *Ernobius*, 1 *Oligomerus*, 1 *Xestobium*, 13 *Xyletinus* and 1 *Dorcatoma*; in addition there are 7 undetermined. However, there is the possibility that there are faulty determinations in both collections.

Bostrychidae do not occur in Klebs' material, and the Copenhagen collection contains only 2 specimens, one of which is a *Rhizopertha* (?). The family is thus rare in amber. In older reports it is mentioned from Baltic amber by Serres (1829), Burmeister (1831, 1832) and Helm (1896c); most probably much of this material belongs to the Scolytidae. Menge (1856, p.23) mentions the find of a *Bostrychus* larva. However, as the Bostrychidae breed in deciduous hard-timber trees of various species, and not in conifers, it must be due to pure chance that a larva should be caught by the amber resin. It may be a case of a larva of an anobiid or, more probably, of a ptinid beetle.

Lyctidae. Of this family, which is very close to the Bostrychidae, only the genus *Lyctus* is mentioned in the literature (Serres, 1829, p.241), a note which demands further confirmation. It is not found in the Copenhagen collection. It breeds in particular in specially starch-containing sapwood of deciduous trees. Its occurrence in the amber forest is highly probable, but it has no relationship to *Pinites.*

Chafers (**Scarabaeoidea**) are extremely rare in amber, and are not found either in Klebs' collection or in the Copenhagen collection. Only few chafers, the most important of them Lucanidae, Dynastinae, Valginae, Trichiinae and Cetoniinae, live in forest vegetation, since they almost all breed in angiospermous woody plants. Since they are in addition almost all large and strong beetles, their absence is easily understandable and does not tell us very much with regard to their real incidence. The generic determinations appear to be somewhat uncertain. Motschulsky describes *Dorcasoides bilobus* (1856, p.27) and Zang describes a total of 5 species of *Systenocerus* and the extinct *Palaeognathus* (1905a, p.197).

Cupedidae. This archaic, highly remarkable group of tropical beetles appears to have been quite common in Central Europe in the older Tertiary. They are found for example in various fossil-bearing layers in England, and are known in amber from the description by Peyerimhoff (1909, p.57) of 3 species belonging to the genera *Cupes* and *Priacma*, as well as 7 specimens of *Cupes* in Klebs. In 1960, Yablokov-Khnzorian described *Cupes rohdendorfi* (Text-fig.55). The family does not appear in the Copenhagen collection. The strongly specialized larva lives in the hardwood of deciduous trees, so the family has not belonged to the actual fauna of the amber tree.

Lymexylonidae, which live in far harder wood than the majority of Anobiidae, are not known from the Copenhagen amber collection. Klebs mentions 5 specimens, of which Edm. Reitter has not determined the generic relationship in the one case. Two individuals are close to *Lymexylon* systematically, two others to the tropical genus *Antractocerus.* The family breeds in deciduous trees. Menge mentions (1856, p.23) the find of a *Lymexylon* larva. Even though the structure of this is very characteristic, the determination is hardly correct.

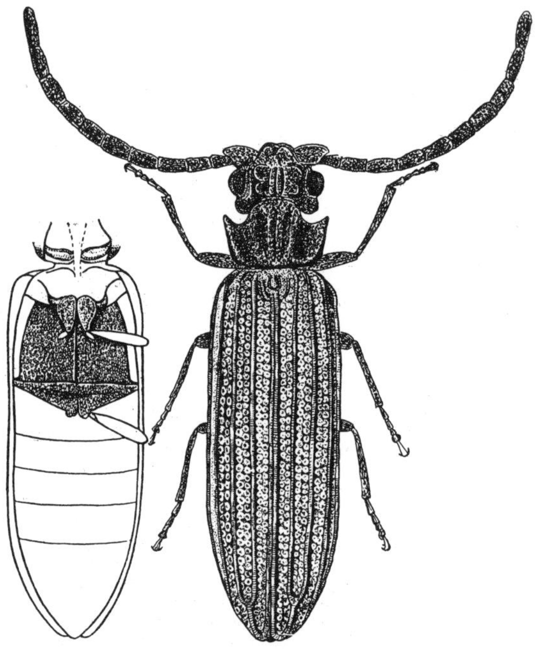

Text-fig. 55. *Cupes rohdendorfi* Yabl. (Cupedidae), ventral and dorsal views. (From Yablokov-Khnzorian, 1960, fig. 1).

Click beetles (**Elateridae**). There are few beetle families which are so uniform and characteristic in their body build, both as imagines and as larvae, and which have at the same time such different biotopes for breeding. The imagines move about freely in the vegetation. The larvae always live concealed, and dig and gnaw their way through the surrounding medium, some in the ground, others in wood or under bark.

A large proportion of them live in fungus-rotted wood.

They are among the most frequently occurring beetles in amber. Klebs' collection includes 286 specimens, while the Copenhagen collection includes 49 specimens as well as a larva. The majority of click beetles have a great tendency to fly as imagines, and this is possibly the reason why many genera are represented which have not had any direct relationship with the forest; for example, in the Klebs' collection there are no less than 34 *Limonius* as well as a number of *Cardiophorus*, *Adrastus* and *Agriotes*. 59 *Hypnoidus* and *Cryptohypnus* must be

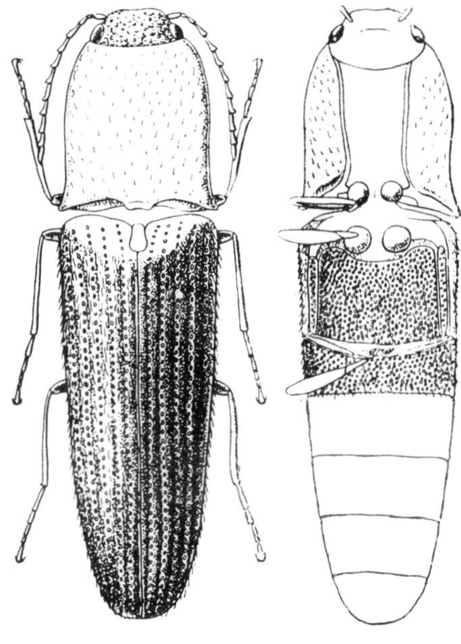

Text-fig.56. *Limonius barovskyi* Yabl. (Elateridae), dorsal and ventral views. (From Yablokov-Khnzorian, 1961b, fig.7).

assumed to originate from meadow-like regions. With regard to the genus *Athous*, the larvae of which for a very great part live in fact in the forest floor itself, there are on the other hand only 10 specimens known. Specially numerous is *Elater* (69 pieces in addition to 42 near relatives), and this has a natural explanation, since its larvae live in the fungus-attacked wood of both deciduous trees and conifers. Four *Adelocera* have

presumably lived as larvae in conifers. Most of these genera are also known from the classic literature. Of the Danish material, part is being studied at the moment by J.N.L. Stibick, U.S.A. It includes various species new to science, for example within the genera *Colaulon*, *Limonius*, *Dipropus*, *Ampedus*, *Anchastus*, *Glyphonyx*, *Silesis*, *Agriotes* and *Cardiophorus*. A taxonomic study of *Elater (Ampedus)* would most likely be of particular interest. It is probable that the *Elater* species, which have most often been rather unhindered in their choice of host plant, used as host plant for their reproduction to a large extent amber trees which were attacked by Polyporaceae.

It is remarkable that *Melanotus* is unknown from amber, and that hitherto only a single *Denticollis* has been found; from a nutritional point of view both these genera have had particularly good conditions for reproduction in the amber forest, but the larvae live preferably in rotting cambium, probably mainly as predators; this biotope, however, characterizes the dead and most often fallen trunks, where all resin production has ceased. The cause may also depend on geographical conditions, as these genera (and this also applies to *Athous)* are fairly strongly associated with boreal conditions.

Only a few species have been described from amber; *Elater naumanni* Giebel (1856, p.91); this species, however, rather belongs to *Limonius* than *Elater* (Text-fig.56). Quite recently, Yablokov-Khnzorian has described a number of species (1961b, p.84-97); Agryptinae: *Plagioraphes fasciatus;* Elaterinae: *Holopleurus succineus, Orthoraphes reichardti, Elater gebleri, Diaraphes kozhantshikovi, Elatron semenovi, Limonius barovskyi* and *Athous olgae;* Cardiophorinae; *Crioraphes rhodendorfi* and *Cardiophorus yatsenkokhmelevskyi;* incertae sedis: *Tetraraphes ebersini.* This list shows how few of the previously known generic names can be reckoned with as persisting after a taxonomic revision.

Of the **Eucnemidae** (Melasidae) there are 34 pieces in Klebs' collection and a few in Copenhagen, all imagines. The larvae of these beetles live in dead wood, often in decaying fungus-attacked trunks or heavy branches, although a few species live in twigs and thin branches. The majority of the species breed in deciduous trees, only *Eucnemis* breeds in spruce, probably also in pine and other conifers; this is the case at any rate for the recent Central European species. More than half of the Eucnemidae found in amber belong in fact to *Eucnemis* (16 specimens in addition to 2 near specimens). In addition, Klebs mentions: *Dirrhagus*, near *Dromaeolus*, *Hypocoelus*, near *Nematodes* and *Xylobius;* 7 samples belong to genera unknown to Reitter. Eucnemid larvae are usually regarded as wood-eaters.

In 1962, Yablokov-Khnzorian described the species *Throscogenius takhtajani* (p.82), which he refers to its own subfamily of Throscidae. According to Cobos (1963, p.346) this genus correctly belongs to the Eucnemidae.

Throscidae. The larvae of some recent species are reported to have a mode of life similar to that of the larvae of Eucnemidae, but there are grounds for not accepting this. It is certainly not the case for the main genus *Throscus*, the larva of which is found in damp meadow ground together with the imago. The larvae are reported to live here on grass roots. Like so many other insects from this biotope, it is relatively common in amber (14 specimens in the Klebs' collection, 12 in Copenhagen, the last-mentioned material at present being studied by Professor Eric Yensen, U.S.A.

Yablokov-Khnzorian in 1962 described *Palaeothroscus sosnovskyi* (p.83) (Text-fig.57) and *Throscites tschitscherini* (p.84), both referred by Cobos (1863) to recent genera, respectively *Throscus* and *Aulonothroscus* (p.348).

Artematopidae (Crowson, 1973, p.228). This family is referred to the very comprehensive group Elateriformia. Its

mode of life is generally speaking unknown. It is relatively common in amber and is represented by 10 imagines in the Danish collection. In the publication mentioned, Crowson has described 2 species: *Electribius oligocenicus* (p.233) (Text-fig.58), which appears to be quite common, and *Protartematopus electricus* (p.234), on a single specimen. Both holotypes are of East Prussian origin, but at any rate the genus *Electribius* occurs also in Danish amber.

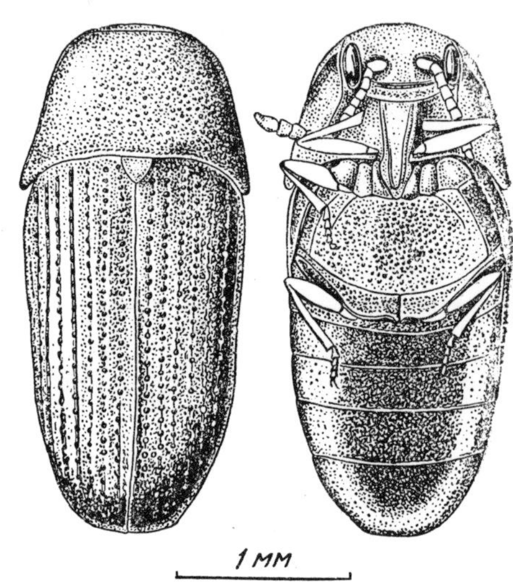

Text-fig.57. *Palaeothroscus sosnovskyi* Yabl. (Throscidae), dorsal and ventral views. (From Yablokov-Khnzorian, 1962, fig.2).

Longicorn beetles (**Cerambycidae**) have without a doubt played an important role in the forest of the amber-land. Nevertheless, pieces of amber with remains of longicorn beetles are quite uncommon. Most of the species are quite large, so it is relatively easy for them to escape from the resin by their own strength, but on the other hand, when captured, they have quickly been eaten by birds. The following is based in particular on Klebs' collection (39 pieces), the determinations in which must be regarded as rather certain (Edm. Reitter det.), but in addition on 2 pieces in the Copenhagen collection. Klebs' material comprises exclusively imagines, the Copenhagen collection only first stage larvae. In addition to these, there are known amber Cerambycidae from Motschulsky (1856), Menge (1856), Giebel (1856), Scudder (1885), Schaufuss (1891), Helm (1896c) and Zang (1905b), in addition to a number of more uncertain determination.

The fact that the sub-family Prioninae is quite unknown from amber is probably due to the often considerable size of these animals; at any rate it is known with reasonable certainty from the upper Oligocene from the Rhineland. The distribution into the other sub-families is quite surprising when compared with the conditions in the present: 25

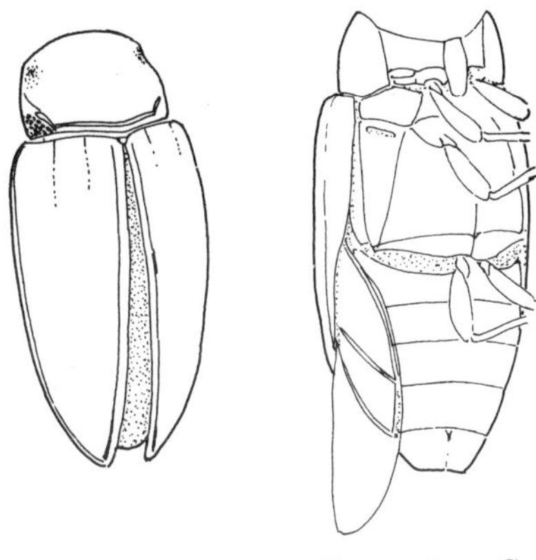

Text-fig.58. *Electribius oligocenicus* Crowson (Artematopidae), dorsal and ventral views. (From Crowson, 1973, fig.5-6).

Aseminae, 7 Cerambycinae, 7 Lepturinae and 6 Lamiinae, in addition to a number not further determined.

It is striking that Aseminae, at present a scantily represented group, as well as being to a very great extent boreal, constitutes individually more than half of the amber material hitherto determined (25 pieces out of a total of 45). In addition, the genus *Nothorrhina* Redtb. seems to dominate very strongly, as apart from this there is only known with certainty *Tetropium* (1 piece). Zang (1905b, p.236) establishes the species *Nothorrhina granulicollis*. According to Helm (1896c, p.229) one of the specimens examined by him is systematically close to the recent *Nothor-*

rhina muricata Schönherr. The species *Spondylus crassicornis* described by Giebel (1856, p.127) is possibly also a *Nothorrhina*.

At the present day, *Nothorrhina* occurs mainly in the Mediterranean countries, but is also found in the adjacent regions of Central Europe. It is thus found considerably more south than the other Aseminae. Only one existing species is known, living on pine. The imago hides in the crannies of the bark, where egg laying also takes place, but it swarms actively in full sunshine. The larva lives under bark in the cambium (Picard, 1929, p.87).

It is not stated how many species have been included in Klebs' material, but it must be reasonable to assume that at any rate one has been a common parasite on the amber tree. This is supported by the observation that the single asemin larva found in the Copenhagen amber is quite young and undoubtedly originates from an egg laid in the depth of a crevice in the bark on the amber tree. Scudder (1885) mentions a *Spondylus* larva; there can be much doubt as to the correctness of this determination, but it is probable that it has been one of the Aseminae, since these larvae are easy to distinguish from the majority of the other larvae of the Cerambycidae.

Aseminae are morphologically an archaic group within the Cerambycidae, and they are exclusively associated with the conifers. It is probable that from a stratigraphic point of view they are older than the other now-living sub-families, whose most flourishing period can only have been reached along with the deciduous trees, which constitute such an important basis for the nutrition of most of the species.

Cerambycinae have 7 representatives in the material mentioned. An imago from the Klebs' collection and probably the one larva in the Copenhagen collection belong to *Callidium*, which at the present breeds in both deciduous trees and conifers, among these some species in *Thuja* and other cypresses; this genus has had rich possibilities for breeding in the amber forest, and can quite well have lived in the amber tree. *Anaglyptus* and *Gracilia* breed today in oak, walnut and other deciduous trees; these genera have had rich possibilities in the Eocene Europe, but they have hardly been breeding in the amber tree. There are 3 specimens of *Obrium* in Klebs. This genus has at the present day both representatives which breed in conifers and species which breed in deciduous trees; whether it has lived in the amber tree is uncertain.

Among Lepturinae, Klebs mentions *Grammoptera*, *Pachyta* and *Strangalia*, Zang (1905b, p.243) describes *Strangalia berendtiana*, Helm (1896c, p.229) mentions *Leptura*, and Menge (1856, p.23) mentions a *Leptura* larva (but the determination of this must be regarded as doubtful). They are all well-known genera in the recent fauna. Lepturinae, which are associated above all with decaying and dead wood, are most often only slightly critical in their choice of host plants, merely demanding moisture in the wood, and are only found near the ground and in fallen trunks. It is therefore most likely that they have lived in all windfalls in the amber forest, also those of amber trees. The adult beetles seek the nectar of open flowers, e.g. of Umbelliferae, so that they have only had a possibility of coming into contact with the resin excretion of living amber trees during their search for a suitable site for egg-laying.

Lamiinae are not represented either in the Copenhagen collection or in the Klebs' collection. The sub-family is mentioned, however, by Berendt (1845, p.56) without any further determination, and other literature contains descriptions of the species *Dorcadionoides subaeneus* Motschulsky (1856, p.27), *Pogonochaerus jaekeli* Zang (1905b, p.233) and *Dorcaschema succineum* Zang (1905b, p.240). Menge mentions the genus *Saperda* (1856, p.21) and Scudder mentions a *Saperda* larva which, however, without doubt must be based on an erroneous determination, as *Saperda* never breeds in conifers. La-

miinae have probably been relatively uncommon in the amber territory, at any rate within the actual region of the amber forest.

Of the longicorn beetles mentioned, *Nothorrhina*, *Tetropium* and *Callidium* live as larvae in the cambium of weakened or recently dead trees, and only when they are fully developed do they form short pupal burrows within the sapwood. Numerous Aseminae likewise burrow in the bast, but only at the commencement of their development, later on they form systems of tunnels deep in the sapwood or heartwood; this developmental picture is also known from many other longicorn beetles. It seems to be the case that the longicorn beetles have primarily attacked weakened trees and lived in the cambium, and this stage is still found among the primitive genera of the various subfamilies. Secondarily, most of the genera have then become specialized as borers in the wood, both in the hardwood (e.g. many Lamiinae), and in heavily decayed wood (to a special degree Lepturinae), but without being dependent on the growth of fungi.

Buprestid beetles **(Buprestidae)**. Most of the genera belonging here breed in the cambium of tree trunks, many in deciduous trees, others in conifers, and both in their mode of existence and appearance, the larvae strongly recall the larvae of longicorn beetles. They often attack while the trees are still alive, and when the dead bark is loosened from the wood, the buprestids have long since left the site.

They are rare in amber. In Klebs' collection there are mentioned 1 *Anthaxia* and 2 *Poecilonota*, in the Copenhagen collection there are no imagines. In his list (1845, p.56) Berendt mentions the genus *Agrilus*. Yablokov-Khnzorian describes (1962, p.88) *Electrapate martynovi*, which he refers to its own family, Electrapatidae. According to Cobus (1963, p.350), however, it simply belongs to the family Buprestidae. Menge (1856, p.23) mentions 2 *Buprestis* larvae, which are not further determined. The Copenhagen collection contains a quite young larva, which with some doubt may be referred to the group of genera around *Phaenops.* Quite possibly it is newly-hatched from an egg laid in the depth of a furrow in the bark. The recent Central·European *Phaenops* species is reported to breed in spruce and pine, and it is within the bounds of possibility that the Eocene specimen has bred in the amber tree. Of the genera mentioned above, *Poecilonota* and *Agrilus* are associated with deciduous trees, while *Anthaxia* has species on both deciduous trees and conifers.

Wood-wasps **(Siricidae),** in their mode of existence, recall very much the wood-boring longicorn beetles, with the difference that the eggs are introduced deep into the wood by means of the ovipositor of the female. Two species of *Sirex* are mentioned by Klebs (1890b).

Weevils **(Curculionidae,** pars) have mainly been discussed earlier, as the great majority of the imagines move around freely on the vegetation. A number of the individuals (in Klebs' collection a total of 13 specimens), however, belong to genera which live in dead wood; most of them breed in deciduous trees, only *Magdalis* is found at the present day just as common in decaying conifers. It is characteristic that none of these genera *(Dryophthorus,* Cossinini, Calandrini, *Choerorrhinus, Mesites, Magdalis, Rhyncolus, Cryptorrhynchus* and *Acalles)* occur with more than 2 specimens in Klebs' collection; none of them have been specially associated with the living amber tree.

In addition to the above, Voss (1972, p.180) has described *Pissodes henningseni* from the Copenhagen amber. Should this species have had a mode of life like the majority of the now-living species, and this is no doubt most likely, it may be assumed to have lived as a larva in the cambium of dying Pinaceae, and the amber tree may have been among these sources of food.

Anthribidae. From earlier times, the genus *Anthribus* has been mentioned from Baltic amber (Berendt, 1845, p.56), but the name must be regarded as a collective term, covering the concept Anthribidae, also mentioned by Helm (1896c, p.228). Voss (1953) has described *Pseudomecorhis simulator.* In Klebs' collection of 10 specimens, Edm. Reitter has referred 7 to *Tropideres* and 2 in systematically close relation to this. The Copenhagen collection of amber contains 3 specimens, not further identified. Today, the larva of *Tropideres* must be sought in decaying wood of various deciduous trees. The Anthribidae found in amber thus do not appear to have had any fixed relationship with the amber tree or with other Coniferae.

The wood-boring **Platypodidae** are apparently uncommon in Baltic amber, but are more common in Quaternary copal from dicotyledonous trees, for example from South East Africa. This is a difference which appears to depend on the fact that this beetle family breed in particular in deciduous hardwood trees and not in conifers. Platypodidae are neither found in the Copenhagen collection nor in Klebs', but finds in the Baltic amber are mentioned in older literature (Guérin, 1825, p.580) and Burmeister (1831, p.1100; 1832, p.635). Schedl (1962, p.1035) has described 3 species of *Cenocephalus* from Mexican *Hymenaea* amber (Oligo-Miocene).

Bark beetles **(Scolytidae).** Wherever there has been a local drying-out or other form of weakening of the bark of the trees, e.g. an unusual exposure to sun or wind, or where there has been persistent sap-sucking by large colonies of aphids, there are good conditions for a number of bark beetle species; others appear later, when the vital layer of the tree, the cambium, has been subjected to a more advancing decay, and often live in the remains from the life activities of the species first present. It is probable that the destruction of the cambium which follows, both in the Palaeogenic as well as at the present day, has been aided by attacks of bacteria and various fungi, and has undoubtedly resulted in loss of the bark.

The bark beetles occurring in amber have been analytically studied by Schedl in two publications, an East Baltic material in 1947 and a Danish material in 1967. A total of 59 pieces had their species determined, while it was only possible to determine the generic relationships in 27 pieces on account of poor quality. The group Hylesinini includes the still living genera *Hylastes* (3 specimens) and *Hylurgops* (44 specimens), as well as the extinct *Hylescierites* (2 specimens), *Xylechinites* (1 specimen), *Charphoborites* (2 specimens) and *Phloeosinites* (21 specimens). The extinct *Taphramites* (12 specimens) is referred to Ipini, as well as the recent *Taphrorychus* (1 specimen). Scolytinae, which often attack perfectly healthy deciduous trees, are not known from amber.

In the present day forest, *Hylastes* breeds in the rootneck of dead or sick trees of the pine family, which means in recent practice in the tree stumps, but generaly each species is very closely associated with a single conifer species or genus. *Hylurgops* attacks mainly dead trunks or exceedingly weakened trees; it is associated with the pine family, but the individual species is much more freely placed in its choice of host plant within this group of plants than was the case for *Hylastes*. As a result of their gnawing out brood galleries in the bark and cambium, the bark beetles were regarded by many earlier amber investigators as an important cause of the flow of resin, which subsequently became amber. However, bark beetles are relatively rare, and *Hylurgops*, which is by far the most dominant genus, does not at the present day attack healthy trees with good capacity for the production of resin, facts which tell strongly against such a concept. The presence of *Hylastes* and *Hylurgops* in amber indicates above all the presence of a relatively rich representation of the pine family in the amber forest, particularly the actual genus *Pinus*, but their direct relationship with the living amber tree has probably been without any practical significance.

Among the beetle larvae he found in his amber material, Menge (1856, p.23) reports 4 *Curculio*, and in the same piece as one of these larvae there was found a *Hylesinus* and a *Dorcatoma* (here it must have been a question of imagines, even though this is not clear from the text). That these *Curculio* larvae should really have been Curculionidae is highly unlikely; undoubtedly they have been Scolytidae, which as a result of woodpeckers hacking on dead or dying sheets of bark have been shaken out and captured by the flow of resin below. It is possible that the larvae belonged to the most common of Schedl's genera, *Hylurgops*.

According to Schedl (1947), *Phloeosinites* is closely related to the recent genus *Phloeosinus*. *Phloeosinites* has been relatively common in the amber forest; no less than 7 species are known. However, none of these species have such an incidence as justifies referring it to the regular parasites of the amber tree. The recent *Phloeosinus* species live on *Thuja* and *Juniperus* and other trees of the cypress family, which were exceedingly common in the amber forest; it is therefore probable that *Phloeosinites* has likewise lived on various cypresses.

Taphramites has the remarkable feature that it is twice as common in the small Copenhagen amber collection (1967) as in the far greater Königsberg material which was available to Schedl prior to 1947. This increases the probability that there is actually a difference in the local origin of most of Danish amber and East Baltic amber. *Taphramites* is related to the recent genera *Thamnurgus*, the species of which breed in various herbaceous plant stalks (in other words an unusual mode of life for a bark beetle), and to *Lymantor*, which breeds in maple and other deciduous trees. Presumably, *Taphramites* has biologically recalled *Lymantor*.

The recent genus *Taphrorychus* is only known from a single piece of amber. In the present, it lives under the bark of a long series of ordinary forest trees, including oak.

Nothing can be deduced with regard to the biology of the rare genera *Hylescierites*, *Xylechinites* and *Charphoborites*, but it is unlikely that they have played any role in the natural history of the amber tree.

One of the most remarkable features of the bark beetles of the amber territory of the Eocene is the total lack of typical Ipini, as for example the genera *Ips*, *Orthotomicus* and *Pityogenes*, which dominate to such a high degree in our present day coniferous forests. Among these, *Ips* at any rate is associated to a far higher degree with spruce than with pine, and spruce was an uncommon tree in the amber forest. Equally striking is the absence of Scolytinae, among whose normal breeding trees is oak, which plays a very big role in the amber forest, but also this group has boreal features. This is a characteristic which encourages a comparison with similar peculiarities in the distribution of caddisflies, ants and aphids, as well as with palaeobotanical conditions of that period.

As a result of the activity of the bark beetles, together with the life functions of micro-organisms and fungi, there is a local break-down of the cambium of the trees, and gradually this break-down works deeper and deeper into the wood, the bark is loosened, and as a result of the hacking by birds during their search for insects, the bark falls off in larger or smaller sheets. This biotope between bark and wood contained a very rich animal life, many of which are known as amber fossils. In agreement with the fact that this fauna lives on dead or dying trees, windfalls and, perhaps in particular, stumps of fallen trees, in other words plants which themselves are not resin-producing, they are all known in only a single or at most in few specimens. Still more species of this category

of amber fauna may be expected, but hardly in large numbers of individuals. Their capture in the resin will usually not have taken place on their natural biotope, but during their stage of spreading, or where birds have hacked them free from the sites of bark beetle attacks on the weakened trees. It is a fauna which both at the present and in the past has included many saprophages, but also phytophages, which have lived on fungi and green algae; in addition there have been many predators, and among these a number which have lived especially on the larvae and pupae of the bark beetles.

A feature of practical significance has undoubtedly been that woodpeckers and other birds with corresponding breeding habits have continued to hack into the depth of the naked parts of the trunk at that time of the year when they build their nest. These nest holes have then for some years changed ownership between birds and small mammals, and have given rise to infections in the wood due to bacteria, fungi, insects and much more. Many of the insects mentioned above can also have gained access in this way. Contributing to a high degree to the destruction of the tree by this means is a fermentation and heat development which takes place in the base of the site, where nesting material, remains of food, excrement and often dead young birds undergo slow combustion in their compressed state. It is not surprising that this unusual biotope has its special fauna. Among the insects already mentioned attention might be drawn for example to species of Phoridae, Silphidae and Siphonaptera. Among the permanent inhabitants there are also many Ptinidae and Dermestidae, of which the former have lived mainly on plant remains, the latter on animal remains.

Ptinidae are first mentioned by Berendt (1845, *Ptinus*) and Helm (1896c). In Klebs' collection, according to Reitter, there were 3 *Niptus* and 16 *Ptinus*. In the Copenhagen collection there are altogether 6 specimens. In the nature of things, none of the scarab-like larvae have been found.

Dermestidae. Of this family, Berendt (1845) mentions *Dermestes* and *Anthrenus*, Helm (1896c) mentions several dermestid species. In Klebs' collection there were 2 *Attagenus*, 1 *Dermestes* and 1 *Globicornis*. The Copenhagen collection contains both imagines and larvae. Of imagines there are 2 *Dermestes*, 3 *Trinodes*, 1 near *Orphinus* and 1 undetermined, and of larvae 1 *Anthrenus* and 1 undetermined. The lively dermestid larvae have had relatively little difficulty in surmounting the edge of the nest.

Among the animals which can originate from the cambium destroyed by the bark beetles, various mites have already been mentioned, also moss scorpions, woodlice and myriapods in the most comprehensive significance of this word, in addition to ground beetles and staphylinids, together with many dipterous larvae. Associated with the dead cambium there are in addition a number of small beetles, among which most of them get their nourishment from dead or living plant cells, at times in the form of the excrement from other animals. As a rule, the plant cells are unicellular green algae or mould fungi, or they may be mycelia of various pore fungi and agarics. Most of these beetles belong systematically to the super-family Cucujoidea, which comprises what is most often designated as Clavicornia + Heteromera. These groups tell us in general nothing special with regard to the natural history of the amber tree, since in agreement with their biology they are most often neutral in their choice of host plant.

Nitidulidae are only very weakly represented, when comparisons are made with the profuseness of species in the present day (10 specimens in Klebs, 2 in the Copenhagen collection). However, many of these small animals may have been overlooked while the amber was being sorted. It is striking that only 2 specimens of *Epuraea* are known, which at the present day are very rich in species and hâve more living representatives in the dead cambium than other genera.

Rhizophagus (**Rhizophagidae**) is mentioned by Berendt (1845, p.56) and by Helm (1896c, p.227), which has escaped Bachofen-Echt's attention, possibly because it does not occur in Klebs' collection. The Copenhagen collection contains only a single specimen. As far as is known, *Rhizophagus* larvae live on other insect larvae and are often encounted in the galleries of the bark beetles.

Among **Cucujidae**, those forms living in cambium are strikingly common in comparison with the other genera. In the Klebs collection, for example, there are 8 specimens of *Silvanus* (or closely related forms), and one *Cucujus* against 8 pieces of other genera (3 *Airophilus*, 1 *Nausibius*, 1 *Platysus* and 3 undetermined).

Among **Cryptophagidae**, a large proportion of the species live on mould fungi; this applies not least to the very small *Atomaria*, of which there are 9 specimens (out of a total of 16) in Klebs' collection. There are 4 Cryptophagidae in the Copenhagen collection.

Lathridiidae are relatively frequent in amber, possibly due to their pronounced tendency to fly. They live mainly on mould fungi. Klebs' collection contains 41 specimens (3 *Corticaria*, 18 *Enicmus*, 1 *Holoparnacus*, 9 *Lathridius*, 7 *Melanophthalma*, 1 near *Rivelieria* and 2 undetermined). In the Copenhagen collection, which has not been worked up, there are 16 pieces of amber with Lathridiidae, mainly *Enicmus*.

In **Colydiidae**, Stein has described *Bothrideres kunowi* and *B. succinicola* (both 1881, p.221), a genus which at the present day occurs particularly as predators in bark beetle tunnels. However, the family is already mentioned by Berendt *(Colydium)* without any further comments. Klebs has had 14 specimens: 1 *Apistus*, 2 *Bothrideres*, 1 *Coxelus*, 2 *Diodesma*, 1 near *Endophloeus*, 1 near *Murmidius*, 2 *Synchita*, 3 *Xylolaemus* and 1 undetermined. It should be mentioned that recent *Xylolaemus* are found in particular under the bark scales on maple. The Copenhagen collection contains 9 pieces with Colydiidae, 7 of them larvae, which might suggest that also among the Eocene Colydiidae there have been species which have lived mainly on the surface of the bark of living trees, and thus also on the amber tree, hidden under the bark scales or in furrows in the bark.

Endomychidae were mentioned already by Berendt and Menge, and Motschulsky has described *Phymaphoroides antennatus* (1856, p.27). The determination of these beetles apparently presents special difficulties, at any rate Reitter hazards a firm determination of the genus in only 2 out of the 11 specimens of Klebs: 1 *Leiestes* and 1 *Mycetina*. In addition, he mentions 1 near *Hyleia*, 1 near *Mycetaea*, 1 near *Symbiotes* and a total of 6 pieces which appear quite unknown to him. The Danish collection contains 10 specimens.

Of **Aspidiophoridae** there are 2 uncertainly determined specimens in the Copenhagen collection.
A few other beetle families, known from the Baltic amber have lived mainly in rotting bark and on similar biotopes: **Clambidae** with *Clambus* (Bachofen-Echt, 1949, p.106) and 1 specimen in Copenhagen. **Hydrophilidae** with *Cercyon* Bachofen-Echt (1949, p.107), none in Copenhagen. **Histeridae** are mentioned by Berendt, Menge and Helm without any other specification than *Hister* and Histeridae. Bachofen-Echt (1949, p.107) mentions the genera *Platysoma*, *Carcinops*, *Abraeus* and *Acritus*, which are all probable. There are no specimens in Copenhagen.
Corylophidae. Klebs' collection contains a single *Corylophus*, the Copenhagen collection contains an imago and a larva, none of which have been determined.
Scaphidiidae are mentioned by Berendt (1845, p.56), but otherwise do not occur in the literature, and are thus not found in Klebs' collection. The Copenhagen collection includes a single specimen, probably belonging to *Scaphisoma* or related genus.

Salpingidae, which are closely related to Pythidae, are rare in the Baltic amber; they are mentioned by Helm (1896c, p.228), but only 4 specimens are known from the Klebs' collection (1 *Lissodema*, 2 *Salpingus* and 1 undetermined), as well as 2 specimens from Copenhagen. The Salpingidae are apparently predators, and at the present day they are found together with their larvae under dead bark, to a very pronounced degree in the tunnel systems of bark beetles. The species lives particularly in deciduous trees, but is by no means uncommon in conifers.

Pyrochroidae, cardinal beetles, live as larvae on the decaying cambium, mainly in deciduous trees. Where they are found at the present day as larvae under the dead bark, they occur most often in quite large numbers. The family is mentioned by Berendt and Helm, and Abdullah (1965, p.40) describes *Palaeopyrochroa crowsoni*.

Gall midges (**Cecidomyiidae**, pars). The Danish material shows that gall midges have played a considerable role in the life of the amber forest, being represented in 174 pieces of amber. However, when the number of gall midges mentioned in the literature is nevertheless very modest (Berendt's collection comprised only 24 specimens, 1845, p.57), the reason must be the lack of interest in these small Diptera, which have been completely submerged in the enormous number chironomids and sciarids. The earliest mention of the family originates from Burmeister, with the genus *Lasioptera* (1832, p.637). In 1850 (p.32) Loew mentions 4 species in addition to a number of genera not further discussed. The most important study to date is that by Meunier (1904b), where more than 30 new species are described.
Biologically, the gall midges vary ex-

ceedingly, but our knowledge in this field is still strongly limited, in particular of the tropical fauna. A considerable number, including the relatively primitive subfamily Lestremiinae, but also many Cecidomyiinae, live as larvae under dead bark, in decaying wood and in other plant detritus, probably mainly as saprophagous species. A very considerable number of Cecidomyiinae obtain their nourishment at the larval stage from living plant tissues, mainly as true gall formers and almost exclusively on angiosperms; also here, however, a deviant biology is found, e.g. some are parasites on other terrestrial arthropods.

In the recent boreal fauna of gall midges, the true gall formers are in the majority, even though also the saprophagous species are quite common, but in the Baltic amber the relationship appears to have been more or less the converse. Not only have Lestremiinae in particular been relatively common, but so have also the saprophagous genera of Cecidomyiinae (*Bryocrypta, Colpodia, Palaeocolpodia* (?) and *Colomyia*).

One could be tempted to the point of view that the original mode of life for the family, as saprophages in the larval stage, has been dominant, because in the Palaeogenic the group was still in the process of development, and had not reached the equilibrium known until the present. In a comparison between the fauna of the amber forest and of the present Baltic region, however, it is a question of the relationship between a subtropical and a typical boreal fauna, and this in fact may be the explanation for the difference. In the present, the gall midges have a very extensive distribution over many climatic regions, but our knowledge of them is based by far on the study of the boreal forms, and among them the the gall formers are as mentioned in the majority. On the other hand, our knowledge of this fauna in the tropics and subtropics is very modest, not least where it concerns species without any economic significance. In these moist warm forest regions, however, the saprophages have quite specially favour-

able conditions for existence on account of the rapid breakdown and conversion of organic material; it must therefore be expected that a more detailed study will show that here also in the present there is a relatively high percentage of saprophages by comparison with the boreal regions. We know in addition that also in the Palaeogenic there has been a boreal flora rich in species north of the amber forest region, both in the old and in the new world. It is reasonable to assume that this flora has had a fauna of gall midges, which has more or less corresponded to the recent boreal gall midge fauna with relatively great frequency for the true gall formers, and that the gall midges of the amber forest, in the same way as appears to be the case within other insect groups, thus represents a geographic section of the total wealth of forms of its epoch, and not a phylogenetic step in the evolution of the group.

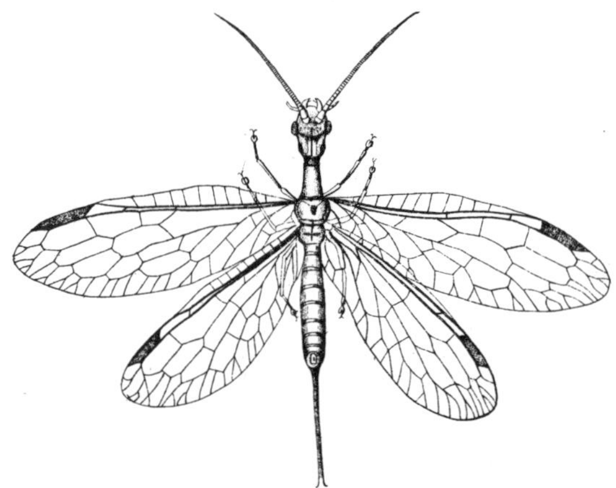

Text-fig.59. *Fibla erigena* (Menge) (Rhaphidioidea), reconstruction. (From Berendt 1856, pl.VIII, fig.14).

Snake flies (**Rhaphidioidea**) occur only rarely in amber, and are not represented at all in the Copenhagen collection. Hagen (1856, p.83) describes a single species *(Rhaphidia (Inocellia) erigena* (Text-fig.59) which is referred by Carpenter (1956, p.79) to the genus *Fibla*. In the same publication, Carpenter himself describes *Rhaphidia baltica* (p.78)

162

and *Inocellia pecularis* (p.80) (Text-fig. 60). Today *Inocellia* is holarctic, while *Rhaphidia* and *Fibla* only occur in the palaearctic; in the Miocene, at any rate, the two last-mentioned genera were also represented in the nearctic fauna (Carpenter, 1936). Together with many similar examples, this example emphasizes the reduction in distribution which has taken place in the course of the deterioration of climate in the Caenozoic.

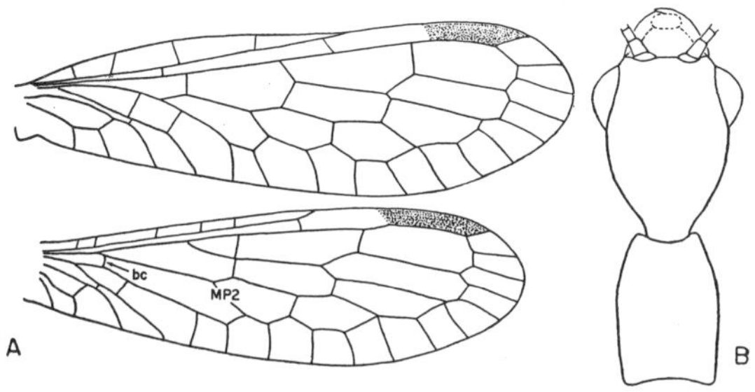

Text-fig.60. *Inocellia pecularis* Carpenter (Rhaphidioidea), A: wings, B: head and prothorax. (From Carpenter, 1956, fig.3).

Imagines must have occurred as free-living predators. The very mobile larvae, which to so high a degree recall staphylinid larvae, have probably had a mode of life with a great similarity to that of most of the Cleridae; they have mainly obtained their food under dead bark on already dead portions of various trees, presumably both deciduous trees and conifers. Just as in the present, they must have been able to change their hunting territory; two amber fossil larvae in Menge's collection (1856, p.15) and one described by Weidner (1958, p.63) suggest this.

Termites **(Isoptera)** (Plate 12) are relatively common in the Baltic amber, and this must be regarded as a confirmation of the subtropical character of the amber territory. The Copenhagen collection contains a total of 39 specimens, 29 of which have been determined by A.E. Emerson. There are 9 pieces of Prussian amber, while the remainder are of Dan-

ish origin. As the Danish collection as a whole contains only small amounts of amber which with certainty originate from Prussia (less than 10%), these figures suggest that the termites have been somewhat more frequent in the eastern region than in the western region; at any rate, the eastern amber seems to contain more termites in relation to other insects than the western amber. Even though the material is weak, it must be regarded as supporting the assumption that there has been some climatic difference between the South Scandinavian and the East Baltic localities of origin of amber.

Some of the known species of amber termites belong to the primitive families Kalotermitidae and Hodotermitidae, others belong to the somewhat less primitive Rhinotermitidae, whereas both the very primitive Mastotermitidae and the family Termitidae, so strongly dominant at the present day and numbering the most specialized forms, are completely lacking. Among the recent Termitidae there are many genera which never reside in the forest, and it is quite natural that these should not be found in amber; but why the numerous forest forms within the family and likewise the xylophagous Mastotermitidae are not represented in the material, is difficult to say. The Mastotermitidae are known from other Eocene deposits (Emerson, 1965), and with regard to the Termitidae Emerson writes (1967, p.276): "- - it is predicted that abundant fossils from tropical regions will prove the existence of numerous advanced genera of Termitidae in Cretaceous time."

Kalotermitidae and Hodotermitidae do not appear to have been very common, *Electrotermes affinis* Hagen and *Termopsis bremii* Heer, with 4 and 3 specimens, respectively, are the most numerous in the Copenhagen collection. Their society has hardly been of outstanding dimensions. Recent species establish their colonies in tree trunks of very varying character: recumbent or erect, in the process of breaking down or completely fresh, many are established in deciduous trees, some in conifers or even in the trunks of palm trees. This has probably also been the procedure in the amber forest, but the living, vital amber tree does not appear to have played any specific role for any of the species.

By far the most common amber termite is *Reticulitermes antiquus* (Germar) (Text-fig.61), described as early as 1813 (p.16); at least 22 of the Danish specimens are of this species, belonging to the Rhinotermitidae. Members of this family typically have their communities in underground nests, established in association with the root neck of the trees, and it must be assumed that this has also been the case in the Palaeogenic species. From the nest, these termites establish earth-covered tunnels on the surface of the bark, often high up in the crowns of giant trees; by this means, the light-shy workers can transport food home to the nest, all the day round and untroubled by their surroundings.

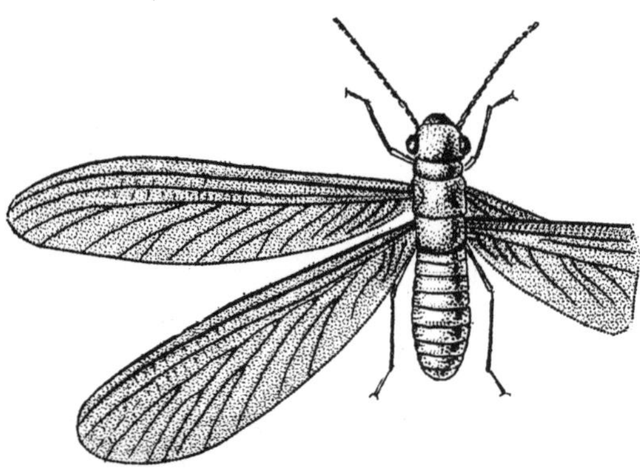

Text-fig. 61. *Reticulitermes antiquus* (Germar) (Isoptera). (From Berendt, 1856, pl.V, fig.6).

All the three termites named here are mentioned by Hagen in Berendt's book (1856). In Berendt's collection they were represented by 38, 15 and 94 specimens, which corresponds rather closely to the numerical relationships in the Danish collection. Among the newer literature with regard to termites from Baltic amber might be mentioned Rosen (1913, p.318-335), Weidner (1955, p.55-74) and Emerson (1969, p.1-57).

There can be no doubt that the termites have played a very important role in the amber forest, even though they do not appear to have had any great significance for the actual amber tree itself. In their attacks on healthy trees, the Kalotermitidae may have contributed to the decay of many old plants, and as a result of their mode of nutrition (the species found here have all been cellulose-eaters) they have together with *Reticulitermes antiquus* been very active demolishers of the quite considerable amounts of dead wood which are always present in tropical or subtropical forests, and which must return to the organic circulation as rapidly as possible.

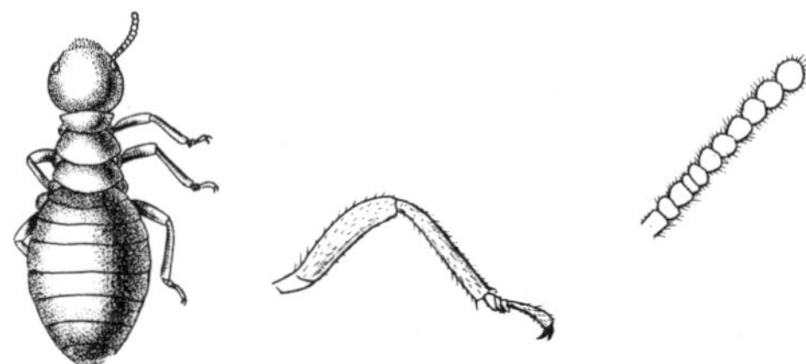

Text-fig.62. Termite-larva (Isoptera), dorsal view, hind leg and antenna. (From Berendt, 1856, pl.V, fig.1).

The great majority of inhabitants of a termite nest lead a hidden existence, which only accident or rather misfortune can interrupt in such a way that they could be captured by the amber resin. The literature also mentions and illustrates only a single worker, or possibly it is a larva (Hagen, 1856) (Text-fig.62). It must therefore be considered that these species in the amber forest have actually been represented by very much larger numbers of individuals than is suggested by the number of fossils found. Practically all known individuals are reproductive males and females, at times we only have their shed wings. Specimens are known with wings and specimens whose wings have been shed just previously, along the characteristic basal suture at the root of the wing. As the termites are very weak flyers, and as their distribution during swarming only stretches over quite short distances, except where the wind has occasionally played a special role, it might be considered certain that at any rate the 3 species mentioned above, which are relatively frequent, must have had their nests within the actual region of the amber forest.

In recent termites, pairing first takes place on the ground after the swarming, which cannot be designated as any real pairing flight. The pairs seek together on the ground or on the tree trunks, often not before the wings have been shed, and in many cases the pieces of amber enclose such a pair, which have been surprised in the course of their search for a suitable site to establish their new colony. At times a piece of amber may contain a large number of individuals (Bachofen-Echt, 1949, p.95); here, the fresh and sticky amber resin must be assumed to have been in immediate neighbourhood of that colony whose young reproductive individuals have swarmed.

References

Abdullah, M., 1964: New heteromerous beetles (Coleoptera) from the Baltic amber of eastern Prussia and gum copal of Zanzibar. — Trans. R. Ent. Soc. London 116, p.329-346.
— 1965: New Anthicidae and Pyrochroidae (Coleoptera) from the Baltic amber (Oligocene). — Entomologist 1965, p.38-42.
Alexander, C.P., 1931: Crane-flies of the Baltic amber (Diptera). — Bernstein-Forsch. 2, p.1-135.
Ander, K., 1942: Die Insektenfauna des baltischen Bernsteins nebst damit verknüpften zoogeographischen Problemen. — Lunds Univ. Årsskr. N.F. (2) 38 (4), 83 pp.
André, E., 1895: Notice sur les fourmis fossiles de l'ambre de la Baltique et description de deux espèces nouvelles. — Bull. Soc. Zool. France 20, p.80-84.
Andrée, K., 1931: Ein merkwürdiger Irrtum. Bemerkungen zu R. Shelfords zweiter Bernstein-Blattiden-Arbeit. — Bernstein-Forsch. 2.
— 1951: Der Bernstein. Das Bernsteinland und sein Leben. — Stuttgart. 96 pp.
Aycke, J.C., 1835: Fragmente zur Naturgeschichte des Bernsteins. — Danzig.
Bachofen-Echt, A.v., 1925: Mikrophotographien von Bernsteininsekten. — Paläont. Zschr. 7, p.162-163.

— 1928a: Insekteneinschlüsse im Bernstein. — Nat. u. Mus. 1928, p.299-302.
— 1928b; Farbenphotographien von im Bernstein und Kopal eingeschlossenen Insekten. — Paläont. Zschr. 9, p.283-284.
— 1928-29: Leben und Sterben im Bernsteinwald (I-III). — Palaeobiologica 1, p.39-50, 2, p.15-18, 264-269.
— 1930: Der Bernstein und seine Einschlüsse. — Verh. Zool. -bot. Ges. Wien 80, p.(35)-(44).
— 1934: Beobachtungen über im Bernstein vorkommende Spinnengewebe. — Biologia Generalis, Wien, 10. p.179-184.
— 1942: Über die Myriapoden des Bernsteins. — Palaeobiologica 7, p.394-403.
— 1944: Einschlüsse von Federn und Haaren im Bernstein. — Ibid. 8, p.113-119.
— 1949: Der Bernstein und seine Einschlüsse. — Wien, 204 pp.
Bagnall, R.S., 1914: On *Stenurothrips succineus* gen. et. sp. nov., an interesting Tertiary Thysanopteron. — Geol. Mag. (6) 1, p.483-485.
— 1923-26: Fossil Thysanoptera I-IV. — Ent. Mon. Mag. (3), 59, p.35-38; 60, p.130-133, 251-252; 62, p.16-17.
— 1924: Von Schlechtendal's work on fossil Thysanopterian in light of recent knowledge. — Ann. Mag. Nat. Hist. (9) 14, p.156-162.
Baker, A.C., 1922: Two new aphids from Baltic amber. — J. Wash. Acad. Sci. 12, p.353-358.
Bakkendorf, O., 1948: A comparison of a Mymarid from Baltic amber with a recent species, *Petiolaria anomala* (Micro-Hym). — Ent. Medd. 25, p.213-218.
Becker, E.C., 1963: Three new fossil Elaterids from the amber of Chiapas, México, including a new genus (Coleoptera). — J. Paleont. 37, p.125-128.
Beier, M., 1937: Pseudoscorpione aus dem baltischen Bernstein. — Festschr. E. Strand 2, p.302-316.
— 1947: Pseudoskorpione im baltischen Bernstein und die Untersuchung von Bernsteineinschlüssen. — Mikroskopie 1, 1946, p.188-199.
— 1955: Pseudoscorpione im baltischen Bernstein aus dem Geologischen Staatsinstitut in Hamburg. — Mitt. Geol. Staatsinst. Hamburg 24, p.48-54.
Berendt, G.C., 1830: Die Insekten im Bernstein, ein Beitrag zur Tiergeschichte der Vorwelt. — Danzig. 38 pp.
— 1945-1856: Die im Bernstein befindlichen organischen Reste der Vorwelt. I-II. — Berlin. 374 pp.
Bervoets, R., 1910: Diagnoses de quelques nouvelles espèces de Cicadines de l'ambre de la Baltique. — Ann. Mus. Nat. Hungar. 8, p.125-128.
Bischoff, H., 1915: Bernsteinhymenopteren. — Schr. phys. ökon. Ges. Königsb. 56, p.139-144.

Boesel, M.W., 1938: Insects and arachnids from Canadian amber. Chironomidae. — Univ. Toronto Stud. Geol. Ser. 40, p.44-55.

Bölsche, W., 1927: Im Bernstein-Wald. — Stuttgart. 78 pp.

Bollow, H., 1940: Die erste Helmide (Col. Dryop.) aus Bernstein. — Mitt. Münch. Entom. Ges. 30, p.117-119.

Brinck, P., 1949: Studies on Swedish stoneflies. — Opusc. Entom. Suppl. 11, Lund.

Brischke, C.G.A., 1886: Die Hymenopteren des Bernsteins. — Schr. Nat. Ges. Danzig, N.F. 6, p.278-279.

Brues, C.T., 1910: Some notes on the geological history of the parasitic Hymenoptera. — J. N. Y. Entom. Soc. 18, p.1-22.

— 1923a: A fossil genus of Dinapsidae from Baltic amber (Hymenoptera). — Psyche 30, p.31-35.

— 1923b: Two new species of Phoridae from the Baltic amber. — Ibid. 30, p.59-62.

— 1923c: Some new fossil parasitic Hymenoptera from Baltic amber. — Proc. Amer. Acad. Arts Sci. 58, p.327-346.

— 1923d: Ancient insects; fossil in amber and other deposits. — Sci. Monthly 17, p.289-304.

— 1926: A species of *Urocerus* from Baltic amber. — Psyche 33, p.168-169.

— 1932: The parasitic Hymenoptera of the Baltic amber. I. — Bernstein-Forschungen 3, p.4-178.

— 1933: Progressive change in the insect population of forests since the early Tertiary. — Amer. Naturalist 67, p.385-406.

— 1938: Insects and arachnids from Canadian amber. Ichneumonoidea, Serphoidea, and Chalcidoidea. — Univ. Toronto Stud. Geol. Ser. 40, p.27-44.

— 1939: Fossil Phoridae in Baltic amber. — Bull. Mus. Comp. Zool. Harvard 85, p.411-436.

Brues, C.T., A.L. Melander & F.M. Carpenter, 1954: Classification of insects. Keys to the living and extinct families of insects, and to the living families of other terrestrial arthropods. — Bull. Mus. Comp. Zool. Harvard 108, 917 pp.

Burmeister, H., 1831-32: Handbuch der Entomologie I-II. — Berlin.

Burr, M., 1911: Dermaptera (earwigs) preserved in amber, from Prussia. — Trans. Linn. Soc. London (2) 11, p.145-150.

Buttel-Reepen, H.v., 1906: Apistica. Beiträge zur Systematik, Biologie, sowie zur geschichtlichen und geographischen Verbreitung der Honigbiene (*Apis mellifica* L.). — Mitt. Zool. Mus. Berlin 3 (2), p.117-201.

Campbell, J.M., 1963: A fossil beetle of the genus *Hymenorus* (Coleoptera: Alleculidae) found in amber from Chiapas, Mexico. — Univ. Calif. Publ. Entom. 31 (1), p.41-42.

Carpenter, F.M., 1934: Fossil insects in Canadian amber. — Univ. Toronto Stud. Geol. Ser. 36.

— 1935a: Fossil insects in Canadian amber. — Ibid. 38, p.69.

— 1935b; Tertiary insects of the family Chrysopidae. — J. Paleont. 9.

— 1936: Revision of the Nearctic Raphidioidea (recent and fossil). — Proc. Amer. Acad. Arts Sci. 71, p.89-157.

— 1938: Insects and arachnids from Canadian amber. Introduction. — Univ. Toronto Stud. Geol. Ser. 40, p.7-13.

— 1954: The Baltic amber Mecoptera. — Psyche 61, p.31-40.

— 1956: The Baltic amber snake-flies (Neuroptera). Ibid. 63, p.77-81.

Carpenter, F.M. & W.L. Brown, 1967: The first Mesozoic ant. — Science 157, p.1038-1040.

Carpenter, F.M. & F.M. Hull, 1939: The fossil Pipunculidae. — Bernstein-Forschungen 4, p.8-17.

Carvalho, J.C.M., 1966: A new fossil genus and species of Miridae from the Baltic amber (Hemiptera). — Revta bras. Biol. 26, p.199-201.

Chopard, L., 1936a: Orthoptères fossiles et subfossiles de l'ambre et du copal. — Ann. Soc. Ent. France 105, p.375-386.

— 1936b: Un remarquable genre d'Orthoptères de l'ambre de la Baltique. — Livre Jubil. Bouvier, Paris, p.163-168.

Christiansen, K., 1963: Notes on Miocene amber Collembola from Chiapas. — Univ. Calif. Publ. Entom. 31 (1), p.46-48.

Cobos, A., 1963: Comentarios criticos sobre alcunos Sternoxia fóssilis del ámbar del Báltico recientemente descritos (Coleoptera). — Eos. Madrid 39, p.345-355.

Cockerell, T.D.A., 1909a: Descriptions of Hymenoptera from Baltic amber. — Schr. phys. ökon. Ges. Königsb. 50, p.1-20.

— 1909b: Some additional bees from Prussian amber. — Ibid. 50, p.21-25.

— 1909c: An apparently new syrphid fly from Baltic amber. — Ibid. 50, p.173.

— 1915: British fossil insects. — Proc. U.S. Nat. Mus. Wash. 49, p.469-99.

— 1916: Insects in Burmese amber. — Amer. J. Sci. 42, p.135-138.

— 1917: Insects in Burmese amber. — Ann. Entom. Soc. Amer. 10, p. 323-329.

— 1919: Two interesting insects in Burmese amber. — Entomologist 52, p.193-195.

— 1920: A therevid fly from Burmese amber. — Ibid. 53, p.169-170.

— 1922: Fossils in Burmese amber. — Nature 109, p.713-714.

— 1923: Insects in amber from South America. — Amer. J. Sci. 5, p.331-333.

Cockerell, T.D.A. & G.E. Clark, 1918: A tipulid fly from Baltic amber. — Canad. Entom. 50, p.115-116.

Cooper, K.-W., 1964: The first fossil tardigrade: *Beorn leggi* Cooper, from Cretaceous amber. — Psyche 71, p.41-48.

Crampton, G.C., 1926: The external anatomy of the primitive tanyderid dipteran *Macro-*

chile spectrum Loew, preserved in Baltic amber. — Bull. Brooklyn Entom. Soc. 21, p.1-14.

Crowson, R.A., 1964: A review of the classification of Cleroidea (Coleoptera), with descriptions of two new genera of Peltidae and of several new larval types. — Trans. R. Entom. Soc. London 116, p.275-327.

— 1965: Some thoughts concerning the insects of the Baltic amber. — Proc. XIIth Intern. Congr. Entom. London 1964, p.133.

— 1973: On a new superfamily Artematopoidea of polyphagan beetles, with the definition of two new genera from Baltic Amber. — J. Nat. Hist. 7, p.225-238.

Crowson, R.A. & M.G. de Viedma, 1964: Observations on the relationships of the genera *Circaeus* Yablok. and *Mycterus* Clairv. with a description of the presumed larva of *Mycterus* (Col. Heteromera). — Eos, Madrid 40, p.99-107.

Dampf, A., 1911: *Palaeopsylla klebsiana* n.sp., ein fossiler Floh aus dem baltischen Bernstein. — Schr. phys. ökon. Ges. Königsb. 51 (1910), p.248-259.

Davis, C., 1939a: Taxonomic notes on the Embioptera. III: The genus *Burmitembia* Cockerell. — Proc. Linn. Soc. N.S.W. 64, p.369-372.

— 1939b: Taxonomic notes on the Embioptera. XII: The genus *Haploembia* Verhoeff. — Ibid. 64, p.561-567.

Demoulin, G., 1955: Remarques critiques sur *Cronicus anomalus* (Pictet). Ephéméroptère de l'ambre oligocène de la Baltique. — Bull. Inst. R. Sci. Nat. Belg. 31(4), p.1-4.

— 1956: *Electrogenia dewalschei* n.gen. n.sp., Ephéméroptère fossile de l'ambre. Bull. Ann. Soc. R. Entom. Belg. 92, p.95-100.

— 1965: Contribution à la connaissance des Ephéméroptères de l'ambre oligocène de la Baltique. — Entom. Medd. 34, p.143-153.

— 1968: Deuxième contribution à la connaissance des Ephéméroptères de l'ambre oligocène de la Baltique. — Dtsch. Entom. Zschr. (N.F.) 15, p.233-276.

— 1970: Troisième contribution à la connaissance des Ephéméroptères de l'ambre oligocène de la Baltique. — Bull. Inst. R. Sci. Nat. Belg. 46 (2), p.1-11.

Domke, W., 1952: Der erste sichere Fund eines Myxomyceten im baltischen Bernstein. (*Stemonites splendens* Rost. fa. *succinifera* nov. foss.). — Mitt. Geol. Staatsinst. Hamburg 21, p.152-161.

Duisburg, H.v., 1863: Beitrag zur Bernstein-Fauna. — Schr. phys. ökon. Ges. Königsb. 3, 1862, p.31-36.

— 1868: 2. Zur. Bernstein-Fauna. — Ibid. 9, p.23-28.

Dybas, H.S., 1961: A new fossil feather-wing beetle from Baltic amber (Coleoptera: Ptiliidae). — Fieldiana, Chicago (Zool.) 44, p.1-9.

Eaton, A.E., 1871: A monograph on the Ephemeridae. — Trans. Entom. Soc. London 1871.

Eckstein, K., 1890: Tierische Haareinschlüsse im baltischen Bernstein. — Schr. Nat. Ges. Danzig. N.F. 7, p.90-93.

Edwards, F.W., 1929: A note on the amber moth-fly *Eophlebotomus connectens* Cockerell. — Ann. Mag. Nat. Hist. 10 (3), p.424-425.

— 1932: Culicidae. — Genera Insectorum 194.

Emerson, A.E., 1933: A revision of the genera of fossil and recent Termopsinae (Isoptera). Univ. Calif. Publ. Entom. 6, p.165-196.

— 1965: A review of the Mastotermitidae (Isoptera), including a new fossil genus from Brazil. — Amer. Mus. Nov. 2236, 46 pp.

— 1967: Cretaceous insects from Labrador 3. A new genus and species of termite (Isoptera: Hodotermitidae). — Psyche 74 (4), p.276-289.

— 1968: A revision of the fossil genus *Ulmeriella* (Isoptera, Hodotermitidae, Hodotermitinae). — Amer. Mus. Nov. 2332, 22 pp.

— 1969: A revision of Tertiary fossil species of the Kalotermitidae (Isoptera). — Ibid. 2359, 57 pp.

— 1971: Tertiary fossil species of the Rhinotermitidae (Isoptera), phylogeny of genera, and reciprocal phylogeny of associated Flagellata (Protozoa) and the Staphylinidae (Coleoptera). — Bull. Amer. Mus. Nat. Hist. 146 (3) p.243-303.

Emery, C., 1891: Le formiche dell'ambra siciliana nel museo mineralogico dell' Universita di Bologna. — Mem. R. Acc. Sci. Ist. Bologna (5) 1, p.141-165.

— 1905: Deux fourmis de l'ambre de la Baltique. — Bull. Soc. Entom. France, p.187-189.

Enderlein, G., 1900: *Epipsocus ciliatus* (Pict.) Hag., eine Psocide des Bernsteins, und die rezente peruanische *Epipsocus nepos* nov. spec. — Berlin. Entom. Zeit, 45, p.108-112.

— 1905: Zwei neue beschuppte Copeognathen aus dem Bernstein. — Zool. Anz. 29, p.576-580.

— 1910: Über die Beziehungen der fossilen Coniopterygiden zu den recenten und über *Archioconiocompsa prisca* nov. gen. spec. — Zool. Anz. 35, p.673-677.

— 1911: Die fossilen Copeognathen und ihre Phylogenie. — Paläontographica 58, p.279-360.

— 1929: Über die Drüsenhaare der Larve des fossilen *Trichadenotectum trigonoscenea* (End. 1911). — Zool. Anz. 83, p.177-180.

— 1930: Die Klassifikation der Coniopterygiden auf Grund der rezenten und fossilen Gattungen. — Arch. Klas. Phylog. Entom. 1 (2), 1929, p.98-114.

Ermisch, K., 1941: Mordelliden und Scraptiiden aus baltischem Bernstein. — Entom. Blätt. 37, p.177-185.

Erwin, T.L., 1971: Fossil tachyine beetles from Mexican and Baltic amber. With notes on a new synonymy of an extant group (Col. Carabidae). — Entom. scand. 2, p.233-236.

Essig, E.O., 1938: Insects and arachnids from Canadian amber. Aphididae. — Univ. Toronto Stud. Geol. Ser. 40, p.17-21.

Ewing, H.E., 1938: Insects and arachnids from Canadian amber. Acarina. — Ibid. 40, p.56-62.

Fennah, R.G., 1963: New fossil fulgoroid Homoptera from the amber of Chiapas, Mexico. — Univ. Calif. Publ. Entom. 31 (1), p.43-48.

Folsom, J.W., 1938: Insects and arachnids from Canadian amber. Order Collembola. — Univ. Toronto Stud. Geol. Ser. 40, p.14-17.

Friedrich, W.L., L.A. Símonarson & O.E. Heie, 1972: Steingervingar í Millilögum í Mókollsdal. (Tertiary fossils from Mókullsdalur, NW-Iceland). — Nátturufræðingurinn 42, p.4-17.

Gautier, Cl., 1947: Les insectes fossiles de l'ambre et l'antiquité. — Bull. Soc. Linn. Lyon 16, p.58-59.

Germar, E.F., 1813: Insecten in Bernstein eingeschlossen, beschrieben aus dem academischen Mineralien-Cabinet zu Halle. — Mag. Entom. 1, p.11-18.

Germar, E.F. & G.C. Berendt, 1956: Die im Bernstein befindlichen Hemipteren und Orthopteren der Vorwelt. (Herausgegeben von H. Hagen, mit Zusätzen von A. Menge). — In Berendt: Die im Bernstein befindlichen organischen Reste der Vorwelt II (1).

Giebel, C.G., 1846: Palaeozoologia. — Merseburg.

— 1856: Die Insekten und Spinnen der Vorwelt mit steter Berücksichtigung der lebenden Insekten und Spinnen; monographisch dargestellt. — Leipzig. 511 pp.

— 1862: Wirbelthier und Insektenreste im Bernstein. — Zschr. Gesamt. Naturwiss. 20, p.311-321.

Gravenhorst, J.L.C., 1835: Bericht über die in Bernstein erhaltenen Insecten der phys. — ökon. Gesellschaft zu Königsberg. — Schles. Gesellsch., Breslau, Übers. 1834, p.92-93.

Gressitt, J.L., 1963: A fossil chrysomelid beetle from the amber of Chiapas, México. — J. Paleont. 37, p.108-109.

Guérin, F.E., 1938: Note sur les insectes trouvé dans l'ambre de Sicile. — Rev. Zool. I, p.168-170.

Hagedorn, M., 1906: Borkenkäfer des baltischen Bernsteins. — Schr. phys. ökon. Ges. Königsb. 47, p.115-121.

— 1907: Fossile Borkenkäfer. — Dtsch. Entom. Zschr. 1907, p.259-261.

Hagen, H.A., 1854: Über die Neuropteren der Bernsteinfauna. — Verh. zool. — bot. Ver. Wien, 4. p.221-232.

— 1856: see Pictet-Baraban & Hagen.

— 1882: Beiträge zur Monographie der Psociden. — Stett. Entom. Zeit. 43.

Hamilton, K.G.A., 1971: A remarkable fossil Homopteran from Canadian Cretaceous amber representing a new family — Canad. Entom. 103, p.943-946.

Handlirsch, A., 1906: Fernand Meunier und seine Arbeiten über die Paläontologie der Insekten. — Wien. 10 pp.

— 1906-1908: Die fossilen Insekten und die Phylogenie der rezenten Formen. — Leipzig. 1430 pp.

— 1921: Paläontologie. — Schröder: Handb. Entom. Jena.

Handschin, E., 1926a: Die Collembolen des baltischen Bernsteins. — Zool. Anz. 65, p.179-182.

— 1926b: Revision der Collembolen des baltischen Bernsteins. — Entom. Mitt. 15, p.161-185, 211-223, 330-342.

— 1926c: Über Bernsteincollembolen. Ein Beitrag zur ökologischen Tiergeographie. — Revue suisse Zool. 33.

Hatch, M.H., 1927: A revision of fossil Gyrinidae. — Bull. Brooklyn Entom. Soc. 22, p.89-97.

Haupt, H., 1950: Die Käfer aus der eozänen Braunkohle des Geiseltales. — Geologica, Berlin 6, 168 pp.

— 1956: Beitrag zur Kenntnis der eozänen Arthropodenfauna des Geiseltales. — Nova Acta Leopoldina (N.F.) 8 (128), 90 pp.

Heer, O., 1865: Die Urwelt der Schweitz. — Zürich.

Heie, O.E., 1967: Studies on fossil aphids (Homoptera: Aphidoidea), especially in the Copenhagen collection of fossils in Baltic amber. — Spolia Zool. Mus. Haun. 26, 274 pp.

— 1968: An aphid identified with *Aphis transparens* Germar et Berendt, 1856, and some other Baltic amber aphids in Germar collections. — Stuttg. Beitr. Naturkd. 184, 7 pp.

— 1969a: Aphids in the Berendt Collection of amber fossils in Berlin (Homoptera: Aphidoidea). — Dtsch. Entom. Zschr. N.F. 16, p.175-183.

— 1969b: The Baltic amber Aphidoidea of the Geologisches Staatsinstitut of Hamburg. — Mitt. Geol. Paläont. Inst. Univ. Hamburg 38, p.143-151.

— 1970a: Lower Eocene aphids (Insecta) from Denmark. — Bull. Geol. Soc. Denmark 20, p.162-168.

— 1970b: Notes on six little known Tertiary aphids (Hem., Aphidoidea). — Ent. scand. 1, p. 109-119.

— 1971: The rediscovered types of the fossil aphids described by Germar and Berendt in 1856. — Dtsch. Entom. Zschr. (N.F.) 18, p.251-264;

— 1972: Some new fossil aphids from Baltic amber in the Copenhagen Collection (Insecta, Homoptera, Aphidoidea). — Steenstrupia 2, (17), p.247-262.

— 1976: Taxonomy and phylogeny of the fossil family Elektraphididae Steffan, 1968 (Homoptera: Aphidoidea) with a key to the species and description of *Schizoneurites obliquus* n.sp. — Ent. scand. 7, p.53-58.

Heie, O.E. & W.L. Friedrich, 1971: A fossil specimen of the North American hickory aphid (*Longistigma caryae* Harris), found in Tertiary deposits in Iceland. — Entom. scand. 2, p.74-80.

Helm, O., 1884a: Mitteilungen über Bernstein VIII. Über einige Einschlüsse im Bernstein. — Schr. Nat. Ges. Danzig (N.F.) 6, p.125-127.

— 1886: Mitteilungen über Bernstein XIII. Über die Insekten des Bernsteins. — Ibid. 6, p.267-277.

— 1896b: Tierische Einschüsse im Succinit. — Ibid 9, p.88-90.

— 1896c: Beiträge zur Kenntnis der Insekten des Bernsteins. — Ibid. 9, p.220-231.

Hennig, W., 1938: Die Gattung *Rhachicerus* Walker und ihre Verwandten im Baltischen Bernstein. — Zool. Anz. 123, p.33-41.

— 1939: Über einen Floh aus der Bernsteinsammlung des Herrn Scheele. — Arb. Morph. Tax. Entom. Berlin-Dahlem 6 (4), p.330-332.

— 1954: Flügelgeäder und System der Dipteren unter Berücksichtigung der aus dem Mesozoikum beschriebenen Fossilien. — Beitr. Entom. 4, p.245-388.

— 1964: Die Dipteren-Familie Sciadoceridae im Baltischen Bernstein. — Stuttg. Beitr. Naturkd. 127, 10 pp.

— 1965: Die Acalyptratae des Baltischen Bernstein und ihre Bedeutung für die Erforschung der phylogenetischen Entwicklung dieser Dipteren-Gruppe. — Ibid. 145, 215 pp.

— 1966a: *Fannia scalaris* Fabricius, eine rezente Art im Baltischen Bernstein? (Diptera: Muscidae). — Ibid. 150, 12 pp.

— 1966b: Dixidae aus dem Baltischen Bernstein, mit Bemerkungen über einige andere fossile Arten aus der Gruppe Culicoidea. — Ibid. 153, 16 pp.

— 1966c: Conopidae im Baltischen Bernstein (Diptera: Cyclorrhapha). — Ibid. 154, 24 pp.

— 1966d: Einige Bemerkungen über die Typen der von Giebel 1862 angeblich aus dem Bernstein beschriebenen Insektenarten. — Ibid. 162, 7 pp.

— 1966e: Spinnenparasiten der Familie Acroceridae im Baltischen Bernstein. — Ibid. 165, 21 pp.

— 1966f: Bombyliidae im Kopal und im Baltischen Bernstein (Diptera: Brachycera). — Ibid. 166, 20 pp.

— 1967a: Die sogenannten "niederen Brachycera" im Baltischen Bernstein (Diptera: Fam. Xylophagidae, Xylomyidae, Rhagionidae, Tabanidae). — Ibid. 174, 51 pp.

— 1967b: Neue Acalyptratae aus dem Baltischen Bernstein (Diptera: Cyclorrhapha). — Ibid. 175, 27 pp.

— 1967c: Therevidae aus dem Baltischen Bernstein mit einigen Bemerkungen über Asilidae und Bombyliidae (Diptera: Brachycera). — Ibid. 176, 14 pp.

— 1968: Ein weiterer Vertreter der Familie Acroceridae im Baltischen Bernstein (Diptera: Brachycera). — Ibid. 185, 6 pp.

— 1969a: Neue Gattungen und Arten der Acalyptrata. — Canad. Entom. 101, p. 589-633.

— 1969b: Neue Übersicht über die aus dem Baltischen Bernstein bekannten Acalyptratae (Diptera: Cyclorrhapha). — Stuttg. Beitr. Naturkd. 209, 42 pp.

— 1969c: Kritische Bemerkungen über die phylogenetische Bedeutung von Bernsteinfossilien: Die Gattung *Proplatypygus* (Diptera, Bombyliidae) und *Palaeopsylla* Siphonaptera). — Estr. Mem. Soc. Entom. Ital. 48, p.57-67.

— 1970: Insektenfossilien aus der unteren Kreide II: Empididae (Diptera, Brachycera). — Stuttg. Beitr. Naturkd. 214, 12 pp.

— 1971a: Neue Untersuchungen über die Familien der Diptera Schizophora (Diptera: Cyclorrhapha). — Ibid. 226, 76 pp.

— 1971b: Die Familien Pseudomyzidae und Milichiidae im Baltischen Bernstein (Diptera: Cyclorrphapha). — Ibid. 233, 16 pp.

— 1971c: Insektenfossilien aus der unteren Kreide III. Empidiformia (Microphorinae") aus der unteren Kreide und aus dem Baltischen Bernstein; ein Vertreter der Cyclorrhapha aus der unteren Kreide. — Ibid. 232, 28 pp.

— 1972a: Beiträge zur Kenntnis der rezenten und fossilen Carnidae, mit besonderer Berücksichtigung einer neuen Gattung aus Chile (Diptera: Cyclorrhapha). — Ibid. 240, 20 pp.

— 1972b: Insektenfossilien aus der unteren Kreide IV. Psychodidae (Phlebotominae), mit einer kritischen Übersicht über das phylogenetische System der Familie und die bisher beschriebenen Fossilien (Diptera). — Ibid. 241, 69 pp.

— 1972c: Eine neue Art der Rhagionidengattung *Litoleptis* aus Chile, mit Bemerkungen über Fühlerbildung und Verwandtschaftsbeziehungen einiger Brachycerenfamilien (Diptera: Brachycera). — Ibid. 242, 18 pp.

Hills, E.S., 1957: Fossiliferous Tertiary resin from Allendale, Victoria. — Proc. R. Soc. Victoria (N.S.) 69, p.15-20.

Hoch, E., 1976: Amniote remnants from the eastern part of the Lower Eocene North Sea basin. — Proc. Centre natl. Rech.

Holland, G.P., 1951: Insects in Canadian amber. — Newsl. Entom. Div. Dept. Agric. Canada, 1951.

Hope, F.W., 1834: Observations on succinitic insects. — Trans. Entom. Soc. London 1, p.133-147.

Horn, W., 1906: Über das Vorkommen von *Tetracha carolina* L. im preussischen Bernstein und die Phylogenie der *Cicindela*-Arten. — Dtsch. Entom. Zschr. p.329-336.

Houlbert, C., 1912: Contribution à l'étude des larves de Cicindélides. — Insecta Rennes 2, 1912, p.1-15.

Hull, F.M., 1945: A revisional study of the fossil Syrphidae. — Bull. Mus. Comp. Zool. 95, p.215-355.

— 1958: Tertiary flies from Colorado and the Baltic amber. — Psyche 64 (1957), p.37-45.

Hurd, P.D., R.F. Smith & J.W. Durham, 1962: The fossiliferous amber of Chiapas, México. — Ciencia (Méx.), 21, p.107-118.

Hurd, P.D., R.F. Smith & R.L. Usinger, 1958: Cretaceous and Tertiary insects in Arctic and Mexican amber. — Proc. 10th Intern. Congr. Entom. Montreal 1956, 1, p.851.

Jakobi, A., 1938: Eine neue Bernsteinzikade. — Sb. Ges. Nat. Freunde p.188-189.

Jerichin, V.V. & J.D. Sukatscheva, 1971: O Melovych Nassekomonossych "Jantarjach" (Retinitach) Severa Sibiri. — Akad. Nauk SSSR, Leningrad, p. 3-48 (in Russian).

Jordan, K.H.C., 1952: *Nabis succini* n.sp. Eine Nabide aus den Bernstein Ostpreussens (Hemiptera Heteroptera). — Beitr. Entom. Berlin 2, p.455-457.

Just, J., 1974: On *Palaeogammarus* Zaddach, 1864, with a description of a new species from western Baltic amber (Crustacea, Aphipoda, Crangonychidae). — Steenstrupia 3, p.93-99.

Karsch, F., 1884: Neue Milben im Bernstein. — Berlin. Entom. Zschr. 28, p.175-176.

Keilbach, R., 1937: Neue Forschungen über samländische Bernsteineinschlüsse. — Naturforsch. 13, p.398-400.

— 1939: Neue Funde des Strepsipterons *Mengea tertiaria* Menge im baltischen Bernstein. — Bernstein-Forschungen 4, p.1-7.

Kelner-Pillault, S., 1970a: L'ambre Balte et sa fauna entomologique avec description de deux Apoides nouveaux. — Ann. Soc. Entom. France (N.S.) 6, p.3-24.

— 1970b: Une Mélipone de l'ambre Balte. — Ibid. 6, p.437.

— 1974: État d'évolution des Apides de l'ambre Balte. — Ibid. 10, p.623-634.

Kerville, H. Gadeau de, 1893: Note sur les Thysanoures fossiles du genre *Machilis* et description d'une éspèce du succin (*Machilis succini* G. de K.). — Ann. Soc. Entom. France 62 (1893), p.463-466.

Kinsey, A.C., 1919: Fossil Cynipidae. — Psyche 26, p.44-49.

— 1938: Insects and arachnids from Canadian amber. Cynipidae. — Univ. Toronto Stud. Geol. Ser. 40, p.21-27.

Klausnitzer, B., 1976: Neue Arten der Gattung *Helodes* Latreille aus Bernstein (Coleoptera, Helodidae). — Reichenbachia 16, p.53-61.

Klebs, R., 1886: Gastropoden im Bernstein. — Jahrb. Preuss. Geol. Landesanst. 1885, p.366-394.

— 1890b: Über die Fauna des Bernsteins. — Versamml. Dtsch. Naturf. 62, p.268-271. (Engl. transl.: Ann. Mag. Nat. Hist. (6) 6, p.486-491).

— 1910: Über Bernsteineinschlüsse im allgemeinen und die Coleopteren meiner Bernsteinssammlung. — Schr. phys. ökon. Ges. Königsb. 51, p.217-242.

Koch, C.L. & G.C. Berendt, 1854: Die im Bernstein befindlichen Crustaceen, Myriapoden, Arachniden und Apteren der Vorwelt (veröffentlicht und mit Zusätzen versehen von A. Menge). — In Berendt: Die im Bernstein befindlichen organischen Reste der Vorwelt. Berlin, 124 pp.

Kolbe, H., 1927: Verfleichender Blick auf die rezente und fossile Insektenwelt Mitteleuropas, und eine Erinnerung an meine Abhandlung über "Problematische Fossilien aus dem Culm". — Dtsch. Entom. Zschr. 1925, p.147-162.

— 1932: Über das Verhältnis der Coleopterenfauna Zentraleuropas der Jetztzeit zur Tertiärzeit. — Entom. Bl. 28, p.147-154.

Kolbe, N.J., 1883: Neue Beiträge zur Kenntnis der Psociden der Bernstein-Fauna. — Stett. Entom. Zeit. 44, p.186-191.

Kolenati, F.A., 1851: Über Phryganiden im Bernstein. — Abh. Böhm. Ges. (5) 6, 15 pp.

Konow, F.W., 1897: Ueber fossile Blatt — und Halmwespen. — Entom. Nachr. 23.

Kornilowitsch, N., 1903: Hat sich die Struktur quergestreifter Muskeln bei Insekten, die wir im Bernstein antreffen, erhalten? — Sb. Naturf. Ges. Dorpat 13, 1902, p.198-206. (In Russian).

Krüger, L., 1923: Neuroptera succinica baltica. Die im baltischen Bernstein eingeschlossenen Neuropteren des Westpreussischen Provinzialmuseums in Danzig. — Stett. Entom. Zeit. 84, p.68-92.

Kuznetsov, N.J., 1941: A revision of the amber Lepidoptera. — Ed. Acad. Sci. USSR, Moscow-Leningrad. 136 pp.

Kühne, W.G., L. Kubig & T. Schlüter, 1973: Eine Micropterygide (Lepidoptera, Homoneura) aus mittelcretazischem Harz Westfrankreichs. — Mitt. Dtsch. Entom. Ges. 32 (3-4), p.61-64.

Larsen, C.S., 1917: Fortegnelse over Danmarks Microlepidoptera. — Entom. Medd. 11 (2), p.28-319.

Larsson, S.G., 1975: Palaeobiology and mode of burial of the insects of the Lower Eocene Mo-clay of Denmark. — Bull. Geol. Soc. Denmark 24, p.65-81.

Lazell, J.D., 1965: An *Anolis* (Sauria, Iguanidae) in amber. [Mexican amber]. — J. Paleont. 39 (3), p.379-382.

Lengerken, H.v., 1913: Etwas über den Erhaltungszustand von Insekteninklusen im Bernstein. — Zool. Anz. 41, p.284-286.
— 1921: Über den Erhaltungszustand von Bernsteininklusen. — Sb. Ges. Naturf. Freunde, Berlin, p.84-86.
— 1922: Über fossile Chitinstrukturen. — Verh. Dtsch. Zool. Ges. 6, p.73-74.
— 1923: Über Widerstandsfähigkeit organischer Substanzen gegen natürliche Zersetzung. — Biol. Zentralbl. 43, p.546-555.
Lindroth, C.H., 1974: On the elytral microsculpture of Carabid beetles (Col. Carabidae). — Entom. scand. 5, p.251-64.
Loew, H., 1850: Über den Bernstein und die Bernsteinfauna. — Program Realschule Meseritz, 1950, p.1-48.
— 1861: Über die Dipterenfauna des Bernsteins. — Amtl. Ber. Vers. Dtsch. Naturf. 35 (1860), p.88-98.
— 1864: On the Diptera or two-winged insects of the amber-fauna. — Amer. J. Sci. (2) 37, p.305-324.
Lucks, R., 1927: *Palaeogammarus balticus*, nov.sp., ein neuer Gammaride aus dem Bernstein. — Schr. Nat. Ges. Danzig (N.F.) 8 (3), p.1-13.
Lühe, M., 1904: Säugetierhaare im Bernstein. — Schr. phys. ökon. Ges. Königsb. 45, p.62-63.
Luna de Carvalho, E., 1961: Révision des Paussides appartenant aux tribus des Pentaplatarthrini et Hylotorini (Col. Carabidae Paussinae). — Rev. Zool. Bot. Afr., Brussels 63, p.1-21.
MacKay, M.R., 1969: Microlepidopterous larvae in Baltic amber. — Canad. Entom. 101, p.1173-1180.
— 1970: Lepidoptera in Cretaceous amber. — Science 167, p.379-380.
MacLeod, E.G., 1970: The Neuroptera of the Baltic amber. I. Ascalaphidae, Nymphidae, and Psychopsidae. — Psyche 77, p.147-180.
Manning, F., 1961: A new fossil bee from Baltic amber. — Verh. XI. Intern. Kongr. Entom. Wien 1960, p.306-308.
Masner, L., 1969a: The geographic distribution of recent and fossil Ambrositrinae (Hymenoptera, Proctotrupoidea: Diapriidae). — Ber. 10. Wandervers. Dtsch. Entom. 1965, Dresden, p.105-109.
— 1969b: A scelionid wasp surviving unchanged since Tertiary. (Hymenoptera: Proctotrupoidea). — Proc. Entom. Soc. Wash. 71, p.397-400.
Mathiesen, Fr. J., 1967: Notes on some fossil Phryganidean larval tubes from the Tertiary of Denmark and Greenland. — Bull. Geol. Soc. Denmark 17, p.90-94.
Mayr, G.L., 1868: Die Ameisen des baltischen Bernsteins. — Beitr. Naturkd. Preuss. Königsb. 1, 102 pp.
McAlpine, J.F., 1970: First record of Calypterate flies in the Mesozoic era (Diptera, Calliphoridae). — Canad. Entom. 102, p.342-346.
— 1973: A fossil Ironomyiid fly from Canadian amber (Diptera: Ironomyiidae). — Ibid. 105, p.105-111.
McAlpine, J.F. & J.E.H. Martin, 1966: Systematics of Sciadoceridae and relatives with descriptions of two new genera and species from Canadian amber and erection of the family Ironomyiidae (Diptera: Phoridae). — Ibid. 98, p.527-544.
— 1969a: Canadian amber. — The Beaver, 1969, p.28-37.
— 1969b: Canadian amber. — a paleontological treasure-chest. — Canad. Entom. 101, p.819-838.
Menge, A., 1855: Über die Scheerenspinnen, Chernetidae. — Schr. Nat. Ges. Danzig 4 (2), p.1-42.
— 1856: Lebenszeichen vorweltlicher, im bernstein eingeschlossener thiere. — Programm Petrischule, Danzig, p.1-32.
— 1866: Über ein Rhipidopteron und einige andere im Bernstein eingeschlossene Tiere. — Schr. Nat. Ges. Danzig (N.F.) 1 (3-4), p.2-5.
— 1869: Über einen Scorpion und zwei Spinnen im Bernstein. — Schr. Nat. Ges. Danzig (N.S.) 2 (10), 9 pp.
Meunier, F., 1892a: Note sur les Leptidae de l'ambre. — Bull. Soc. Entom. France (1892), p. LXXXIII.
— 1892b: Aperçu des genres de Dolichopodidae de l'ambre suivi du Catalogue bibliographique des Diptères fossiles de cette Résine. — Ann. Soc. Entom. France, p.377-384.
— 1894a: Note sur quelques Tipulidae de l'ambre Tertiaire (Dipt.). — Bull. Soc. Entom. France 1894, p.CLXXVII-CLXXVIII.
— 1894b: (Dolichopodidae). — Bull. Soc. Entom. France 1894, p.111.
— 1894c: Note sur quelques Mycetophilidae, Chironomidae, et Dolichopodidae de l'ambre Tertiaire. — Ann. Soc. Entom. France 1894, p.22.
— 1894d: Note sur les Mycetophilidae de l'ambre Tertiaire. — Wien. Entom. Zeit. 13, p.62-64.
— 1895a: Observations sur quelques Diptères Tertiaires et catalogue bibliographique complet sur les insectes fossiles de cet ordre. — Ann. Soc. Sci. Brux. 19, p.1-16.
— 1895b: Note sur quelques Empidae et Mycetophilidae et un curieux Tipulidae de l'ambre Tertiaire. — Bull. Soc. Entom. France, p.XII-XV.
— 1899a: Notes sur les Dolichopodidae de l'ambre Tertiaire. — Bull. Soc. Entom. France, 1899, p.322.
— 1899b: Etude de quelques Diptères de l'ambre Tertiaire. — Ibid., 1899, p.334-335, 358-359, 392-393.
— 1899c: Révision des Diptères fossiles types

de Loew conservés au Musée Provincial de Koenigsberg. — Entom. Misc. 7, p.161-165, 169-182.

— 1900: Etudes de quelques Diptères de l'ambre Tertiaire. — Bull. Soc. Entom. France 1900, p.111.

— 1901: Nouvelles recherches sur quelques Cecidomyiidae et Mycetophilidae de l'ambre. — Ann. Soc. Sci. Brux. 25.

— 1902a: Etudes de quelques Diptères de l'ambre. — Ann. Soc. Sci. Nat. Zool. Paris (8) 16, p.395-406.

— 1902b: Déscription de quelques Diptères de l'ambre. — Ann. Soc. Sci. Brux. 26, p.96-104.

— 1903a: Les Pipunculidae de l'ambre. — Rev. Sci. Bourbon, p.148-151.

— 1903b: Un nouveau genre de Sciaridae de l'ambre. — Ibid. 16, p.165-167.

— 1904a: Contribution à la faune des Helomyzinae de l'ambre de la Baltique. — Feuille Jeun. Nat. 35, p.21-27.

— 1904b: Monographie des Cecidomyiidae, Sciaridae, Mycetophilidae et Chironomidae de l'ambre de la Baltique. — Ann. Soc. Sci. Brux. 28, p.13-276.

— 1904c: (Scatopsidae, Simuliidae). — Jahrb. Preuss. Geol. Landesanst. Berlin, 24, p.392-395.

— 1905a: Sur un curieux Psychodidae de l'ambre de la Baltique. — Misc. Entom. 13, p.49-51.

— 1905b: Contribution à la faune acalyptères Agromyzinae de l'ambre. — Ann. Soc. Sci. Brux. 29, p.89-94.

— 1905c: Monographie des Psychodidae de l'ambre de la Baltique. — Ann. Mus. Hungar. 3, p.235-255.

— 1906a: Un nouveau genre de Psychodidae et une nouvelle espèce de *Dactylolabis* (Tipulidae) de l'ambre de la Baltique. — Naturaliste 28, p.103-104.

— 1906b: Les Tipulidae de l'ambre de la Baltique. — Ann. Soc. Sci. Brux. 36, p.213-215.

— 1906c: Monographie des Tipulidae et des Dixidae de l'ambre de la Baltique. — Ann. Sci. Nat. (Zool.) (9) 4, p.349-401.

— 1906d: Les Dolichopodidae de l'ambre de la Baltique. — C.R. Acad. Sci., Paris, 143, p.617-618.

— 1907-08: Monographie des Dolichopodidae de l'ambre de la Baltique. Naturaliste, 29-30.

— 1908a: Monographie des Empidae de l'ambre de la Baltique et catalogue bibliographique complet sur les Diptères fossiles de cette résine. — Ann. Sci. Nat. (Zool.) (9) 7, p.81-135.

— 1908b: Les Asilidae de l'ambre de la Baltique. — Bull. Soc. Entom. France (1908), p.18-20.

— 1908c: Les Phoridae et les Leptidae de l'ambre de la Baltique. — Comptes Rend. Acad. Sci. Paris 147, p.1362-1363.

— 1909a: Sur quelques Diptères (Xylophagidae, Therevidae, Arthropidae, Stratiomyidae, Tanypezinae et Ortalinae) de l'ambre de la Baltique. — Ann. Soc. Sci. Brux. 32, p.258-266.

— 1909b: Les Phoridae et les Leptidae de l'ambre de la Baltique. — Ibid. 33, p.97-101.

— 1910a: Sur un Cyrtidae de l'ambre de la Baltique. — Bull. Soc. Entom. France (1910) p.177-179.

— 1910b: Les Stratiomyidae de l'ambre de la Baltique. — Ibid. (1910) p.199-201.

— 1910c: Un Bombyliidae de l'ambre de la Baltique. — Ibid. (1910), p.349-350.

— 1912a: Monographie der Leptiden und Phoriden des Bernsteins. — Jahrb. Preuss. Geol. Landesanst. Berlin, (2) 30 (1909), p.63-90.

— 1912b: Coup d'oeil rétrospectif sur les Diptères du succin de la Baltique. — Ann. Soc. Sci. Brux. 36, Mém., p.160-186.

— 1916a: Sur quelques Diptères (Bombyliidae, Leptidae, Dolichopodidae, Conopidae et Chironomidae) de l'ambre de la Baltique. — Tijdschr. Entom. 59, p.274-286.

— 1916b: Beitrag zur Monographie der Mycetophiliden und Tipuliden des Bernsteins. — Zschr. Dtsch. Geol. Ges. 68, p.477-498.

— 1917: Über einige Mycetophiliden und Tipuliden des Bernsteins nebst Beschreibung der Gattung *Palaeotanypeza* derselben Formation. — N. Jahrb. Mineral. Geol. Paläont. 1917, p.73-106.

— 1922-23: Nouvelle contribution à la monographie des "Mycetophilidae" (Ceroplatinae, Mycetophilinae et Sciophilinae) de l'ambre de la Baltique. — Rev. Sci. Bourbon. 1922 (4), p.114-120; 1923 (1), p.14-34.

Meyer, A.B., 1887: Notiz über in Ostsee-Bernstein eingeschlossene Vogelfedern. — Schr. Nat. Ges. Danzig, N.F. 6, p.206-208.

Motschulsky, V. de, 1856: Lettre à Ménétriés, No. 4. — Etudes Entom. 5, p.21-38.

Muesebeck., 1960: A fossil Braconid wasp of the genus *Ecphylus* (Hymenoptera) [From Chiapas amber]. — J. Paleont. 34, p.495-496.

— 1963: A new Ceraphronid from Cretaceous amber (Hymenoptera: Proctotrupoidea). — Ibid. 37, p.129-130.

Murdoch, J., 1934: Amber in California. — J. Geol. 42 (3), p.309-310.

Olfers, E.v., 1908: Die "Ur-Insekten" (Thysanura und Collembola) im Bernstein. — Schr. phys. ökon. Ges. Königsb. 48, 1907, p.1-40.

— 1911: Ein neuer Thysanure im Bernstein. — Berlin. Entom. Zschr. 56, p.151-152.

Paramonow, S.J., 1938: Zur Entomofauna des Bernsteins. — Trav. Inst. Zool. Biol. Acad. RSS. Ukraine 18 (1936), p.53-64.

— 1939: Kritische Übersicht der gegenwärtigen und fossilen Bombyliiden-Gattungen der ganzen Welt. — Rep. Inst. Zool. Biol. Kiew, 23, (1938), p.23-50.

Parsons, C., 1939: A ptiliid beetle from Baltic amber in the Museum of Comparative Zoology. — Psyche 46, p.62-64.

Patch, E.M., 1938: Food-Plant catalogue of aphids of the world. — Maine Agric. Exp. Sta. Bull. 393. 430 pp.

Petersen, A., 1924: Bidrag til de danske Simuliers Naturhistorie. — D. Kgl. Danske Vid. Selsk. Skr., Nat. — math. Afd. (8) 4, p.235-341.

Peterson, B.V., 1975: A new cretaceous Bibionid from Canadian amber (Diptera: Bibionidae). — Canad. Entom. 107, p.711-15.

Petrunkevitch, A., 1942: A study of amber spiders. — Trans. Conn. Acad. Arts Sci. 34, p.119-464.

— 1946: Fossil spiders in the collection of the American Museum of Natural History. — Amer. Mus. Nov. 1328, 36 pp.

— 1950: Baltic amber spiders in the Museum of Comparative Zoology. — Bull. Mus. Comp. Zool. Harv. 103, p.259-337.

— 1957: *Eohelea stridulans* n.gen., n.sp., a striking example of paramorphism in an amber biting-midge. — J. Paleont. 31, p.208-214.

— 1958: Amber spiders in European collections. — Trans. Conn. Acad. Arts Sci. 41, p.41-400.

— 1963: Chiapas amber spiders. — Univ. Calif. Publ. Entom. 31 (1), p.1-60.

Peus, F., 1968: Über die beiden Bernstein-Flöhe (Insecta, Siphonaptera). — Paläont. Z. 42, p.62-72.

Peyerimhoff, P.de, 1909: Le *Cupes* de l'ambre de la Baltique. — Bull. Soc. Entom. France, 1909, p.57-60.

Picard, F., 1929: Conditions de la répartition géographique des Coléoptères Cérambycides en France. — C.R. Soc. Biogéogr. Paris 6, p.85-88.

Pictet, F.J., 1854: Traité de Paléontologie ou histoire naturelle des animaux fossiles considérés dans leur rapports zoologiques et géologiques. — Paris.

Pictet-Baraban, F.J. & H. Hagen, 1856: Die im Bernstein befindlichen Neuropteren der Vorwlt (mit Zusätzen von A. Menge). — In Berendt: Die im Bernstein befindlichen organischen Reste der Vorwelt II (2), p.41-125.

Pietrzak, Teresa, 1972: Baltic Sea amber in Polish literature (1534-1967) (in Polish). — Polska Akad. Nauk, Muz. Ziemi, Warsz.

Ping, C., 1931: On a Blattoid insect in the Fushun amber. — Bull. Geol. Soc. China 11, p.205.

Presl, J.S., 1822: Additamenta ad Faunam protogaeam, sistens descriptiones aliquot animalium in succino inclusorum. — Deliciae Prag. Hist. Nat. Spect. 1.

Priesner, H., 1924: Bernstein-Thysanopteren. — Entom. Mitt. 13, p.130-151.

— 1929: Bernstein-Thysanopteren II. — Bernstein-Forsch. 1, p.111-138.

Protescu, O., 1937: Etude géologique et paléobiologique de l'ambre roumain; les inclusions organique de l'ambre de Buzau; première partie. — Soc. Romana Geol. B, 3, p.65-110.

Quate, L.W., 1961: Fossil Psychodidae in Mexican amber. Part. 1. — J. Paleont. 35, p.949-951.

— 1963: Fossil Psychodidae in Mexican Amber, part 2. — Ibid. 37, p.110-118.

Quiel, G., 1911: Bemerkungen über Coleopteren aus dem baltischen Bernstein. — Berlin. Entom. Zeit. 55, p.181-192.

Rebel, H., 1934a: Microlepidopteren aus dem baltischen Bernstein. — Forsch. Fortschr. 10, p.372-373.

— 1934b: Bernstein-Lepidopteren. — Paläobiol. 6, p.1-16.

— 1935: Bernstein-Lepidopteren. — Dtsch. Entom. Z. Iris, 49, p.162-186.

— 1936: Mikrolepidopteren aus dem baltischen Bernstein. — Naturw. 24, p.519-520.

— 1937: Zur Systematik der Bernstein-Lepidopteren. — Z. Österr. Entom. Ver. 22, p.1-3.

Reis, O.M., 1890: Über eine Art Fossilisation der Muskulatur. — Mitt. Ges. Morph. Physiol. München, p.1-6.

— 1893-98: Untersuchungen über die Petrifizierung der Muskulatur. — Arch. Mikr. Anat. Entw. Gesch. 41, 44 & 52.

Richards, W.R., 1966: Systematics of fossil aphids from Canadian amber (Homoptera: Aphididae). — Canad. Entom. 98, p.746-760.

Romer, A.S., 1946: Vertebrate Paleontology. 2. edit. — Chicago. 687 pp.

Rosen, K.v., 1913: Die fossilen Termiten: eine kurze Zusammenfassung der bis jetzt bekannten Funde. — Trans. II. Intern. Congr. 1912, p.318-335.

Ross, E.S., 1956: A new genus of Embioptera from Baltic amber. — Mitt. Geol. Staatsinst. Hamburg 25, p.76-81.

Ross, H.H., 1958: The Cretaceous caddisfly, *Dolophilus praemissus* Cockerell. — Proc. 10th Intern. Congr. Entom. Montreal 1956, 1, p.849.

Saalas, U., 1923: Die Fichtenkäfer Finnlands II.)-Ann. Acad. Sci. Fenn. (A), 22.

Sabrosky, C.W., 1963: A new Acalyptrate fly from Tertiary amber of México (Diptera: Milichiidae). — J. Paleont. 57, p.119-120.

Salt, G., 1931: Three bees from Baltic amber. — Bernstein-Forsch. 2, p.136-147.

Sandberger, F.v., 1886: Bemerkungen über einige Heliceen im Bernstein der preussischen Küste. — Schr. Nat. Ges. Danzig, N.F. 6, p.137-141.

Sanderson, M.W. & T.H. Farr, 1960: Amber with insect and plant inclusions from the

Dominican Republic. — Science 131. p.1313.

Saunders, W.B. et al., 1974: Fossiliferous amber from the Eocene (laiborne) of the Gulf Coastal Plain. — Geol. Soc. Amer. Bull. 85, p.979-984.

Schaufuss, C., 1891: Preussens Bernstein-Käfer. Neue Formen aus der Helm'schen Sammlung im Danziger Provinzialmuseum. — Berlin. Entom. Zschr. 36, p.53-64.

— 1896: Preussens Bernsteinkäfer II. Neue Formen der Helm'schen Sammlung im Danziger Provinzialmuseum. — Ibid. 41, p.51-54.

Schaufuss, L.W., 1888: Einige Käfer aus dem Baltischen Bernstein. — Ibid. 32, p.266-270.

— 1890a: Die Scydmaeniden des baltischen Bernsteins. — Nunquam otiosus 3, p.561-586.

— 1890b: Systemschema der Pselaphiden, ein Blick in die Vorzeit, in die Gegenwart und die Zukunft. — Tijdschr. Entom. p.101-162.

Schedl, K.E., 1947: Die Borkenkäfer des baltischen Bernsteins. — Zentralbl. Gesamtg. Entom. 2 (1), p.12-45.

— 1962: New Platypodidae from Mexican amber. — J. Paleont. 36, p.1035-1038.

— 1967: Bernsteinborkenkäfer aus dem Zoologischen Museum der Universität Kopenhagen. — Entom. Medd. 35, p.85-87.

Schlechtendal, D.v., 1887: Physopoden aus dem Braunkohlengebirge von Rott am Siebengebirge. — Zschr. Naturwiss. 60, p.551-592.

— 1888: Mitteilungen über die in der Sammlung aufbewahrten Originale zu Germar's: "Insekten im Bernstein eingeschlossen" mit Rücksicht auf Giebel's "Fauna der Vorwelt". — Zschr. Naturwiss. 61, p.473-491.

Schlee, D., 1970: Insektenfossilien aus der unteren Kreide. — 1. Verwandtschafts-forschung an fossilen und rezenten Aleyrodina (Insecta, Hemiptera). — Stuttg. Beitr. Naturkd. 213, 72 pp.

Schlee, D. & H.G. Dietrich, 1970: Insektenführender Bernstein aus der Unterkreide des Libanon. — N. Jahrb. Geol. Paläont. Mh. 1970, p.40-50.

Schlotheim, E.v., 1820: Die Petrefactenkunde. — Gotha.

Schlüter, T., 1975: Nachweis verschiedener Insecta-Ordines in einem mittelkretazischen Harz Nordwestfrankreichs. — Entom. Germanica 1 (2), p.151-161.

Schubert, K., 1939: Mikroskopische Untersuchungen pflanzlicher Einschlüsse des Bernsteins. I. Holz. — Bernstein-Forschungen 4, p.23-44.

Scudder, S.H., 1885: Systematische Übersicht der fossilen Myriopoden, Arachnoideen, und Insekten. — Zittel: Handb. der Paläontologie 2 (1).

— 1890: The Tertiary insects of North America. — Rept. U.S. Geol. Surv. Terr. 13, p.103-116, 624.

— 1891: Index to the known fossil insects of the world including myriopods and arachnids. — Bull. U.S. Geol. Surv. 71, 744 pp.

Seevers, C.H., (in press): Fossil Staphylinidae in Tertiary Mexican amber (Coleoptera). — Univ. Calif. Publ. Entom. 31 (2).

Seidlitz, G., 1896: Naturgeschichte der Insecten Deutschlands. Coleoptera.

— 1898: Naturgeschichte der Insecten Deutschlands. Coleoptera.

Selberg, Inga-Britt, 1963: Preliminär bestämning av insekter och övriga landarthropoder i Gyllenhalska Bärnstenssamlingen, Uppsala. — Unpublished.

Sellnick, M., 1919: Die Oribatiden der Bernsteinsammlung der Universität Königsberg in Preussen. — Schr. phys. ökon. Ges. Königsb. 59, p.21-42.

— 1927: Rezente und fossile Oribatiden (Acar. Orib.). — Ibid. 65, p.114-116.

— 1931: Milben im Bernstein. — Bernstein-Forsch. 2, p.148-180.

Sendel, N., 1742: Historia succinorum corpora aliena involentium et naturae opere pictorum et caelatorum ex regiis Augustorum cimeliis Dresdae condictis aeri insulptorum conscripta. — Leipzig 328 pp.

Serres, 1829: Géognosie des terrains Tertiaires. — Paris.

Sharov, A.G., 1957: The first record of a Cretaceous Aculeata (Hymenoptera) (in Russian). — Dokl. Acad. Nauk SSSR, Moscow 112, p.943-944.

— 1966: Basic Arthropodan stock, with special references to insects. — Pergamon Press. 271 pp.

Shelford, R., 1910: On a collection of Blattidae preserved in amber from Prussia. — J. Linn. Soc. Zool. 30, p.336-355.

— 1911: The British Museum collection of Blattidae enclosed in amber. — Ibid. 32, p.59-69.

Silvestri, F., 1913: Die Thysanura des baltischen Bernsteins. — Schr. phys. ökon. Ges. Königsb. 53, 1912, p.42-66.

Skalski, A.W., 1973a: Studies on Lepidoptera from fossil resins. Part II. *Epiborkhausenites obscurotrimaculatus* gen. et sp. nov. (Oecophoridae) and a tineidmoth discovered in the Baltic amber. — Acta Palaeont. Polon. 18 (1), p.153-160.

— 1973b: Notes on the Lepidoptera from fossil resins. — Pol. Pismo Ent. 43, p.647-654.

— 1973c: Studies on the Lepidoptera from fossil resins. Part VI. *Tortricidrosis inclusa* gen. et sp.nov. from the Baltic amber. — Dtsch. Entom. Zschr. (N.F.) 20, p.339-344.

— 1974: Zwei neue Gattungen und Arten der Familie Tineidae aus dem baltischen Bernstein. Studien an Lepidopteren aus fossilen Harzen. — V.) Beitr. Entom. 24, p.97-104.

— 1975: Notes on present status of botanical and zoological studies of ambers. Atti della cooperazione interdisciplinare Italo-Polacca. — Studi e Ricerche sulla problematica dell' ambra 1, p.153-175.

— 1976a: Studies on the Lepidoptera from fossil resins. Part III. Two new genera of Tortricidae from the Baltic amber. — Field Zool.

— 1976b: Les Lépidoptères fossiles des l'ambre. — Etat actuel de nos connaissances (1-3). — Linneana Belgica 6, p.154-169, 195-208, 221-233.

— unpublished: Composition and ecological problems of Oligocene Microlepidoptera from the Baltic amber in the light of observations of cases these insects are drowned in recent resins.

— unpublished: Micropterygidae from the Baltic amber with redescription of taxon *proavitella* Rebel, 1935.

Snyder, T.E., 1928: A new *Reticulitermes* from Baltic amber. — J. Acad. Sci. 18, p.515-517.

— 1960: Fossil termites from Tertiary amber of Chiapas, México (Isoptera). — J. Paleont. 34, p.493-494.

Stach, J., 1922: Eine neue *Sminthurus*-Art aus der Bernsteinfauna. — Bull. Acad. Polon. Sci. Lettr. (B) 1922, p.53-61.

Stannard, L.J., 1956: Two new fossil thrips from Baltic amber (Thysanoptera; Terebrantia). — Fieldiana, Zool., Chicago, 34, p.453-460.

Steffan, A.W., 1968: Elektraphididae, Aphidinorum nova familia e sucino baltico (Insecta: Homoptera: Phylloxeroidea). — Zool. Jb. Syst. 95, p.1-15.

Stein, J.P.E., 1877: Drei merkwürdige Bernstein-Insekten. — Mitt. Münch. Entom. Ver. 1, p.28-30.

Stitz, H., 1939: Ameisen oder Formicidae. — Tierwelt Deutschlands 37.

Strassen, R. zur, 1973: Fossile Franzenflügler aus mesozoischem Bernstein des Libanon. — Stuttg. Beitr. Naturkd. 256, 51 pp.

Stuckenberg, B.R., 1974: A new genus and two new species of Athericidae (Diptera) in Baltic amber. — Ann. Natal Mus. 22 (1), p.275-288.

— 1975: New fossil species of *Phlebotomus* and *Haematopota* in Baltic Amber (Diptera: Psychodidae, Tabanidae). — Ibid. 22(2) p.455-464.

Sturtevant, A.H., 1963: A fossil Periscelid (Diptera) from the amber of Chiapas, México. — J. Paleont. 37, p.121-122.

Stys, P., 1969: Revision of fossil and pseudofossil Enicocephalidae (Heteroptera). — Acta Entom. Bohenoslov. 66, p.352-365.

Takhtajan, A., 1969: Flowering plants. Origin and dispersal. — Edinburgh. 284 pp.

Teskey, H.J., 1971: A new soldier fly from Canadian amber (Diptera: Stratiomyidae). — Canad. Entom. 103, p.1659-1661.

Théobald, N., 1937: Les insectes fossiles des terrains oligocènes de France. — Bull. Mens. Soc. Sci. Nancy (N.S.) 2 bis, 473 pp.

Tillyard, R.J., 1926: Insects of Australia and New Zealand. — Sidney.

Townsend, C.T., 1942: Manual of myiology, part XII. (Fossil forms and a list of fossil localities: p.7-22). — Itaquaquecetuba 1942;

Tryapitsyn, V.A., 1963: A new Hymenopteron genus from Baltic amber. (In Russian). — Paleont. Zhurn. no. 3, p.89-95.

Türk, E., 1963: A new Tyroglyphid deutonymph in amber from Chiapas, Mexico. — Univ. Calif. Publ. Entom. 31 (1), p.49-51.

Uhmann, E., 1939: Hispinen aus baltischem Bernstein. — Bernstein-Forsch. 4, p.18-22.

Ulmer, G., 1912: Die Trichopteren des baltischen Bernsteins. — Beitr. Naturkd. Preussens, 10, 380 pp.

Ulrich, W., 1927: Über das bisher einzige Strepsipteron aus dem baltischen Bernstein und über eine Theorie der Mengeinenbiologie. — Zschr. Morph. Ökol. Tiere 8, p.45-62.

Voigt, E., 1936: Die Lackfilmmethode, ihre Bedeutung und Anwendung in der Paläontologie, Sedimentpetrographie und Bodenkunde. — Zschr. Dtsch. Geol. Ges. 88, p.272-292.

— 1937a: Paläontologische Untersuchungen an Bernsteineinschlüssen. — Paläont. Zschr. 19, p.35-46.

— 1937b: Der Erhaltungszustand der tierischen Einschlüsse im Bernstein. — Chem. Zeitg. nr. 83.

— 1938: Eine neue Methode zur mikroskopischen Untersuchungen von Bernstein-Einschlüssen. — Forsch. Fortschr. 14, p.55-56.

Voss, E., 1953: Einige Rhynchophoren der Bernsteinfauna (Col.). — Mitt. Geol. Staatsinst. Hamburg, 22, p.119-140.

— 1957: *Archimetrioxena electrica* Voss und ihre Beziehungen zu rezenten Formenkreisen (Col. Curc.). — Dtsch. Entom. Zschr. 4, p.95-102.

— 1972: Einige Rüsselkäfer der Tertiärzeit aus baltischem Bernstein (Coleoptera, Curculionidae). — Steenstrupia 2 (11), p.167-181.

Wagner, H., 1924: Ein neues *Apion* aus dem baltischen Bernstein. — Dtsch. Entom. Zschr. p.134.

Wasmann, E., 1926a: Die Paussidengattungen des baltischen Bernsteins. — Zool. Anz. 68, p.25-30.

— 1926b: Die *Arthropterus*-Formen des baltischen Bernsteins. — Zool. Anz. 68, p.225-232.

— 1927: Die Paussiden des baltischen Bernsteins und die Stammesgeschichte der Paussiden. — Tijdschr. Entom. 70, p.LXII-LXIX.

— 1928: Die Bernsteinpaussiden und die

Stammesgeschichte der Paussiden. — Verh. X. Intern. Zool. Kongr. Budapest, 1927, p.1497-1515.

— 1929: Die Paussiden des baltischen Bernsteins und die Stammesgeschichte der Paussiden. — Bernstein-Forsch. 1, p.1-110.

— 1932: Eine ameisenmordende Gastwanze (*Proptilocerus dolosus* n.g., n.sp.) im baltischen Bernstein. — Ibid. 3, p.1-3.

Watson, J.C., 1925: Fossil resins (retinite) from Yallourn, Allendale and Lal lal. — Rec. Geol. Surv. Vict. 4 (4), p.438-485.

Weidner, H., 1952: Insektenleben im Bernsteinwald. (Ein Bericht über die Bernsteinsammlung des Geologischen Staatsinst. Hamburg). — Entom. Zschr. 62, p.62-72, 88.

— 1955: Die Bernstein-Termiten der Sammlung des Geologischen Staatsinstituts Hamburg. — Mitt. Geol. Staatsinst. Hamburg 24, p.55-74.

— 1956: Die Bernstein-Heuschrecken (Saltoptera) der Sammlung des Geologischen Staatsinstituts Hamburg (Orthopteroidea). — Ibid. 25, p.82-103.

— 1958: Einige interessante Insektenlarven aus der Bernsteininklusen-Sammlung des Geologischen Staatsinstituts Hamburg (Odonata, Coleoptera, Megaloptera, Planipennia). — Ibid. 27, p.50-68.

— 1964: Eine Zecke, *Ixodes succineus* sp.n., im baltischen Bernstein. — Veröff. Überseemus. Bremen (A) 3 (3). p.143-151.

— 1968: Über die im deutschen Tertiär gefundenen Termiten-Arten. — Beih. Ber. Naturhist. Ges. 6, p.13-20.

Wheeler, M.R., 1963: A note on some fossil Drosophilidae (Diptera) from the amber of Chiapas, México. — J. Paleont. 37, p.123-124.

Wheeler, W.M., 1915: The ants of the Baltic amber. — Schr. phys. ökon. Ges. Königsb. 55, 1914, p.1-142.

Wille, A., 1959: A new fossil stingless bee (Meliponini) from the amber of Chiapas, Mexico. — J. Paleont. 33, p.849-852.

Wille, A. & L.C.A. Chandler, 1965: A new stingless bee from the Tertiary amber of the Dominican Republic (Hymenoptera: Meliponini). — Revta Biol. Trop., San José, 12, 1964, p.187-195.

Williams, R.M.C., 1968: Redescriptions of two termites from Burmese amber. — J. nat. Hist. 2, p.547-551.

Williamson, G.C., 1932: The book of amber. — London. 268 pp.

Willmann, C., 1931: Moosmilben oder Oribatiden. — Tierw. Dtschl. 22.

Wilson, E.O., F.M. Carpenter & W.L. Brown, 1967: The first Mesozoic ants, with the description of a new subfamily. — Psyche 74, p.1-19.

Wittmer, W., 1963: A new Cantharid from the Chiapas amber of Mexico. — Univ. Calif. Publ. Entom. 31 (1), p.53.

Wygodzinsky, P., 1959: A new Hemipteran (Dipsocoridae) from the Miocene amber of Chiapas, Mexico. — J. Paleont. 33, p.853-854.

Yablokov-Khnzoryan, S.M., 1960: New Coleoptera from Baltic amber (in Russian). — Paleont. Zhurn. Akad. Nauk. SSSR, 1960 (3), p.90-101.

— 1961a: Helodidae (Coleoptera) from Baltic amber (in Russian). — Ibid. 1961 (1). p.107-116.

— 1961b: New Elateridae from the Baltic amber (in Russian). — Ibid. 1961 (3), p.84-97.

— 1961c: Circaeidae. — a new family of Coleoptera from amber (in Russian). — Dokl. Akad. Nauk. SSSR, 136, p.209-210.

— 1962: Representatives of Stenoxia (Coleoptera) from the Baltic amber. Family Throscidae Bach, 1849 (Trixagidae sensu Crowson). (in Russian). — Paleont. Zhurn. Akad. Nauk. SSSR, 1962 (3), p.81-89.

Zaddach, G., 1864: Eine Amphipode im Bernstein. — Schr. phys. ökon. Ges. Königsb. 5, p.1-12.

Zang, R., 1905a: Über Coleoptera Lamellicornia aus dem baltischen Bernstein. — Sb. Ges. Nat. Fr. Berlin, 1905, p.197-205.

— 1905b: Coleoptera Longicornia aus der Berendtschen Bernsteinsammlung. — Ibid. 1905, p.232-245.

Zeuner, F., 1936: The recent and fossil Tympanophoridae (Tettig., Salt.). — Trans. Entom. Soc. London 85, p.287-302.

— 1939: Fossil Orthoptera Ensifera. — Brit. Mus. (Nat. Hist.), London. 321 pp.

Žerichin: see Jerichin.

Review

Extensive forests have covered large areas of the earth far back in the ages of the cryptogams, but it was not before the arrival of the coniferous plants towards the end of the Palaeozoic that plants evolved which both spontaneously and as a defence against damage secreted resin in large amounts. Those plants which secreted specially large amounts of resin all seem to have belonged to the tropic or subtropic regions of the earth. At an early stage, there appears to have developed both retinite-producing stocks and succinite-producing ones. The diterpenes in the last-mentioned stocks were originally of the labdane type (type I in text-fig.2), which on selenium dehydrogenation mainly produced agathalene, a state of affairs very similar to that in recent Araucariaceae. This stage was found in the extinct pinacean *Pinites* and occurs also in some few now-living Pinaceae. In the majority of Pinaceae, however, a further evolution has probably taken place, and here are the diterpenes of the abietane type (type III). These last-mentioned resins, retanes, are mainly found in recent boreal species, while resins of type I in particular have been retained in tropic and subtropic species; the abietane group seems far less resistant to breakdown, it is not known from any Palaeogenic fossil, and its resin occurs in far smaller amounts. This is presumably an important reason why fossil resins are not known which have been formed in boreal regions, apart from the very warmest regions here.

By far the greater part of the amber known today originates from younger Cretaceous and older Tertiary. During the Neogenic deterioration of climate, the plants which were strong producers of resin, and among which there were now in addition numerous angiosperms, were steadily forced towards the equator. The formation of amber in the Baltic region appears to have stopped early in the Oligocene, somewhat later in the Rhineland and in Roumania, while it was still functioning in the Miocene on Sicily and in other regions at corresponding latitude.

Baltic Tertiary amber which is dated with certainty is known today from the earliest Eocene (Mo-clay amber) and from the earliest Oligocene (Samland amber), with a time interval of about 20 million years, but most of the Danish amber must have been formed during the intervening period. It is probable that the East Baltic material from Samland should merely be regarded as the final phase of a much longer period, which has undoubtedly even comprised the Bornholm amber from the Jurassic and unknown ambers of Cretaceous age.

By amber forest must be understood these parts of a much greater forest region (extending high northwards in the purely boreal regions), in which the amber tree *Pinites* has appeared in such large numbers that it has left its traces in posterity. The Baltic amber forest must in all probability have had a very great distribution from west to east, it must have had conditions of growth from the North Sea of that epoch and possibly right east to the Sea of Ob east of the Urals. Northwards, the forest type to which the amber tree has belonged has reached till about 60° North (Text-fig.6). From here, it has presumably in its more western section spread southwards over the Baltic and down to those parts of Central Europe which at that time were land, in particular the Hercynian regions, where it has in fact been possible to trace amber direct. In Russia, the Sea of Tethys must presumably have been the southern boundary. Biologically, this great region would appear to fall into two sections: a western and an eastern, but there has hardly been any sharp boundary between them. As only very little is known about the West Baltic amber, the differences between the two faunas are still difficult to define. In addition, the difficulties are increased by the fact that a quite considerable mixing of western and eastern amber has undoubtedly taken place in the south-western Baltic regions.

The core of the region of formation in Scandinavia must have been Southern Sweden. One may picture the most northern forests in which *Pinites succinifer* can have grown, as having covered the southern slopes of Värmland. From here and southwards, comprising Västergötland and the intervening lowland, there has been a rich variation of landscape, perhaps richer than anywhere else in the extensive eastern amber regions. From the South Swedish water-shed, brooks and streams have possibly collected together to form that river which can have been the predecessor of the Götaelv, and in that case have had presumably the same bed as now. The stream has continued over those parts of the Kattegat which at that time consisted of drained areas, and has poured its content out into the North Sea, in the direction of the present day eastern estuary of the Limfjord (Text-fig.7). Even today, considerable amounts of amber are released each time the entrance to the Limfjord is dredged up at Hals Barre; this is an indication of the great amounts which are possibly still hidden in the sea floor. Also the Mo-clay amber must originate from here. Gradually, as Jutland was raised above sea level in the course of the Tertiary, and the North Sea was displaced towards south-west, the continually changing river bed must presumably have released previously deposited materials and carried these down to still younger deposits. This has happened for example in the lignite deposits of Mid-Jutland, and it has happened along the coast of West Jutland especially from Ringkøbing to Fanø, where much amber is still released today and washed ashore. The southgoing movement of the amber has continued, right up to the time that the new Kattegat interrupted the old connection between Sweden and North Jutland.

Whether part of the amber of Skåne and East Zealand originates from the same Scandinavian amber forest region, is very doubtful; the most recent investigations (Nilsson, 1973) appear to confirm the old theory by Holst that the Alnarp Valley in Skåne (Text-fig.7) has received its rich content of amber direct via the river Vistula in the Quarternary. In the course of the Eocene, those portions of East Baltic amber which were carried to the North Sea of that day (much has presumably been led southwards to the coasts of the Sea of Tethys), have presumably in the main been deposited further south than the Scandinavian deposits, in the neighbourhood of the present-day Rügen or west of this, right out to the Bay of Helgoland.

In the Palaeogenic, the earth is presumed to have had a geographical distribution into plant zones which were more or less as they are today, except that the single parts of the zone ring were displaced towards the poles in relation to the present, corresponding to a warmer climate. The tree or group of trees which produced by far the greatest part of the amber resin, and which probably with reason is identified with *Pinites succinifer* Conwentz, has in the main grown in the subtropical section of the original, widely spread primeval forests, and in the warmest parts of the adjacent boreal region. On morphological grounds, one must conclude that this plant has belonged to the Pinaceae, but its resin had the special chemical constitution (type I), which deviated from that of the great majority of the now-living Pinaceae (type III). We are unable to recognize *Pinites* in any recent plant species.

If one combines the modest knowledge available of the flora of the amber forest in the form of plant remains embedded in amber, with the far greater knowledge of the Eocene flora in Europe one has obtained from other sources, it is possible to form a limited picture of what may have grown of plant genera, but this does not provide much information as to how this flora was composed in the individual plant societies. In that respect, wind-spreading of pollen, spores and other relatively characteristic constituents have played too great a role. Possible wind-spreading also reduces the significance which can be attributed to those parts of the fauna which have had a greater chance of being introduced from afar. A considerable part of the fauna, however, has been bound to one place, and for biological or physical reasons has had only a slight possibility of being taken up in any air plancton, and in this way the picture of the biological societies of the amber forest can be amplified to some degree. First and foremost the arthropod fauna is of significance, since we are in possession of tens of thousands of fossils of this fauna, in the finest possible state of preservation from a taxonomic point of view.

Pinites has undoubtedly been a characteristic tree in this type of primitive forest. It has grown in extensive Eocene forests, from which rich amounts of succinite have been obtained, but it has probably not been particularly dominant. It has been a tree which has grown high in stature, of a type of growth resembling that of the most vigorous *Pinus* species of today, with which it must have had great external resemblance. It must have been a tree which produced only inconsiderable shade, and which together with oak and many plants with corresponding physical characteristics has provided a soil for the rich and luxuriant undergrowth, which we can see has had a very wide distribution.

An evaluation of the ecological demands which *Pinites* has made on its surroundings is difficult — apart from the heat requirements of the tree, which

must have been about the same order of magnitude as those of the Araucariaceae in the present. If the content of insect fossils of amber is taken as an indicator for the moisture requirement of the tree, one must come to the conclusion that it has required a considerable air humidity, and that it has been able to grow quite close to open water; no amber fossils are known which can be taken as a sign of a dryer climate, e.g. on mountain slopes, in such amounts that they can be ascribed real significance. However, with the picture of the wandering of all amber by waterways as a background, it is not possible to regard these considerations as decisive. Because of its relatively slight power of resistance to various destructive forces, Eocene amber is never found in its primary stratum, but only where it has been placed secondarily under specially favourable circumstances leading to natural preservation. It has always been transported to these secondary deposits by small and large rivers, which have first of all torn it away from its primary deposit in the forest floor. This means that the present existence of the amber depends on the producing tree having grown in the neighbourhood of running water, but it reveals nothing as to the actual tolerance of the amber tree to water and to water vapour. We have no possibility for finding amber from trees which might have grown under more arid conditions, and the transport must have taken place within a relatively short period of years (geologically speaking) after the excretion of the resin.

The entomofauna bears evidence of both running water and stagnant water, and as the transport of the more or less fossilized resin, the amber-to-be, has taken place with the aid of running water, it is natural that just this fauna should be encountered particularly frequently; it is generally accepted that the amber forest has been highland forest. What is true in this belief is that insects, among which many breed in rivers and mountain streams, are represented in the amber by abnormally many alpine species in proportion to insects from stagnant water, this being true not only for the number of species, but also for the number of individuals. This becomes very pronounced, for example, in Plecoptera, Ephemeroptera and not least in Trichoptera (Plate 4,A). In many cases, however, the numbers of individuals for a single species are very modest, and may bear evidence of relative rarity. This is the case to a striking degree for the Simuliidae, which at the present day are exclusively associated with water of this nature, and which generally occur in enormous swarms, wherever they live on their true biotope.

There is a further element of the fauna which provides a telling piece of evidence for the character of the amber landscape. This is the Chironomidae, undoubtedly that group of insects which is most numerous in the Baltic amber, at any rate in the Scandinavian amber. These thousands and thousands of fossils belong almost all to the true Chironomidae, which to a great extent breed in stagnant water, in addition often very oxygen-poor water, e.g. at the bottom of large lakes. Their presence in such large amounts is evidence that in the amber forest there have been plentiful numbers of dams and lakes of very varying character. In the landscape, a lake normally represents a larger or smaller surface which is relatively uniform, traversed by one or more incoming streams, which are in turn connected with an outgoing stream. If amber trees have grown along the banks of such a lake, the resin which has been produced has remained lying in most places in the terrestrial vegetation of the bank, and has therefore been destroyed in the course of the years; only the small amounts formed near the inlet and outlet have had a chance of being caught by the stream and taken to a place with possibilities for preservation. Year after year the Chironomidae have occurred in the amber forest in swarms of milliards, and this fact, together with the fact that the most important time of swarming has coincided with that time

of the year when the production of resin has been most active, is a condition for their relative frequency in the amber today. It is definitely necessary to reckon with the fact that the probability of the single individual becoming a fossil has been considerably smaller than was the case with the Plecoptera, for example. The real difference in the incidence of the two groups in the amber forest has been even more strongly pronounced than appears from the amber fossils.

This correction of the values found leads to considerations of the study by Ulmer on the amber Trichoptera. It was Ulmer who first emphasized the significance of mountain streams for the character of the amber fauna, and identified the amber territory as a mountainous territory. The percentage of Trichoptera breeding in running water was in his East Prussian material quite abnormally high, while the percentage of species from stagnant water was correspondingly low. If, as is done above, adequate attention is paid to the possibility of the fauna not only being trapped in the resin, but also "surviving" and becoming carried down as fossils to sites which are functionally capable of preservation, then there will be an equalization between the two elements of the fauna, and the Trichoptera fauna of the amber territory will in no way be distinguished significantly from normal.

A further factor with regard to the waterways and lakes of the forest can be recognized by means of the fauna. In the course of time, a number of finds have been made of pronounced freshwater fauna, the occurrence of which in amber is very difficult to understand. In particular, one might mention finds of *Corixa* larvae, *Nepa* and several specimens of *Palaeogammarus* (Text-fig.35), but also finds of a may-fly larva (Text-fig.18), a trichopterous larva (its shelter was lost), *Gerris* larvae, *Heterocerus* larva, a few dytiscids and some byrrhids (species living on sandbanks), can be referred to this category. Here it is probably a case of the consequences of quite brief violent floodings, where the fauna of the banks has been carried further inland by sudden torrents, and where the freshwater fauna has later become stranded by the water level once again falling. It is most likely a spring phenomenon, where considerable volumes of water are carried out to sea from the fells or from more northerly lying parts of the territory; this may also have to do with specially heavy amounts of rain associated with the time of the year.

Between the crowns of the trees of the high wood and the forest floor, there has been a soil for a rich and luxuriant undergrowth, and it is probably here that the subtropical vegetation, so rich in species, has been particularly prominent in between the growing young trees of the giant tree species. It is here that so many species of Cupressaceae, which we know from amber, have had their crowns, Lorantaceae and other epiphytes have been common, while shoots of *Smilax* and other lianas have been entwined among them. This vegetation has formed a roof over the forest floor, where at all times there has been a twilight illumination, and a herbaceous vegetation has been typically lacking. The forest floor has been covered by a thick layer of old rotting leaves, and fallen tree trunks and branches under rapid decay in the permanently humid air have formed an impenetrable confusion. Moss and liverwort have spread a thick carpet practically everywhere and a good distance up the trunks of the old trees, broken through by a multitude of the fruiting bodies of various fungi.

In this forest floor the organic conversion has been considerable, and it must have had room for a very rich animal life, but our knowledge of this is slight — apart from the micro-fauna which has spread from the moss cushions of the forest floor up into those growing on the basal parts of the tree trunks. Also here it is a fact that our special knowledge is limited to that region of the forest which has been close to the foot of the amber trees, and where in addition a brook or a larger river in the immediate neighbourhood has been able to carry away the amber resin over a period of time which from a geological point of view is by no means boundless.

However, among the amber fossils, there are very large numbers of one biological type from that animal life which has been so widely distributed in the forest floor. This is the type of animal which lives its larval life in the mould-forming plant material, but has its adult existence free in the vegetation of the forest undergrowth. Its most characteristic representatives are the Sciaridae, which are among the fossils most commonly found, and known in many species (Plate 5,B); but examples could also be mentioned from other insect groups, for example Scydmaenidae among the beetles, a family whose imagines swarm under quite definite ecological conditions. The larvae of the Sciaridae have undoubtedly been saprophages just like the larvae of recent species under similar conditions, and they have presumably been hunted by the larvae of Empididae and Dolichopodidae, whose imagines are just as frequent in the amber. During their hopping flight from twig to twig in the undergrowth, the restless imagines of all these Diptera have landed on the amber tree and have been an easy prey to the resin. There is no doubt that insects of this type are very strongly over-represented in amber, by comparison with other faunal elements of the forest floor, as for example Carabidae, Oniscidae, Myriapoda and Pulmonata. The Sciaridae are such a characteristic and central element in this fauna, that it would seem justified to call the deepest-lying, continually twilight-lit, humid region of this type of amber forest, for the *Sciara* zone.

It is not easy to say how common this type of forest has been, but probably it has been widespread. It has hardly been the major type of forest in any mountainous region, but it has more been a case of forest in a level country, often more or less morass-like, where overshadowed streams have found their way through the landscape. This has been a land suitable for the majority of Helodidae (Text-fig.19). In the case of the Scandinavian amber territory, it brings to mind particularly the extensive lowland, which presumably also at that time has lain between Värmland and Västergötland. The fact that the forest has not covered over the streams everywhere, however, can be seen from one element of the fauna which admittedly is not specially frequent, but is nevertheless obvious. This consists of a number of Curculionidae and some Thysanoptera, characteristic for herbaceous vegetation and quite particularly for various types of monocotyledons. This has probably been a case of a river bank fauna. It has not been associated with the continuous, dark forest, but rather with the free shores, from where at any rate Thysanoptera as low-floating air plankton can have driven into the forests, possibly along the waterways. The relatively many *Phyllobius* which are known from Baltic amber, have most probably lived in adjacent forest edges, as these weevils today are in fact found particularly on the low hanging branches of oak and in freely exposed parts of young trees.

The fauna has also been of a different type in the lightly shadowed, airy, partly sunny regions in the upper crowns of the undergrowth, where the mosaic of leaves has played such a large part for the distribution of the light, and where the humidity of the air must have been much lower than near the floor of the forest. Here too there have been many Diptera. It was here that Rhagionidae (Plate 5,C) and other brachycerous flies have lain in wait for their prey, often Dolichopodidae and Empididae, which had been flying up to the light from the forest deep. This is the zone which has contained many spiders (Text-figs.49-51), and where a large number of parasitic wasps (Plate 1) have sought their victims, mostly different kinds of insect larvae, whether they have been free-living on the surface of the plants or hidden in the bark and wood. It is likewise at this level in the forest that the majority of cicadas (Plate 3,A,B) have

lived, hardly particularly common on the amber tree itself, but mainly in the leafage of the many different angiosperms, whose branches had blended with those of the amber trees. In this way, apparently many free-living insects have been captured by the fresh amber resin. None of these species, however, had achieved such a great frequency as many of the ordinary inhabitants of the *Sciara* zone. The reason for this is in part that none of the animals in this zone have been represented by such large numbers of individuals as many in the *Sciara* zone; the total amount of food has simply been too small for this. In addition, however, the amount of resin has increased from the top of the tree to the forest floor; everywhere, it has amounted to the sum of what was locally produced and what had come gliding down from above, and this must have given increased opportunities of capture downwards. Fossils which with certainty must have originated from the crown of the amber tree are unknown.

Quite a lot is known about that animal life which has been busy on the bark of the amber tree at and above the level of the undergrowth. It is here among lichens and algae that Psocoptera (Plate 9,B,C) have had their most important biotope, hunted by small spiders, *Malthodes* larvae and many other diminutive robbers.

It is extremely likely that certain aphids and scale insects have been specially associated with the trunk of the amber tree; in fact bearing recent conditions in mind, one would consider it unlikely if such parasites were not found. That aphid which to a special degree has had the chance of playing a role particularly for the fate of the amber tree in the forest, is *Germaraphis* (Text-fig.10). We know that its "honeydew" has been of significance as a source of nourishment for the ant *Iridomyrmex*, and that it has probably played a similar role for *Formica flori* and other Formicinae. Most important of all, no doubt, have been the biological effects on the local state of health of the bark, following the persistent sucking of sap by a total population of aphids. We know from recent investigations that such a sucking of sap can damage the bark of the tree so much that it results in attacks of a far more serious kind. First it opens up a possibility of various types of fungal infections, among others by Polyporaceae. Hereby, the tree acquires an inevitably fatal disease, which over the course of years will slowly kill it from within. The weakened bark, however, also gives rise to attacks by bark beetles and other insects, which implies local destruction of the cambium, and from a short term point of view this can be even more serious than attacks by fungi; the bark becomes loosened and gradually falls off in larger and smaller flakes. In this connection, woodpeckers are particularly active; they hack industriously and strongly during their seeking for food. That this has also been the case in the Eocene forest is seen from numerous chips of hacked wood which are found in amber. It is quite possible that these sites have also been preferred for attack where woodpeckers and other forest birds which hatch in holes in the tree trunk have sought a nesting place. Also in the case of xylophagous termites like *Reticulitermes antiquus* (Text-figs. 61, 62; Plate 12), the dead patches on the tree trunks have been of significance and undoubtedly often the goal for their covered-over transport routes.

The influence of such damage on the amount of amber resin secreted has often been discussed. It appears to be quite definite and well documented that the healthy *Pinites* plants have freely secreted large amounts of resin, corresponding to the conditions in the present-day *Agathis* species. However, just as it is possible by means of artificial wound production to tap recent resin-producing plants for larger amounts of resin than they would have yielded voluntarily, so has the amber tree undoubtedly also as a result of local damage produced more than normal — as a kind of self-defence. That this actually has been the case, can be seen from the amounts of detritus which together with resin have been pressed out from the larval burrows in the bark, with a content of gnawed chips, larval excrement and the little mite *"Acarus" rhombeus* Koch & Berendt. As the flow of resin apparently has been especially dominant in the annual growth period of the forest, it is by far most probable that this voluntary flow of resin has been normal, and that the flow after damage has been of purely secondary significance. The possibility cannot be refuted that trees which have been subjected to attack have at the beginning produced particularly large amounts of resin, until they became noticeably weakened.

Both the attack of the bark beetles and the attacks by fungi have provided possibilities to thrive for various microfloras and for a fauna of phytophagous animals and their predators. It appears as if these two routes of entry have been used by two biologically quite sharply distinguished types of fauna. *1.* The bark beetles seem at first sight to have been accompanied by Cerambycidae with a corresponding mode of life, e. g *Nothorrhina*, and perhaps by Buprestidae. Thereafter, a rich fauna of main saprophages has followed, which has consisted especially of small forms, including other bark beetles, for example *Hylurgops*. *2.* The Polyporaceae have been followed by species of Serropalpidae, Mordellidae (Plate 11,C) and Alleculidae. These forms appear primarily to have lived in the fruiting bodies of fungi, secondarily as xylophages in the wood which has been partly destroyed by mycelia. Among the last-mentioned, Anobiidae have played a very great part. This combined activity of Polyporaceae and insects must have involved a considerable weakening of the physical resistance of the tree trunks against storms and other forms of bad weather. In the fallen tree trunks, life has continued to exist in the form of continually changing fauna of insects, myriapods, oniscids and scorpions, a fauna which gradually, under the advancing destruction of the wood, has been changed into a pure forest floor fauna. With regard to this fauna, we have as mentioned only little knowledge, for here there was no flowing amber resin to produce relatively imperishable preparations.

Even if one could reckon that by far the majority of the amber fossils have lived in those amber forests where they became fossils, it is nevertheless probable, in some cases almost certain, that wind-spreading has played a part. Spreading as air plankton has had no practical significance for that part of the fauna belonging to the *Sciara* zone, but it has been of significance mainly for the winged small insects from the higher layers of the forest and from the herbaceous vegetation along the banks of ponds and rivulets. It is most striking for those Thysanoptera which have lived on the monocotyledonous vegetation of the bank, but also among aphids probable examples can be mentioned; presumably this has also been the case among scale insects, Aleurodidae, gall-forming midges and wasps and other small fliers.

That fauna with which we are presented via the amber fossils, is a selection of the dominant animal life in the amber forest, but it is very far from giving a numerically truthful impression of its composition. There is a very great preponderance of mobile, small flying insects, for example Sciaridae and Dolichopodidae, which in fact move around at the base of the trees, where the collections of resin have been greatest. There is a great deficit of insects which have lived a hidden life, such as Carabidae, Myriapoda and other animals which have never or rarely been flying. The time of the year at which a species has had its distribution period has played a very great role, as well as the duration of this period; this has undoubtedly been a factor determining the great incidence of the Chironomidae and the total lack of winged specimens of *Germaraphis*. The "converse" proportion between males and females among the Chironomidae is most probably due to the fact of the much shorter lifetime of the males and

thereby the more limited possibility of being caught by the resin; the real numerical relationship has undoubtedly been about the same as today. Statistical treatment of the fossils has therefore only a very limited value by comparison with life in the present, and it can even be directly misleading.

It has often been emphasized (for example by Ander, 1942) that the amber fauna has a very great similarity with the palaearctic and the nearctic in the present, while others (for example Petrunkevitch, 1958) have emphasized its dominating relationship with recent tropical fauna, particularly that of South East Asia, and both sides have to some extent been correct. The zoogeographic problems correspond very closely to the botanical problems (see section III). On transition to the Oligocene, a fauna was found rich in species and genera, of which part, e.g. the cockroaches, had a very long prehistory, as was also the case with the conifers. A very important part of this fauna, namely specialized phytophages, was not developed until some time in the course of the Mesozoic, in step with the angiosperms to which they were adjusted, and just like the flora they had never been exposed to serious biological overload in form of severely deteriorated conditions. In the warm period of the Mesozoic and early Tertiary, the temperature differences were also less pronounced from north to south, and it must be assumed that the individual species have correspondingly had a greater distribution in the north-south direction. Both flora and fauna had in the main in the course of the warm period developed into their morphological status as it is today, only apart from the warm-blooded birds and mammals which are subjected to an enormous evolution during the Tertiary.

In the climatic deterioration of the Neogenic, temperature gradients became sharper from north to south, particularly the boreal and the subtropical belt being considerably reduced, and an arctic shield spread further and further south from the north. The climate-belts became displaced towards the equator, and the difference between them became more and more pronounced. Fauna and flora spread southwards, but along the northern boundary of their distribution, plants and animals disappeared in a somewhat more rapid tempo. In the Baltic region, almost all life from the amber period was destroyed, because the ice on the young alpine folded chains developed into an insurmountable barrier for the fauna and flora, which from the north was pressed by the Scandinavian ice. In a corresponding manner, life was destroyed all over the world, the regions of distribution of the genera were split up by seas and mountains during their spread towards the equator, while in the intermediate regions they became extinct, or their remains persisted in isolated refuges.

The result of these biogeographical wanderings so far can be seen in the present state of affairs. It is a result of the post-glacial filling-up of those empty regions which the Ice Age had brought about during its maximum state. The Ice Age had exerted its weakest effect on the equatorial regions, its most severe effect on the boreal regions. As already mentioned (section III), the starting point was a uniformity of life forms which in the subtropics of the Eocene were widespread from the Pacific to the Pacific. Between the two northern continents, however, there has been a considerably greater similarity in the north (at a latitude corresponding to England and the Baltic countries) than further south, and this is in good agreement with the continental drift theories of among others Dietz & Holden (1969), according to which it was not until late in the Mesozoic that the atlantic rift became effective in the region Canada-Greenland-England-Scandinavia (Text-fig.6).

The Baltic amber fauna contains a large subtropical element. To a pronounced degree, these animals have their closest recent relatives in equatorial forests in South America, Africa and quite particularly frequently in East Asia. They are often found only in one of these regions of the world, often in several, but no matter the extent of their present distribution, they must be regarded as the remains of groups which in the Eocene have had a much greater, often circumpolar distribution.

Another great element in the Baltic amber is the boreal. It is much more difficult to evaluate from climatological points of view than the subtropic. Considered strictly, we only know that the southern boundary for the distribution of these animals has been within the region where *Pinites* has grown more or less generally; but we know nothing as to their full distribution in those plant societies which have grown north of the amber forest. At the present day, many species are found, not least insects, which without any hesitation must be designated as being absolutely boreal, but which nevertheless can be encountered far south in the subtropics, provided merely that the other conditions for existence are in agreement with their needs. Such has undoubtedly also been the case in the Baltic region in the Eocene. It has never been possible to draw a sharp boundary between subtropical and boreal regions, and the conditions have probably been even more blurred in the Eocene, with the relatively weak temperature differences, than in the present day. It is impossible to see how far south and how far north the individual boreal species have had their boundaries.

What we have less knowledge of at all is the purely boreal fauna (and flora), which quite definitely must have existed north of the amber forests. It may be said with great conviction (but consequently without proof), that a very rich representation of the amber forest fauna has also lived here, and many of its species have probably had their main distribution here. In addition, the great contingent of life forms which we "lack" in the amber must have lived here; one can almost say, from morphological and phylogenetic points of view, that we know they must have existed here. As examples among many others, this is the case for several dominating aphid families (Adelgidae, Lachnidae, etc.), bark beetle genera (for example *Ips* and *Scolytes)* and Limnophilidae among Trichoptera. These boreal groups have most probably constituted from west to east a very uniform, but in its detailed content, richly varied holarctic fauna.

It is this holarctic fauna which more than any other has had to pay for the cold wave which culminated in the Ice Age. What was saved in Europe **north** of the Alps, was little. On the other hand, South East Asia functioned as an enormous refuge. In North America, there was a somewhat similar difference between west and east, although the conditions were less destructive for the Westerly fauna, thanks to the course of the mountain chains in the north-south direction. What was saved of that fauna which had its representatives in the amber forest, depended on many accidents. What we refind today in both the northerly world regions, we call the holarctic fauna of the amber, what is refound only in the old world, is called its palaearctic fauna, and what is found only in North America, is called its nearctic fauna. In reality, they are all merely fragments of the same original fauna from the Palaeogenic age.

That fauna which postglacially has replaced the Baltic amber fauna, consists mainly of purely boreal elements, which during the Ice Age were "preserved" in Asia; it has a great similarity to part of the amber fauna, but the taxonomic relationships between the extinct and the recent Baltic species are relatively moderate, and it has obvious continental features in its biology. A far smaller part of the fauna has immigrated from South Europe and western North Africa, particularly along the coast of the Atlantic. It belongs to a relatively less requiring fauna, which to some degree can also live in subtropical climates, and it has pronounced oceanic features. But the original Baltic amber fauna appears to be totally extinct.

... in amber in
Copenhagen

Collectio[...] natural h[...] long esta[...] are far m[...] Copenhag[...] tion is in[...] century o[...] entomolo[...] its conten[...] material [...] from the [...] collections [...] years mus[...] formation [...] Tertiary a[...] which is p[...] Baltic amb[...] good reaso[...] content. T[...] approximat[...] contained i[...]

[illegible]	1	Cantharoidea	5
[illegible]	2	Melyridae	11
[illegible]	3	Helodidae	71
[illegible]	2	Heteroceridae	1
[illegible]	3	Cleridae	23
[illegible]	223	Temnochilidae	8
[illegible]	636	Anobiidae	37
[illegible]	188	Ptinidae	6
[illegible]	1	Bostrychidae	2
[illegible]	11	Sternoxia	96
[illegible]	14	Dermestidae	9
[illegible]	2	Byrrhoidea	5
[illegible]	353	Clavicornia	73
[illegible]	120	Heteromera	146
[illegible]	1	Cerambycidae	1
[illegible]	2	Chrysomelidae	7
[illegible]	23	Anthribidae	3
[illegible]	156	Curculionidae	8
[illegible]	2	Scolytidae	24
[illegible]	269	Nematoda	4
[illegible]	7	Oligochaeta	2
[illegible]	139	Mammalia (hairs)	2

Oniscoidea	4	Aleuroidea	16
Amphipoda	1	Aphidoidea	207
Diplopoda	13	Coccoidea	111
Chilopoda	15	Heteroptera	48
Pseudoscorpionida	15	Neuroptera s.l.	11
Opilionida	28	Mecoptera	1
Araneida	503	Trichoptera	152
Acarina	424	Lepidoptera	58
Collembola	357	Mycetophilidae	464
Thysanura	40	Sciaridae	371
Ephemeroptera	13	Scatopsidae	26
Plecoptera	16	Bibionidae	4
Blattoidea	31	Tipuloidea	84
Mantoidea	1	Culicidae	3
Saltatoria	11	Psychodidae	156
Phasmoidea	1	Chironomidae	912
Dermaptera	3	Ceratopogonidae	275
Isoptera	39	Simuliidae	9
Psocoptera	92	Cecidomyiidae	190
Thysanoptera	72	Xylophagoidea	2
Auchenorrhyncha	103	Rhagionoidea	30
Psylloidea	2	Athericidae	1

Copenhagen amber material in preparation 1977:

Carpenter, G.H.: Neuroptera, Mecoptera.
Heie, O.E.: Aphidoidea.
Illies, J.: Plecoptera.
Klausnitzer, B.: Helodidae.
Lawrence, J.F.: Ciidae (one specimen).
Mackauer, M.: Aphelinidae, Aphidiidae.
MacLeod, E.G.: Neuroptera.
Masner, L.: Proctotrupoidea.
Moure, J.S.: stingless bees.
Ogaza, B.: Coccoidea.
Remington, C.L.: Thysanura (pars).
Skalski, A.W.: Lepidoptera.
Stibick, J.N.L.: Elateridae.
Suter, W.R.: Scydmaenidae.
Yensen, E.: Throscidae.